Thinking about the Earth

Studies in the History and Philosophy of the Earth Sciences

Volumes in the series include:

David R. Oldroyd *Thinking About the Earth: A History of Ideas in Geology*

THINKING ABOUT THE EARTH
A History of Ideas in Geology

DAVID R. OLDROYD

ATHLONE
London

First published 1996 by
THE ATHLONE PRESS
1 Park Drive, London NW11 7SG

British Library Cataloguing in Publication Data
A catalogue record for this book is available from the British Library

ISBN 0485 11432 1

Library of Congress Cataloging in Publication Data

Oldroyd, D. R. (David Roger)
Thinking about the earth: a history of ideas in geology / by David R. Oldroyd.
p. cm.
Includes bibliographical references (p. –) and index.
ISBN 0-674-88382-9
1. Geology—History. I. Title.
OE11.043 1996
551'.09—dc20 95—4823
CIP

The author and publishers gratefully acknowledge a grant from the Australian Academy of the Humanities towards the publication of this book.

Typeset by WestKey Limited, Falmouth, Cornwall
Printed and bound in Great Britain by
the University Press, Cambridge

To Monty

CONTENTS

ILLUSTRATIONS

ACKNOWLEDGEMENTS

I am indebted to The University of New South Wales for a period of study-leave in 1994 that gave me the time needed to complete this book. Also, The University of Lancaster provided me with much appreciated facilities as a visiting scholar during a portion of the six months leave. My sister-in-law, Mrs Sarah Carrier, nobly provided accommodation in London for an extended period, without which the project certainly would not have been completed in 1994.

In preparing *Thinking About the Earth*, I have received invaluable help from the libraries of The University of New South Wales, the Geological Society, Cambridge University, the Sedgwick Museum, University College (London), Imperial College (London), the British Library, the British Museum (Natural History), the Royal Society, the British Geological Survey and Lancaster University.

Permission to reproduce illustrations is gratefully acknowledged as follows: 1.2, the State Library, Vienna; 1.3, 1.4, 1.5, 2.1, 2.2, 2.3, 3.1, 3.2, 3.3, 3.4, 10.1, the President and Council of the Royal Society; 3.5, 3.6, 5.2, 7.2, 7.5, 8.1, 8.3, 10.3, the British Library; 3.7, the Verona Public Library, 4.1, the British Geological Survey; 4.3, Sir John Clerk of Penicuik; 5.1, 5.3, 7.4, the British Museum (Natural History); 8.2, 8.4, 8.5, 9.1, 9.5, 9.8, 10.2, 10.4, 10.6, 10.7, 11.1, 11.4, 11.8, 11.11, the Geological Society; 9.2, the Geological Survey of Finland; 11.6, 11.7, the library of the Sedgwick Museum; 11.9, the late Stanley Keith Runcorn and Academic Press Inc.; 11.10, Geological Society of America; 11.13, Herr Klaus Vogel; 11.14, S. Warren Carey and Stanford University Press; 12.1, W.W. Norton & Co.

I am grateful to Rowland Hilder and Jane Oldroyd for assistance with computer graphics. And in particular, I am indebted to three referees for their careful reading of the text and their numerous helpful comments and suggestions, and for putting me right on a number of matters about which I should have known more and better.

GLOSSARY

Acidic/basic	Igneous rocks may be classified according to their content of silica (silicon dioxide). Those composed of more than two thirds silica are said to be 'acidic'. Those low in silica (less than half) are said to be 'basic' or 'ultrabasic'. The others may be classified as intermediate.
Agonistic field	Literally 'a field of contest'. A term used by some writers on the sociology of scientific knowledge to describe, metaphorically, the nature of the scientific community and the way it functions.
Albite	One the plagioclase feldspars (sodium/calcium aluminium silicates), consisting of sodium aluminium silicate ($NaAlSi_3O_8$).
Alpine Orogeny	A series of diastrophic earth movements beginning in the late Triassic and continuing to the present. Manifest particularly in the European Alps.
Amphiboles	An important family of dark-coloured ferromagnesian minerals, of variable composition, the most common being hornblende. The amphiboles characteristically have two cleavage systems, approximately at 120° to one another, a fact useful in diagnosis of hand specimens. (Cf. pyroxenes.)
Analyser	The polarizing device (formerly a Nicol prism, but nowadays a Polaroid filter) that resolves the polarized light after it has passed through a thin section of a rock or mineral under microscopic examination.
Andesine	One of the plagioclase feldspars, which can be regarded as a solid solution of albite ($NaAlSi_30_8$) and anorthite ($CaAl_2Si_2O_8$), with between 50 and 70 per cent albite.

Andesite A volcanic rock made chiefly of andesine and mafic minerals, such as pyroxene, hornblende or biotite.

Anorthite One of the plagioclase feldspars, consisting of calcium aluminium silicate ($CaAl_2Si_2O_8$).

Antecedent drainage A drainage pattern established in a district before earth movements occur there, and retained during and after the movements owing to erosion proceeding at a rate comparable to the earth movements.

Anticline A fold in strata such that the convex side of the folded strata is uppermost.

Aragonite One of the crystalline forms (orthorhombic) of calcium carbonate ($CaCO_3$).

Archaean Formerly used as a term (meaning 'ancient') to name the rocks more ancient than those of the Cambrian system (namely Precambrian), but now used to refer to one of the early subdivisions of the Precambrian.

Asthenosphere One of the outer shells of the earth, immediately below the lithosphere, where the material is weak and susceptible to plastic movement, and makes isostatic adjustments possible. Magmas may be generated in the asthenosphere.

Augite A common green, brown or black member of the pyroxene family of ferromagnesian minerals.

Azoic rocks Literally 'without life'. Formerly used to refer to the stratified Precambrian rocks; later restricted to the Archaean or early Precambrian. Now largely an obsolete term.

Basalt A fine-grained, dark-coloured igneous rock, commonly forming dykes or lava flows. Basalts contain calcic plagioclase and pyroxenes (commonly augite); olivine is also present in many basalts.

Base level The level below which a land surface cannot be reduced by running water: principally sea level. Base levelling is the reduction of a land surface to a base level by erosion.

Batholith A large mass of igneous rock, often consisting of granite. Generally emplaced at depth, but may be exposed by erosion.

Bathymetric survey A survey of the depths of lakes, seas, or oceans.

Benioff–Wadati zone A large fault zone at which subduction is occurring, detectable by the earthquakes generated at the plane of the fault.

Biogeochemistry A branch of geochemistry that deals with the effects of life processes on the distribution and fixation of chemical elements in the biosphere.

Biogeography — The science of the geographical distribution of living organisms.

Biosphere — The zone at or close to the earth's surface where life is found; or all living organisms of the earth.

Boulder clay — An unstratified or little stratified and unsorted deposit of silt and clay in which may be embedded particles ranging in size from sand grains to boulders; produced by glacial action, being left by retreating glaciers.

Bowen reaction series — A series of minerals formed as a molten magma cools and solidifies. The series arises as a consequence of a mineral phase reacting with the melt later in the differentiation process, yielding a new mineral further down in the series.

Calcite — One of the crystalline forms (hexagonal–rhombohedral) of calcium carbonate ($CaCO_3$).

Caledonian Orogeny — A large-scale series of earth movements that occurred during Lower Palaeozoic times and led to the formation of a mountain system extending from Ireland through Scotland to Scandinavia, as well as mountains along the Atlantic coast of North America.

Cambrian — The oldest of the Palaeozoic systems (beginning about 570 million years ago, or possibly less), containing the first surviving fossils of organisms with hard parts.

Carboniferous — The Palaeozoic system between the Devonian and the Permian, beginning about 362 million years ago.

Catastrophe — A sudden violent change in the conditions of the earth's surface.

Catastrophism — The doctrine that some past geological changes have occurred as a result of sudden catastrophes. Loosely linked with the notion that the present conditions on the earth are not necessarily the same as those of the past: that the present is not the key to the past. In an extreme version, the doctrine may take the form that the laws of nature are not always the same.

Chemical geology — The science which seeks to establish the earth's history by considering the chemical reactions which may be supposed to have occurred during its past.

Chlorite — A common group of greenish flaky minerals, being hydrous silicates of aluminium, iron, and magnesium.

Chondrites — Stony meteorites containing small rounded bodies of various materials (chondrules), chiefly iron and/or nickel and peridotite, embedded in a fine-grained matrix of olivine and pyroxene.

Cladistics	A taxonomic procedure used by evolutionary biologists or palaeontologists according to which organisms are grouped according to the number of shared derived characters they have in common.
Cleavage	A tendency to split along definite, parallel, closely spaced planes. Found in both minerals (such as micas) and rocks (such as slates).
Coal Measures	Strata containing coal beds, particularly those of the Carboniferous System.
Consequent drainage	A drainage pattern arising on a land surface as a consequence of earth movements of that surface.
Continental drift hypothesis	The concept of the continents slowly moving over the surface of the earth; made possible because of the weakness of the suboceanic crust, as ice might drift through water.
Controversy-ladenness of observations	The notion that observations may be made in the context of a scientific controversy and be influenced thereby.
Craters of elevation theory	A theory espoused by von Buch and others in the early nineteenth century, which supposed that mountains were formed rather like blisters on the earth's surface by intrusion of material from the interior.
Craton (or Kraton)	A large relatively immobile part of the earth's surface (such as the Canadian Shield).
Cretaceous	The youngest of the Mesozoic systems, following the Jurassic and beginning about 208 million years ago. Contains the well-known chalk unit in Europe.
Crust	Outer layer of the earth. Originally thought to overlie a molten interior, it is now defined in various ways, e.g. lithosphere, sial, material above the Mohorovicic discontinuity, tectonosphere.
Cyclothem	A series of beds of various kinds formed during a cycle of sedimentary deposition.
Deism	Belief in the existence of God, but God not known by revelation; 'natural religion' or religion based on reason. Sometimes referred to as the 'absentee landlord' view of God.
Devonian	The Palaeozoic system lying between the Silurian and the Carboniferous and beginning about 408 million years ago.
Diapir	A piercement fold; or the material that moves upward so as to pierce through overlying strata as a result of differences in density of different crustal materials.

Diapirism	A process in which a mobile rock (e.g. rock-salt) rises up and pierces through a more dense and brittle overlying rock, but also producing arching or doming of that rock.
Diastrophism	The process or processes whereby the earth's crust is deformed, producing continents and ocean basins, plateaux and mountains, folds and faults.
Diluvialism	The doctrine that the earth's surface has been significantly affected by large (catastrophic) floods, especially the Noachian flood.
Diorite	A coarse-grained igneous rock, formed at depth and consisting typically of intermediate plagioclase feldspar and hornblende, with quartz, biotite (brown mica) or a pyroxene in small amounts.
Dip	The maximum inclination of a stratum or other planar feature to the horizontal.
Directionalism	The doctrine that geological changes have proceeded in a 'directed' manner (i.e. in some direction), rather than in a cyclic pattern, or in a wholly random fashion. The notion of some external (perhaps divine) directing agent is not, however, necessarily implied.
Dolerite	A family of dark, medium-grained, intrusive rocks containing minerals such as calcic plagioclase, augite and olivine, and in many cases with ophitic texture. Dolerite is equivalent to basalt in composition.
Dolomite	A sedimentary rock consisting predominantly of calcium magnesium carbonate (the mineral dolomite).
Druse	An old German term for a cavity in a rock which has minerals encrusting the cavity walls.
Dyke (or dike)	A tabular body of igneous rock that cuts across the structure of adjacent rocks or cuts massive rocks. So called because it may appear as a 'wall' of rock after the surrounding strata have been eroded away. If the material of the dyke is weathered more easily than the surrounding rock, a trench may be formed.
Dynamo-metamorphism	Metamorphism caused by rock deformation, chiefly folding and faulting.
Eclogite	A granular metamorphic rock, composed chiefly of garnet (particularly almandine–pyrope), hornblende and pyroxene (particularly omphacite). The bulk composition is similar to that of gabbro, the rock being thought to form at depth under very high pressure.

Embryotectonics	A term used by Argand to refer to the tectonic development of the earth's crust, and particularly its mountain ranges.
Enstatite	A form of pyroxene, varying from greyish white to olive green and brown, consisting of iron/magnesium silicates.
Eozoon	A name given to what was for a time in the nineteenth century believed to be the remains of life in Archaean rocks.
Euhemerism	The view that the gods of mythology were deified humans; or the method of interpreting myths which regards them as traditional accounts of real historical events.
Eustacy	Supposed worldwide changes in sea level.
Eutectic point	The minimum melting temperature of a mixture, the components being in such proportions as to produce that minimum melting point, provided that the components do not form solid solutions.
Facies	The general appearance of a sedimentary rock body, with respect to type of composition, bedding, fossil content, etc., indicative of its conditions of formation. Also used in relation to igneous rocks where some part of the rock mass differs from the rest. For example, a granite might be said to show a porphyritic facies at its margins. The term 'metamorphic facies', denoting particular temperature–pressure regimes in metamorphism, is also used.
Fault	A fracture of part of the earth's crust along which there has been relative displacement of the ground or crust on the two sides of the plane of fracture.
Feldspars (formerly often written *felspars*)	A common group of rock-forming minerals, being aluminosilicates of calcium, sodium and potassium. The main types are orthoclase (potassium aluminium silicate) and plagioclase (a series from calcium aluminium silicate to sodium aluminium silicate, divided, arbitrarily, into the minerals anorthite, bytownite, labradorite, andesine, oligoclase and albite).
Fixism	A doctrine, associated for example with Stille, asserting that the positions of the continents remain fixed relative to one another through geological time.
Floetz rocks	The name give by Werner to parallel-layered, approximately horizontal rocks. Roughly equivalent to the sedimentary strata from the Old Red Sandstone upward.
Fluvialism	The doctrine that the major features of the earth's topography have been produced by the erosive action of running water.

Foliation — The laminated or banded structure found in metamorphic rocks arising from the segregation of minerals into layers of different composition. The term is also used more loosely to refer to the parallel fabrics (schistosity) of metamorphic rocks).

Foreland and backland — Two terms used by Suess. In folded mountain ranges, one may distinguish between a rigid unyielding unfolded mass (the 'foreland'), and the zone of folding and the zone of diminishing action, where the folding gradually dies away or ends in a fault (the 'backland'). The term 'foreland' can also refer to the block toward which geosynclinal sediments move when compressed; or the region in front of a series of overthrust sheets.

Formation — A unit in lithostratigraphy consisting of a succession of strata useful for mapping or description. Lyell defined a formation as any assemblage of rocks having some character in common, whether of origin, age or composition. Dana defined a formation as a group of related strata formed in a geological period.

Fossil — Anything dug from the ground; but in its modern meaning the term refers to the remains of animals or plants preserved in the earth's crust by natural causes.

Foyaite — A type of nepheline-syenite.

Gabbro — A dark, coarse-grained, basic igneous rock, equivalent in composition to the fine-grained basalt. Contains calcic plagioclase, and minerals such as augite and olivine.

Gaia — One of the ancient Greek gods. Also a name used by Lovelock as a personification of his theory/hypothesis that in some respects the earth behaves like a living, self- repairing, organism.

Genetic theories of the earth — Theories which envisage the earth changing over time in some predetermined way, or according to certain determinable process operating in accord with chemical and physical principles.

Geodesy — The branch of applied mathematics used to determine the figures and areas of large portions of the earth's surface, and the figure of the earth as a whole.

Geognosy — The term favoured by Werner for the scientific study of the earth. It was intended to be used in contradistinction to the term geology, which was considered more speculative. For Werner's school, it indicated an emphasis on the study of

	rocks and minerals, and attempts to develop stratigraphy on lithological rather than palaeontological principles.
Geomagnetic reversals	Reversals of the earth's magnetic field which have apparently occurred from time to time.
Geosyncline	A large and long 'trough', subsiding over a long period of time, in which sediments have accumulated. The strata of many geosynclines have become subsequently folded, forming mountain ranges.
Gneiss	A coarse-grained metamorphic rock, roughly granitic in composition, with layers or lenticles of granular minerals alternating with layers or lenticles of platy materials. Some gneisses are marked by alternating light-and dark-coloured layers, respectively rich in quartz and feldspar and in dark minerals such as biotite (brown mica) or hornblende (amphibole).
Gnosticism	The system of beliefs of Gnostics: certain heretical sects amongst early Christians who claimed to have superior knowledge of spiritual things, and sought to interpret sacred writings by means of mystical philosophy.
Gondwána or Gondwánaland	A hypothetical ancient continent, including India, Australia, Antarctica, and parts of South America and southern Africa, which supposedly fragmented and drifted apart in post-Carboniferous time.
Graded river/slope/profile	A river has a graded profile when the slope of its bed at every point is just sufficient to enable the water to carry its load of sediment, neither depositing material nor eroding the river bed.
Granite	A family of generally light-coloured, coarse-grained igneous rocks, formed at a depth; rich in quartz and potassic/sodic feldspar; usually mica is also present.
Granitization	The conversion of sediments into granite by some, often unspecified, process.
Graptolites	A group of extinct invertebrates, found as fossils in Palaeozoic rocks. The organisms were colonial animals, with individual polyps living in small cups (thecae) arranged along a common stem. In appearance, the fossils may be like part of a fretsaw blade. They were originally so called because they sometimes resemble writing on a slate.
Greenalite	Iron silicate, occurring in Precambrian ironstones.
Greywacke	A loose term derived from the German word *Grauwacke*, referring to a dark sandstone or grit with angular fragments of quartz and feldspar in a more fine-grained 'clay' matrix.

Guyot	A submarine mountain, the top of which is a comparatively smooth platform.
Gypsum	A common monoclinic mineral, hydrated calcium sulphate, used in the manufacture of plaster of Paris.
Haematite	A red, reddish brown or blackish oxide of iron, being an important form of iron ore.
Hercynian Orogeny	A series of late Palaeozoic diastrophic movements, beginning perhaps in the late Devonian and continuing to the end of the Permian. Produced mountain ranges in the wooded regions of Middle Germany (Saxony).
Historicist accounts of the earth	Accounts based on attempts to determine the history of the contingent events and processes whereby the earth has reached its present state. Hence geological phenomena are explained in historical terms.
Humboldtian science	A term suggested by Susan Cannon for the kind of work undertaken by Alexander von Humboldt. Focusing on astronomy and on the physics and biology of the earth from a geographical standpoint, it sought to discover quantitative relationships between many variables. Typically, data were presented on maps with (iso)lines representing equal values of some variable. Emphasis was placed on quantitative determinations, using instruments.
Hypabyssal rocks	Igneous rocks that have risen from the depths of the earth as magma, but have solidified chiefly as minor intrusions before reaching the surface.
Ichor	A term proposed by Sederholm for a fluid (juice or liquor) capable of converting rocks into granite, being itself derived from a granitic magma.
Igneous rocks	A rock 'formed by fire' – i.e. by solidification from a molten or partly molten state.
Isoseismal lines	Lines drawn on a map, showing places on the earth where an earthquake shock is of the same intensity.
Jurassic	The highly fossiliferous Mesozoic system, between the Triassic and the Cretaceous, beginning about 213 million years ago.
Laccolith	A mass of igneous rock, approximately mushroom-shaped in section, thrust up through sedimentary beds, and giving a dome-like form to the overlying strata.

Laurasia	A hypothetical continent in the northern hemisphere, which supposedly broke up at about the end of the Carboniferous to form the present northern continents.
Limestone	A sedimentary rock, consisting chiefly of calcium carbonate.
Limnological objection	An objection to the idea that valleys are carved by the rivers that flow in them, based on the occurrence of lakes in valleys.
Liquation	The separation of the components of a mixture when that mixture is melted, two immiscible liquids separating from a common solution, for example a magma.
Lithology	The scientific study of rocks and stones (not their fossil contents); or the study of rocks according to their appearance in hand specimens, not thin sections.
Lithosphere	The earth's crust; the outermost portion of shell of the globe.
Loess	A fine-grained, homogeneous, yellow-grey, wind-blown deposit, consisting chiefly of silt.
Love waves (Q waves)	A transverse wave propagated along the boundary between two rigid media; or one of the possible surface seismic waves.
Loxodromism	The tracing of or movement in a loxodromic course; sailing to a rhumb line so that a constant direction is maintained in the presence of a steady wind. Von Humboldt used the term to refer to mountain ranges that maintained a constant direction on the curved surface of the earth.
Mafic	A term used to refer to rocks composed chiefly of minerals containing magnesium and iron (magnesium and ferrum); or to the dark, basic, ferromagnesian minerals.
Magma	The molten fluid or pasty mass generated in the earth's interior from which igneous rocks are formed by crystallization. Includes volatile fluxes and residual liquors which may have escaped during or after consolidation. So the chemical composition of a rock may be different from the magma from which it developed.
Magmatic differentiation	The formation of different igneous rocks from a parent magma.
Magmatic stoping	The emplacement of magmatic masses involving shattering of rocks of the roof or walls of a magma chamber, the sinking of the shattered blocks into the liquid magma, and the consequent enlargement of the chamber.
Magmatist	An upholder of the idea that much of the world's granite has, at some time, existed in the form of molten magma.

Marine terrace	A shore platform, produced by marine erosion.
Mechanical philosophy	The view that all phenomena are ultimately explicable in terms of matter and motion; that mechanical models or analogies are more intelligible and scientifically useful than other forms of explanation; and that the universe may be regarded as a giant mechanism.
Mesozoic	An era of geological time, made up of three systems: Triassic, Jurassic and Cretaceous.
Metamorphism	Changes in the texture or composition of a rock produced by heat, pressure, moisture or the chemical environment, without the occurrence of complete melting. Schists and gneisses are important examples of products of metamorphic action on a regional scale.
Metascience	The study of science from a historical, philosophical or socio-political perspective, or some combinations of these.
Metasomatism	The processes by which one mineral or component of a rock is replaced by another of different chemical composition owing to reactions set up by the introduction of material from external sources.
Mica	A group of minerals with sheet-like structures arising from perfect cleavage in one plane; particularly biotite ('brown mica') and muscovite ('white mica').
Migma	A material having many or all the characteristics of magma, but derived from the process of granitization.
Migmatist	An upholder of the idea that much granite has been produced by granitization rather than by direct crystallization from a molten magma.
Migmatite	Composite rocks such as gneisses produced by the injection of granitic magma between the foliae of a schistose rock.
Mobilism	The doctrine that the positions of the continents relative to one another have changed over time.
Mohorovicic discontinuity	A discontinuity which separates the earth's crust and mantle, and which produces seismic discontinuities.
Mosaic geology	A type of geology which seeks to reconcile observations of the earth's crust with the account of earth's origin and early history as described in the Old Testament, and supposedly by Moses.
Natural theology	A theology based on reason and/or observation rather than revelation.
Nepheline	A rock-forming mineral: sodium potassium aluminium silicate; hexagonal system.

Neptunism	The doctrine associated with the theories of Werner that all the rocks of the earth's crust were originally deposited from water, either by chemical precipitation, or by erosion and deposition of sediment from the supposedly precipitated rocks.
New Red Sandstone	A series of red sandstones formed under desert conditions, constituting much of the Permian and Triassic systems in Britain and elsewhere in Europe.
Nicol prism	A device invented by William Nicol to produce plane-polarized light. A crystal of calcite is cut and then cemented together again in such a way that only the 'extraordinary' ray arising from the double refraction of the calcite is transmitted; and this ray is plane polarized. The nineteenth-century petrographic microscopes used two Nicol prisms, one as 'polarizer' and the other as 'analyser'.
Normal fault	A fault in which the displacement of strata is downward on the fault plane. This arises when the overall maximum stress is vertical rather than horizontal. (Cf. reverse fault.)
Obsidian	A dark volcanic glass with conchoidal fracture and chemical composition similar to that of rhyolite.
Old Red Sandstone	A succession of conglomerates, red shales, and sandstones forming the continental facies of the Devonian system in Europe.
Olivine	A group of green, brown-green or yellow-green orthorhombic minerals (iron magnesium silicates), with glossy fracture, commonly found in basic igneous rocks.
Oolite	A limestone mode of small cemented spherical granules of calcium carbonate, somewhat resembling fish roe in appearance.
Ophitic texture	A term applied to the texture of dolerites in which the feldspar (plagioclase) crystals are enclosed within grains of pyroxene, chiefly augite.
Ordovician	The Palaeozoic system between the Cambrian and the Silurian that began about 505 million years ago.
Orocline	A mountain arc owing its form to horizontal crustal displacement after the mountain range was initially formed.
Orogeny	The formation of mountain chains, particularly by folding and thrusting rather than by erosion. A period or process of mountain building.
Overdetermination of theories	A state of affairs in which the empirical evidence demands that a hypothesis be true, but no explanation is forthcoming

as to why it is true. For example, geological evidence might suggest the hypothesis that there has been an Ice Age, but a satisfactory theory as to why the Ice Age occurred might not be available.

Palaeontology The scientific study of fossils.

Palaeozoic The period of geological time following the Precambrian, encompassing (in Europe) the Cambrian, Ordovician, Silurian, Devonian, Carboniferous, and Permian systems, and beginning about 570 million years ago.

Palaetiological A term coined by Whewell in reference to sciences which seek to explain past phenomena in terms of existent principles of cause and effect; and to explain present states of affairs in terms of anciently acting causes.

Pallasite A type of meteorite consisting of approximately equal parts of metal and silicates.

Pangaea A hypothetical great continent, which, according to Wegener's theory of continental drift, fragmented to produce the present continents.

Panthalassa A hypothetical great ocean.

Pantocrator The supposed divine measurer and creator of the cosmos.

Peridotite A general term for ultrabasic plutonic rock, rich in olivine and other mafic minerals, but deficient in feldspars.

Permanentism The doctrine that the relative positions of the continents have remained largely unchanged through geological time.

Permian The youngest of the Palaeozoic systems, following the Carboniferous and beginning about 286 million years ago.

Petrogenesis The branch of petrology which studies the origin of rocks, especially igneous rocks.

Petrography The branch of science dealing with the systematic description and classification of rocks.

Petrology The branch of science dealing with the study of rocks, including their classification, origin, and alteration.

Physico-mythology The study of myths which regards them as mythical or allegorical records of actual physical events; for example, a volcanic eruption might be remembered in a mythopoeic culture as a battle between rival gods.

Physico-theology A theology interwoven with a science; or vice versa. For example: Mosaic geology.

Plagioclase The series of sodium/calcium aluminium silicates forming an important group of feldspars: albite, oligoclase, andesine, labradorite, bytownite and anorthite.

Planetesimal — A small solid body orbiting the sun like a mini-planet. According to some theorists, planets were formed by the aggregation of planetesimals.

Plate tectonics — Theory of the broader structural features of the earth which supposes that there are large crustal 'plates' on the earth's surface, capable of lateral movement by virtue of the plasticity of the underlying material and convection currents in the earth's mantle.

Pleistocene — The geological system immediately preceding the recent period when much of the earth was glaciated; beginning about 1.6 million years ago.

Plutonic — Relating to the earth's depths. Plutonic rocks are thought to have been formed by the slow cooling and consolidation at depth, with the consequent formation of coarsely crystalline rock.

Plutonism — The doctrine that rocks such as granite have been formed by the cooling and consolidation of magma deep under the earth's surface.

Polarized light — Light in which the electromagnetic vibrations are 'constrained'. Thus, in plane-polarized light, the vibrations are confined to a single plane.

Polarizer — The Nicol prism (or Polaroid filter) of a petrographic (polarizing) microscope through which light passes before reaching the thin section being examined. It is that part of the microscope that produces the polarized light required for the examination of rocks or minerals in thin section.

Polarizing microscope — A microscope used in petrography, provided with polarizer and analyser, and with a rotating stage on which thin sections can be mounted for examination. Minerals capable of doubly refracting light display characteristic colours and appearances when viewed in this way, and these features can be diagnostic.

Porphyry — A medium-or fine-grained igneous rock, consisting of larger, well-formed crystals in a matrix or groundmass of finer grain.

Positivism — A system of philosophy which regards facts and observable phenomena as the basis of knowledge.

Precambrian — The immense period of geological time prior to the Cambrian; also the rocks formed in that time. Precambrian rocks were formerly thought to be without fossils and so antedated the beginning of life on earth; but fossil remains without hard parts have now been found in the Precambrian.

Precession of the equinoxes — The movement of the equinoctial points, namely the apparent positions of the sun against the background of the stars at the two times of the year when night and day are of equal length. The movement is due to the slow cyclic movement of the earth's axis of rotation relative to the stars, over thousands of years. A single cycle was called the 'great year' by the ancients and was thought to take about 26,000 years.

Primary — An obsolete term, formerly used to designate what is now called the Precambrian; then extended to include the Palaeozoic; and then used to refer to what is now called the Palaeozoic. (Thus at one time the Palaeozoic, Mesozoic and Tertiary eras were called Primary, Secondary and Tertiary.)

Primitive rocks — An obsolete term, formerly used to refer to rocks that were believed to be those first formed on the earth – such as gneisses.

Propylite — A term used to refer to hydrothermally altered andesites, etc., where the feldspars are replaced by such minerals as albite, epidote, calcite, chlorite, and the mafic minerals by chlorite, epidote, pyrite, and iron ores.

Punctuated equilibrium theory — A version of evolutionary theory proposed by Gould and Eldredge, which holds that species may remain unchanged for long periods (stasis), and then evolution may proceed more rapidly at times – or in 'bursts'.

P waves — Seismic 'pressure waves', in which the vibratory motion is parallel to the direction of transmission of the wave.

Pyrites — Iron pyrites ('fool's gold'). Brass yellow, cubic crystals: FeS_2. An important ore of sulphur.

Pyroxene — A family of generally dark-coloured ferromagnesian minerals, the commonest being augite, found in basic igneous rocks. The pyroxenes typically have two cleavage systems, approximately perpendicular to one another, which feature is useful for recognition in hand specimens. (Cf. amphiboles.)

Quartz — The commonest naturally occurring form of silica (silicon dioxide). Well-shaped clear crystals are called 'rock crystal'. Quartz is an essential component of acidic igneous rocks, and sand grains are commonly fragments of the mineral.

Quaternary — The geological time of the post-Tertiary era. Subdivided into the Pleistocene and Recent (Holocene) epochs.

Raleigh waves — *R* waves. A surface seismic wave propagated along the plane surface of a homogeneous elastic solid.

Remanent magnetism — The magnetization retained after the magnetizing field is reduced to zero.

Reverse fault — A fault arising from the action of compressive forces, such that the movement of strata on one side of the fault has been upward. (Cf. normal fault.)

Rhyolite — A fine-grained acidic igneous rock equivalent in composition to a granite.

Rift valley — Valley produced by subsidence of a strip bounded by two parallel rifts or strike-slip faults.

River terrace — A plain and escarpment in a river valley, together forming a terrace. There may be a series of such structures in a river valley, resulting from periodic uplifts of the land surface.

Rock basins — Depressions or basin-like excavations in the solid rock, sometimes occupied by lakes, and in many cases formed by the action of glaciers.

Salic — A term used to refer collectively to the 'acidic' minerals, rich in silica and/or alumina, such as quartz and the feldspars. In rocks, such minerals are called felsic.

Salt, Sulphur, and Mercury theory — A theory popular in the sixteenth century and early seventeenth century that matter contains, or is perhaps made up of, three property-conferring 'principles'. Sulphur, for example, was supposed to be responsible for flammability.

Saving the appearances — A medieval term for a view of scientific theories, subsequently referred to as 'instrumentalism'. On this view, a theory need not describe reality, or the way the world actually is. It is sufficient that the theory should allow prediction, and the intelligible coordination of observations. Ptolemy's system of epicycles and deferents, used to describe and predict planetary motions, offers a good example in that it did not purport to describe real physical entities. It was a kind of mathematical (geometrical) calculating device.

Schist — Medium-to coarse-grained metamorphic rock with strong foliations, fissile from the preferred orientation of abundant equidimensional mineral grains such as micas (flakes) or amphiboles (rods).

Schistus — A loose term, used by Hutton, to refer to a variety of rocks from schists to greywackes.

Sea-floor spreading — The process whereby, according to plate tectonic theory, basaltic matter rises from the earth's interior along mid-oceanic ridges and spreads laterally forming additions to the intervening sea floor as continents move apart from one another.

Secondary rock — At one time used as an alternative term for Werner's *Floetz* series (the stratified rocks). Also used for Mesozoic rocks. Now obsolete.

Sedimentary cycling — The notion that the material of sediments is cycled between marine and terrestrial environments, sometimes in part via the atmosphere or living organisms.

Seismic — Pertaining to earthquakes or earth vibrations.

Seismogram — A record made by a seismograph.

Seismograph — An instrument which records seismic waves.

Seismometer — An instrument that detects seismic waves.

Shield area — A continental block of the earth's crust that remains stable over a long period of time.

Sial — The upper zone of the earth, underlying continents, composed chiefly of silica and alumina. Ranges from granitic at the top to gabbroic at the base.

Siderite — Chalybdite. A mineral made of iron carbonate, with some magnesium and manganese.

Silica — Silicon dioxide, SiO_2.

Silurian — The Palaeozoic system between the Ordovician and the Devonian, beginning about 438 million years ago.

Sima — The lower part of the earth's crust, underlying the sial of the continents, with rock of basaltic or peridotitic composition, containing particularly silica and magnesia.

Soret's principle — If the temperature varies from one point to another in a solution, the concentration of the solute is higher where the temperature is lower. The system will only achieve equilibrium when the concentration is everywhere appropriate to the temperature.

Stoicism — A philosophy founded by Zeno about 300 BC. Emphasized acquisition of happiness through virtue, which was regarded as the prime good. The Stoic doctrine of the universe was pantheistic, and the cosmos was held to be regular in its behaviour by virtue of a 'world-reason'.

Stoping — The ascent or descent of blocks of crust or country rock into an enclosing magma.

Stratigraphy — That part of geology which studies the formation, composition, sequence and correlation of the stratified rocks of the earth's crust.

Strike	The direction or bearing of a horizontal line on an inclined bed or other structural plane, such as a fault, at 90° to the direction of the dip. Equivalent to a contour line on a bed, etc.
Strike-slip fault	Transcurrent fault. A fault in which the overall movement is in the direction of the strike of the fault.
Subduction	The descent, according to plate tectonic theory, of material from the ocean floor beneath the continents and into the earth's interior.
Superposition, principle of	The principle that in a sequence of strata the lower ones are older and the upper ones younger. This principle may not always apply where strata have undergone deformation and thrust faulting.
S waves	Seismic 'shear waves', in which the vibratory motion is perpendicular to the direction of transmission of the wave.
Syenite	A coarse-grained igneous rock of intermediate acid/base composition, containing alkalic feldspar such as orthoclase, and some dark minerals such as hornblende or brown mica. Small amounts of quartz and plagioclase may also be present.
Syncline	A fold in strata such that the concave side of the folded strata is uppermost.
Tectonosphere	The zone within the earth where crustal movements occur.
Terrane	The area or surface over which a particular rock or group of rocks is prevalent; a connected series, group or system of rocks or formations; or a subsection of a landmass which has been joined to other parts of the mass as a result of plate-tectonic activity, i.e. a 'micro-plate' according to plate tectonic theory.
Tertiary	A name that has survived from the old division of rocks into Primary, Secondary and Tertiary, now used to refer to the following epochs: Palaeocene, Eocene, Oligocene, Miocene and Pliocene. Coming after the Cretaceous, the Tertiary period began about 65 million years ago, and lasted about 63 million years.
Tethys	An elongated east–west seaway that supposedly lay between Europe and Africa and extended across southern Asia in pre-Tertiary time.
Theory-ladenness of observations	The notion that what is observed is dependent on the theories of the observer: there is more to seeing than meets the eyeball.
Thrust fault	A low-angle reverse fault.

Till — Unsorted, non-stratified sediment carried or deposited by a glacier.

Tomography — A term from medicine used in geology to refer to the technique of building up a three-dimensional picture of the earth's interior by analysis of seismic waves by means of seismographs located at different positions on the earth's surface; or the mapping of the three-dimensional structure of the earth from the analysis of seismic data.

Trachyte — An extrusive volcanic rock composed of alkalic feldspars, with biotite, hornblende or pyroxenes. The extrusive equivalent of a syenite.

Transcurrent fault — See strike-slip fault.

Transform faults — A set of strike-slip faults developed in association with the extrusion of magma from an oceanic ridge. (See Figure 11.12e.)

Transition series — In the nomenclature of Werner, rocks (now designated as Cambrian, Ordovician and Silurian) made of limestones, traps and greywackes, and constituting the first orderly deposits, supposedly formed world-wide from a universal ocean; lying between the crystalline 'primitive rocks' such as granite and gneiss and the lowest *Floetz* rock (Old Red Sandstone).

Trap — A term derived from German and Swedish words meaning 'step' or 'stair', used to refer to dark-coloured dyke and flow rocks, chiefly basalt. (In some places, successive lava flows may give the appearance of steps or stairs.)

Triassic — The first system of the Mesozoic era, between the Permian and the Jurassic systems, and beginning about 248 million years ago.

Unconformity — A surface of erosion separating younger from older strata, with the older beds generally dipping more steeply than the younger ones.

Underdetermination of theories — The state of affairs when more than one theory (or hypothesis) can be used to account for some observations, and the observations are insufficient to determine which theory (or hypothesis) is to be preferred.

Uniformitarianism — A term with various meanings, including the methodological principle that 'the present is the key to the past', the assumption that the laws of nature are uniform, the hypothesis that conditions and processes in the past have been much the same as at the present, and that geological changes occur slowly and steadily, without sudden discontinuities or 'catastrophes'.

Vicariance biogeography — The study of the separation of populations of organisms by the formation of geographical barriers and their subsequent differentiation by evolution; or the study of speciation arising from the formation of geographical barriers (a process called vicariance).

Vulcanist — An upholder of the doctrines of Hutton regarding the origin of plutonic rocks; and of the view that basalts were igneous in origin, not deposited from solution as envisaged by Werner.

Walther's law — Contiguous sedimentary facies in a vertical sequence were laterally contiguous in the area of deposition. That is, the vertical sequence at any given locality is made up of lithofacies that have migrated laterally. (For further explanation, see p. 335.)

Whig historiography — Type of history written by certain nineteenth-century historians who were supporters of the Whig party, and saw history as essentially progressive, and culminating in the present. Used for history of science writing that regards present knowledge as correct, and makes judgements about past science by comparison with modern knowledge. Associated, therefore, with historiographical anachronism.

INTRODUCTION

In 1991, at the age of 55, I saw my first active volcano. It was in the remote island of Tanna, in the archipelago of Vanuatu in the South Pacific. Well, I had seen other volcanoes before: in New Zealand, Japan and Italy, and had climbed a few. And there were plenty of other hills and mountains I had walked over – such as those of the Massif Central in France or Arthur's Seat near Edinburgh – that geologists, guidebooks, maps, and geological theory decreed were extinct or dormant volcanoes. After seven years' sojourn in New Zealand, boiling mud pools, all smelly with sulphurous fumes, were quite familiar, even homely. But Tanna was something utterly new to me. Only one word seemed even remotely adequate: the sight was awful. Literally, it filled me with awe.

Visitors to Mount Yasur make an easy climb up the side of the lava cone as dusk is falling. Some of the rock is so spongy that one can easily hold quite a large boulder above one's head with one hand. The stench becomes overpowering if the wind blows towards one, but unexpectedly it is the ears that receive the most astonishing sensation. It is as if one is hearing a huge surf breaking against a line of cliffs – but actually it is molten lava sloshing around in the interior of the mountain. At the rim of the crater, there is not so much to see while there is still light: just some small glowing spots at the bottom of the abyss below, and an occasional loud rumble accompanying an obscurely visible fountain of lava. But as darkness falls, the whole scene changes. After the short tropical twilight, a glowing cauldron of molten rock is seen which lights up the huge amphitheatre of the mountain. Then, from time to time, as the ground trembles and the sloshing sound intensifies, there appears a great fountain of red-hot matter that makes the world's finest fireworks display seem as nothing. But the thought, feeling or sensation that was most central to my mind was that I was able to peer into, or just catch a glimpse of, the earth's interior – apparently a seething mass of molten matter. And really I knew nothing about why it all was so. Perhaps I knew something in an abstract sense, but not the way one knows one's own personality, one's home, how a bicycle works, the way to the office or how to write a letter. Mount Yasur revealed a deep mystery at the heart of things, the nature of our world and of our universe.

Tanna is one of the last fertile and comfortable spots in the world that is hardly touched by Western 'civilization'. In 1991, there were just eleven European residents. The island is fertile and green, and the coral sea teems with fish. Most of the locals have some contact with the Western world, through small-scale trading, a daily plane service, radio, schools and a small hospital. But some choose to live a traditional life, with nothing but grass skirt or penis guard by way of clothing, and pigs and yams as currency. We saw a tribal dance, with chanting and stamping of the ground in a special rhythmic way. The whole village joined in, and it was interesting to watch the children seeking to emulate their elders in the dance. They were being acculturated. The dancing, and no doubt other rituals that we were not privileged to observe, would have served to weld the villagers into a cohesive and successful social unit.

Among the forces bringing about that cohesion would have been the forces of myth. Specifically, the Tannese believe the following story about the origin of Mount Yasur:

> A homeless wanderer looked all over Tanna for somewhere to settle down. Near Sulphur Bay he found two women, Savai and Mounga, making laplap and weaving baskets, as women are inclined to do. Seeing an empty hole nearby the [site of the] volcano, he asked if he might rest his weary self there. Once in the hole he writhed and twisted to keep warm; as he did [so] ash began to form and the frightened girls ran off. When they returned to retrieve their laplap the newly created ash enveloped them. They remain to this day in the volcano as two of the smaller craters (Douglas and Douglas, 1990, p. 91).

Obviously, such a legend does not strike us as plausible. But I suppose it made sense to the villagers in the jungles: at least it explained the unknown in terms of things that were known to them.

Tribal societies typically have hundreds of such legends and myths, which have the vital cognitive and social role of helping to make the unintelligible intelligible, sometimes imparting information of a practical or moral kind, and generating social cohesion. No little boy with his penis guard could say that he thought the legend absurd. This would invite ostracism. Worse, he might be thought mad for trying to think the unthinkable, and would be treated accordingly.

I mention these matters at some length, partly because (as the reader will have gathered) I was profoundly impressed by my experiences of a 'live' volcano at Tanna. (The notion of living and dead volcanoes survives, it may be noted, in our everyday speech.) In the context of a remote South Sea island, my experiences at Tanna gave me, I think, just a glimpse of what life might have been like, and how people may have thought, in the days of prehistory, in a primal culture, before the invention of writing, before science, before nearly all the technology we use today was devised.

The earth, of course, is the stage for all human activity, and myths and legends about the earth abound. Myths continue to be of interest to geologists, in that they may sometimes offer clues about ancient events of a geological nature that are otherwise lost to the historical record (Vitaliano, 1973; Greene, 1992). For the historian and philosopher of earth science, however, they are of even greater interest, for it is in the change from mythopoeic culture(s) to that of science that the first stages in the history of ideas about the earth are to be found. So this will be the first step of our enquiry.

Thinking About the Earth is a book whose title expresses its subject matter rather exactly. It is about ideas concerning the earth, or rather the history of ideas about the earth. This is not quite the same as saying that it is a book about the history of geology, although it is geologists' ideas that chiefly concern me. One could write a history about cosmologists' thinking about the earth; or agriculturists' ideas; or poets' thoughts. I leave such tasks to other authors. Even with such limitations, a book of the present dimensions cannot present a comprehensive history of earth science. As my remarks about Tanna indicate, there are ways of thinking which should not be forgotten in a survey of what people have thought about the earth. Nevertheless, most of what I have to say will be concerned with the geological tradition of Western science. (Regrettably, I am able to write on the tradition of Western thought only.)

So I shall not attempt to provide a technical history of geology *per se*. But, as my task is chiefly to provide an account of ideas within a particular science, technical matters necessarily form a significant part of the whole. In the course of the historical account, we encounter numerous questions. What is the earth? How did it come into being? What is it made of? How old is it? Has it always been the same in the past as it appears at the present? Does it have a history? If so, how can that history be known? What is a science of the earth all about? Is it a form of cosmogony? Or physics? Or chemistry? What are the important things to observe, for the person who wishes to study the earth? What makes for geological analysis? And synthesis? What role can scientific instruments and experimentation play in thinking about the earth?

There seems to be no end to the questions that might be asked. I do not hope to answer them all satisfactorily. What I offer is a broad-brush contribution to the study of the development of ideas and theories about the earth. It is the product of extended personal study of the history of geology,[1] but is also intended as a general synthesis of what historians of science and scientist-historians have written and thought about the history of that science. This is reflected to some extent in the attention given to different topics. Rather more has been written on the early history of geology than on recent topics, and this inevitably shows through in what I have written. Even so, I have attempted to bring matters round to some interesting recent 'holistic' ideas that are emerging in geology, and attracting public attention; and I include discussion of some topics that have received little attention in most histories of geology.

Published information on much of the history of geology is patchy. Little has been written on the later histories of mineralogy and petrology and not a great deal on geophysics. There is little on the history of sedimentology. Geomorphology and palaeontology are somewhat better served, but studies of the history of stratigraphy and map-work are very scattered, as dictated by the nature of the subject. It is difficult indeed to write a unified account of all the many separate investigations that have been carried out over the years on the strata of different parts of the globe, and we know rather little in the West of the history of researches in countries such as China or Russia. Nevertheless, I have attempted to take examples from as many parts of the world as possible.

All sciences have their special features and particular intellectual problems. So far as geology is concerned, the major problem is that it endeavours to understand the past, but can only do so through the lens of the present. It is, to use a word coined by the nineteenth-century historian, scientist and philosopher, William Whewell, a palaetiological science. That is, it deals with causes that operated in the past, rather than the present. Whether those causes are the same as those operating now or different is a constant problem, both for the geologist, and for the philosopher of geology. Thus, in a certain sense, the problems of the geologist are analogous to those of the general historian or even the anthropologist.

So although geologists do not have to deal with the vagaries of human nature as the object of enquiry, their task is a kind of hermeneutic exercise. It has been a common metaphor that strata are similar to a book, which has to be 'read'. Inevitably, many of the pages are missing. Reading always involves interpretation, and the geologist has to become skilled at this task. But the history of geology should be a constant warning to those who think that the task is simple or that we have got it right now. If the history of ideas about the earth can teach us one thing, it is that interpretations are constantly changing. So to suppose that the interpretations we presently favour are correct for all time is an act of hubris.

Partly for this reason, in the later chapters I consider some ways of thinking about the earth that seem to me to be interesting, but which may be incorrect. However, their consideration has a great advantage for me as I approach my task as an author, for I am thereby able to bring my account full circle, so to speak. For, even if we do not end with myth, we find 'respectable' modern ways of thinking that regard the earth as a quasi-organic entity, as did people of mythopoeic cultures, or the philosophers of the ancient world, and as did writers well into the seventeenth century.

Thinking more particularly about the modern period (nineteenth and twentieth centuries), and leaving aside economic geology (and funding), the general problem for the geologist is fundamentally threefold: first to represent the earth, its structure, and behaviour by some means or other; then to tell its history in some way; and finally to get other people to agree with one's representation, history or theory. These have not always been the ways in which people have thought about the earth, and they are not the ways of the poet or the farmer. Formerly, the earth may chiefly have been regarded as an object of awe, reverence and worship, or a plaything of the gods. Or it was simply a source of interesting things that might be classified and placed in a cabinet as objects of aesthetic gratification. Certainly, it has not always been thought of as having a history such that it was formerly different from the present. But, in the modern period, those three goals seem to have been to the fore.

Representation of the earth and its constituent objects typically involves the preparation of geological maps and sections, or other kinds of visual aid that convey large amounts of information in a small compass. The first 'geological' maps simply entered the locations of different kinds of rocks or minerals – usually ones of economic importance such as metalliferous ores, coal or limestone – on topographic maps. But at the beginning of the nineteenth century there began the process of the classification of the earth's strata according to their fossil contents where possible, or sometimes according to their lithologies, depicting the different divisions in different colours or by means of other signs, and entering all the information on a map. The preparation of a map might seem a fairly routine and atheoretical enterprise, but this was far from the case.[2] The divisions or classifications that were employed in mapping did not come provided with some transcendentally given badge of truth. They were human creations, and as such were fought over in some cases with something approaching military zeal.

The earth is, of course, a huge place, and obviously it cannot be squeezed into a meeting room for discussion. In a lecture room, it is representations of the earth that are considered; and the same goes for verbal or pictorial representation in books. Maps, sections, diagrams, photographs, or specimens are displayed and debated, and from them ideas about the geological history of the earth can be educed. The maps and sections are human artefacts; but in a certain sense so also are the specimens. They are abstracted – or extracted – from their original situations, and are commonly presented so as to substantiate a particular point of view. Ideas or theories about the earth and its history are commonly deployed in the making of the maps or sections.

Readers of this book will no doubt observe that it is illustrated rather freely for one dealing with the history of ideas. This is because geologists typically help to represent their thoughts with the help of what Rudwick (1976) has called a 'visual language'. Geological ideas are indeed conveyed particularly effectively with the help of diagrams, and the same is true of geography. It is no coincidence, I suggest, that visual signals form so great a part of the language of the earth sciences. The earth is, as I say, a large place. It is often thought about most easily with the help of diagrams which reduce the large to the small or great lengths of time to manageable representations. So figures are as essential for my purposes as they are for geologists.

Much of what has been written on the history of geology in the twentieth century has to do with the establishment of the 'plate tectonics paradigm' (see Chapter 11). Like other historians of the earth sciences, I treat this topic rather liberally, although perhaps at the expense of other matters no less interesting. My justification for treating matter in Chapter 11 that is quite readily available elsewhere is twofold, apart from the obvious fact that no history of ideas about the earth that comes into the twentieth century can ignore the plate-tectonics revolution. First, it allows me to consider briefly in Chapter 13 the question of truth in science. Second, it allows me to look towards even grander theoretical syntheses, which may or may not turn out to be acceptable to science in the long run, but which are none the less of the highest interest. It is scientists' task to determine whether the ideas are or are not correct. But we, as bystanders (metascientists), can interest ourselves in what is going on currently, and see it as part of a much longer and broader sequence of thoughts. I therefore crave readers' indulgence in taking them into somewhat insecure terrain towards the end of the book, where once again we may perhaps be approaching the realm of myth. Elsewhere, I play a straighter bat on an easier wicket, and 'tell the story'. It purports to be a 'standard account', although I have endeavoured to look at some topics that I judge to be important, such as aspects of the history of petrology, which general histories of geology have largely bypassed.

Thinking About the Earth does not assume previous knowledge of geology/earth science, and only a little chemistry is needed. But knowledge of geography – or an atlas – may be useful! When I started writing this book, I thought it would be a relatively straightforward task, as it would chiefly synthesize work that I had undertaken previously or which was already available in the literature. In the event, the task proved considerably more complicated than I anticipated. The materials on which I have drawn, or could draw, are exceedingly diverse, and in many cases excessively complicated. I hope, however, that I have succeeded in some measure in pleasing all my audiences: fellow historians of science, historians of ideas, geologists, students of earth science, and that mythical being in whom all authors must believe – the general reader. May they find this book easier to read than I have found it to write.

CHAPTER 1

A Mythical and Living World: Ideas About the Earth in Antiquity, the Middle Ages and the Renaissance

Of the numerous myths[1] that have come down to us, those that deal with creation are in many ways the most potent. They tell of where mankind may have originated, and in this way they make some sort of sense to us as to why we are here and why we are the way we are. In a certain sense modern science has a mythic character too so far as the question of origins is concerned. We may think we originated from animals by the Darwinian process of natural selection; or that the universe was formed the way physicists tell us in their 'big bang' theory. Such ideas are nurtured within our education systems and are thereby transmitted from one generation to the next. And, although scientific ideas are constantly exposed to criticism within the scientific community, 'big' ideas are notoriously difficult to shift; and, for all but a tiny fraction of the human population, we have no way of making independent judgements of the truth of scientific pronouncements about such matters as the origin of human beings, the earth or the cosmos. We have to take them on trust; and we do so, trusting our 'wise men'. In that sense, for most people there is a faith in such matters that is not wholly removed from the quality of religious belief.[2] Or we may be said to be dealing in myths. Paul Feyerabend (1970), for example, has contended that 'classical empiricism' has mythic aspects.

Since we have been acculturated with these myths they make sense to us, more or less; and in so far as they don't, we tend to put them out of our minds. Yet if the stage is reached where one is in a position to make an informed judgement about complex issues such as evolutionary theory, it appears that the science is a seething mass of argument. Thus what is mythical for the populace at large is not necessarily so for scientists and philosophers, engaged in their critical discussion. This is one of the main factors that distinguish modern scientific myths from those of the more traditional variety.[3]

Of course, a modern scientific 'creation myth' originates in a mass of empirical and theoretical work. It certainly isn't drawn out of a hat like a rabbit. It is possible also, indeed more than likely, that some creation myths of the ancient world relate to actual physical events, or processes of change, knowledge of which has survived in an imperfect way, being handed down through oral traditions and eventually set down in writing many years after they first arose. The analysis of such processes is, however, a task that is fraught with difficulty. For example, there is a book of notorious complexity by de Santillana and von Dechend entitled *Hamlet's Mill* (1969/1977), which endeavours to show that many myths from different parts of the world are 'talking about' the precession of the equinoxes in cryptic allegorical ways. Needless to say, this enterprise involves subjective exegesis and, while I acknowledge that the authors have some arguments in their favour, a great deal of their case seems dubious. With such an example before us, one must, I think, be cautious before concluding that the many flood myths round the world all represent some worldwide catastrophe, and, like the precession of the equinoxes, a single event or process. Yet this, of course, has been perhaps the most important point of contact between geological science and mythography – in the matter of the Noachian Flood, described in Genesis.

Creation myths commonly relate to important features of local existence. Some years ago, travelling in northern Thailand, I enquired through an interpreter of the local villagers where they supposed their world had come from. They said that it emerged from a 'great gourd'. This is entirely understandable, given the importance of the gourd to the economy in that part of the world. More recently, an informant in Bali told me that the earth originated by God meditating.[4] In the Bible, one of the creation myths has it that the world began in a garden. I suppose this to refer to some kind of oasis, which is obviously a place of immense importance and delight to those who live in arid regions.

One of the oldest creation myths that has come down to us is the *Enuma Elish* (*la nabu shamanu*)[5] from Babylonia, inscribed on clay tablets found in the ruins of the library of Ashurbanipul at Nineveh. It tells of the emergence to power of Marduk, the mighty god of Babylonia, who, following a fearsome battle with the primeval waters (represented by the goddess Tiamut), used the body of his slain foe to make the sky, placed the constellations in their places, measured the year and established the pole of the heavens. The sun and moon were set in their courses, and part of the body of the defeated goddess was used to construct the earth and the topography of the Euphrates/Tigris region. But Marduk himself was not the first god. Marduk, it seems, was conceived in the heart of the ocean, Apsu. So the ocean and the primeval waters apparently came first in time (Sproul, 1979: 91–113).[6] The myth is interesting from a political perspective, in the sense that it seems to state in mythopoeic terms the rise to power of the Babylonian kingdom, by war and conquest. Also, it had an answer to a possibly awkward question: where did Marduk himself come from? As to whence Apsu and Tiamut arose – well, the myth seems to assume that they were 'just there'.

The Bible contains a number of independent creation myths: the story of the creation in six days in Genesis 1; the story of Adam and Eve in the Garden of Eden of Genesis 2; and further accounts in Psalms 33 and 104, the Book of Job, and the St John gospel. Although the Genesis 2 account is placed second in the Bible, it is universally regarded by scholars as older than the creation myth of Genesis 1, and is generally taken to be nothing more than a myth by all but the most literally minded of fundamentalist Christians.

Scholars mostly date the creation myth of Genesis 1 to about 400 BC (the time of the Babylonian exile), and it is associated with the so-called Priestly school.[7] That is, it is supposed that the account was put together by Jewish priests, perhaps the six days of creation and the seventh day of rest signifying the seven-day cycle of worship in the community, organized by the priests in accord with the requirements of the lunar cycle. The myth is interestingly different from the *Enuma Elish* story in that the divine being, the Creator (*Elohim*), is simply postulated to exist. 'He' was just there, and set to work to create the world out of primeval chaos – also simply assumed to have prior existence. The Creator did not emerge as the most powerful god by might of arms: he was the most mighty *ab initio*.

At least until the middle of the nineteenth century, a good deal of effort was made to reconcile the geological record, displayed in the rocks, with the stages of creation in six days, as described in Genesis 1. It was, I fear, a forlorn and losing battle. Today, people of the Christian or Jewish faiths mostly hold that Genesis 1 conveys an allegorical truth, and is not saying anything about the geological history of the globe. The claimed 'truth', stated allegorically by means of the creation myth, is that there is one God; that He created the universe at His own decision, bringing order out of chaos; and that mankind was the pinnacle of creation so far as living beings are concerned. The whole process involved rational intent.

This exegesis is, I suggest, nothing more than wishful thinking, clinging to the tattered shreds of an ancient myth. Indications of descent from the horrific legend of the *Enuma Elish* to the myth of Genesis 1 can be discerned. If this is its ancestry, which seems not unlikely, given that Genesis 1 was probably constructed at the time of the Babylonian Exile, it seems to say little that is edifying from the point of view of human understanding or conduct. But that it should have arisen as and when it did is perfectly understandable.

The emergence of the great pantheon of Greek gods is described in the *Theogony* (or *Descent of the Gods*) of Hesiod (probably eighth century BC), the first Greek poet known to us who spoke in his own name.[8] First there was chaos. Then came 'mother earth': 'wide-bosomed Earth, the ever-sure foundation of all the deathless ones who hold the peaks of snowy Olympus and dim Tartarus in the depth of the wide-pathed Earth' (Hesiod, 1914: 87). It was from mother earth that the gods were born. She bore also the starry heaven 'to cover her on every side', 'long Hills, graceful haunts of the goddess-Nymphs', 'the fruitless deep with his raging swell, Pontus' and 'deep-swirling Oceanus' (ibid.: 89). Like the *Enuma Elish*, Hesiod's story was not always edifying.[9]

Such myths may seem fanciful. But it has been cogently argued by Greene (1992) that parts of Hesiod's *Theogony* in fact describe major geological events in mythopoeic terms. Two major battles are described in the poem: one between Zeus and his followers and the Titans (those born from the 'wily Cronos') – a truly Titanic struggle for power; the other between Zeus and the giant Typhoeus. By considering the elements of these myths, Greene argues, plausibly I believe, that they represent respectively the huge volcanic eruption of the island of Santorini (or Thera) in the Mediterranean, which destroyed the Minoan civilization some time in the second millennium BC; and an eruption of Etna, probably the one that took place in 735 BC.[10]

The volcanic events such as that which took place in Santorini in ancient times are comparable to the disaster that hit the island of Krakatoa in the Dutch East Indies in 1883. In such an event, the magma chamber of the volcano blows up as its roof collapses, and sea water rushes in, only to vaporize with a sudden explosion. In the case of Krakatoa, the noise was such that it was heard nearly three thousand miles away in Australia; and as is well known there were brilliant sunsets around the world for the next few years due to the dust from the explosion. (So it was also in Australia in 1991 after the eruption of Mount Pinatubo in the Philippines.) How could people of the ancient world – those living close enough to Santorini to have seen and heard the catastrophe, but not so close as to have been killed – have made sense of the event? I suggest that they would almost inevitably have been driven to an explanation in anthropomorphic terms – in terms of sentient beings doing violence to one another. As for the volcanic activity of Etna, both major and minor, this was ascribed by Homeric myth to the occasional stirrings and rumblings of Typhoeus, entombed by Zeus under the giant volcano.[11]

Not all myth can be interpreted successfully in terms of recorded memories of actual physical occurrences, and there are various theories of the nature and character of myths that have little regard to geology or astronomy.[12] But, in agreement with Greene, I think it plausible that some parts of myths do give a clouded record of physical events that actually happened. This mode of interpretation of myth I have called 'physico-mythology' (Birkett and Oldroyd, 1991). We shall encounter this approach in Chapter 3 when we consider the beginnings of geological science.

Myths such as Hesiod's *Theogony* were initially transmitted orally. There would, perhaps, have been minor variants in the words used to convey the stories, but the basic order of the elements would have been retained. And once a storyteller had launched into recounting the legend, one would not have been able to stop him in midstream, so to speak. One cannot imagine him telling the tale, and then saying, as at the end of a modern lecture: 'Any questions please?' Thus the poetic form would have had the effect of preserving the myth. Indeed, it would have been one of the principal means of preserving a cultural memory or a cultural heritage; for parts of the myths, besides making some sort of sense of the cosmos, also carried moral meaning. And they spoke with great power. They are published, read and studied to this day.

Considering such matters, scholars in recent years have suggested that the invention of alphabetic writing by the Greeks (in perhaps the seventh century BC) gradually undermined the oral tradition, and brought about major intellectual changes as a result. Robert Logan, for example, in an interesting book, *The Alphabet Effect* (1986), suggests that the invention of alphabetic writing encouraged the codification of knowledge and the 'invention' of logic. It may be further suggested that the writing down of the myths and legends of Homer and Hesiod, and no doubt others that are now lost to us, encouraged a different attitude to the 'knowledge' that they transmitted. The transition to literary transmission permitted the critical discussion of the texts, and the postulation of new accounts of the nature of the universe in terms other than those of the old deities. As is well known, particularly from a famous essay of Sir Karl Popper (1958–9), entitled 'Back to the Pre-Socratics', there was established in Milesia in the period before the time of Socrates (*c*.470–399 BC) a succession of thinkers who were the founders of philosophy as the subject is known today.[13]

What was remarkable about these men was first their willingness to make a distinction between phenomenal appearance and physical reality – to recognize that things may not be as they seem. Second, they recognised the possibility that there might be some underlying 'stuff' or 'stuffs', the changes of which could give rise to the constant flux and change that we see in appearances. Third, they put forward different suggestions as to what the underlying material 'stuff' might be. And, fourth, the different ideas were, it seems, exposed to critical discussion.[14]

The earliest of the Milesians of whom we have knowledge, Thales (*c*.624–548/545 BC), suggested that the fundamental 'stuff' of which the world was made was water.[15] This was not an unreasonable candidate for a fundamental substance, given that water may be converted into steam or ice. And the suggestion that there might be some naturalistic physical explanation to account for all the variety of things seen in the world, rather than one couched in terms of the doings of gods, was surely a remarkable accomplishment. The theory could be extended too in a useful way. Thales supposed that the earth floated on a great ocean, like a huge floating island, and earthquakes might be caused by occasional wobbles of the mass.

But the idea had obvious problems; and we find that instead of being accepted uncritically, it was held up to scrutiny by Thales' own pupil or disciple, Anaximander (*c*.611–547 BC). (So far as I am aware, this is the first record we have of a pupil taking issue with his teacher, instead of simply seeking to learn and transmit the master's teachings.) Anaximander asked the following question: if the earth is held up by the water, what holds up the water?

Clearly, there was something radically wrong with Thales' hypothesis. And so Anaximander proposed his own conjecture, namely that the earth floated free in space, so to speak, being equidistant from all other things; and in consequence it had no cause to move in one direction or another. This was a fine argument, based on reason. But Anaximander also had an empirical component to his thoughts about the earth. Our senses suggest that it has an irregular but essentially flat surface. So Anaximander proposed that it was drum shaped. And in principle there might be people living on the other side of the drum (the *perioeci*). Thus there was no privileged 'up' or 'down' in the universe. This was a remarkable transcendence of common sense.

Further, Anaximander rejected the idea that the basic 'stuff' of the universe was similar to any of the things with which we are familiar in our everyday lives – like Thales' water. Instead, Anaximander suggested that there was an underlying substratum, which he called the *apeiron*. This term cannot be translated satisfactorily, but it is usually construed as the 'boundless', the 'unbounded', the 'formless' or the 'unformed'. The merit of Anaximander's suggestion was that it recognized that one cannot explain something in terms of that which is to be explained – say water in terms of water or matter in terms of matter. To do so really provides no explanation at all. So even if the *apeiron* was a mysterious entity, never seen as such, it had one of the main requirements of a satisfactory scientific explanation. Anaximander's answer to the question of matter theory had something in common with his argument against Thales concerning the earth. If water held up the earth, what held up the water? Likewise, if everything was made of water, what was water made of?

Further interesting philosophical speculations were due to Heraclitus (*c*.540–480 BC), Parmenides (born *c*.515 BC), and the atomists Leucippus (fifth century BC), Democritus (*c*.460–*c*.370 BC), and Epicurus (341–270 BC). Any philosophy of matter should attempt to give an adequate account of change. Heraclitus took the problem to be so important that he proposed that the fundamental 'stuff' of the universe was fire. It was, we must suppose, a 'philosophical' kind of fire. It was a hypothetical 'stuff' that possessed the essential quality of changeability. Thus the changes perpetually occurring in the world were explained by suggesting that everything was constituted of a 'stuff' that constantly changed.[16] Fire was the nearest known approximation to this 'stuff'. So one might call it fire. But it was a 'philosophical' entity, not just the 'thing' we see burning in the grate. In our terms, fire is more like energy or a process than a kind of 'stuff'. And so it was with Heraclitean 'fire': it was a 'principle' of change.

Parmenides' emphasis was the exact opposite of that of Heraclitus. Parmenides took his stand on the basis of the purest of pure reason. He argued first that there is a 'link' between thought and reality, and that one cannot think of nothing. (That is, all the objects of our thought must exist in some manner, otherwise we could not think of them.) Following this line of argument, then, we are forced to conclude that non-being cannot be. We cannot think of non-being. To assert or think about the being of non-being involves a logical contradiction.

Further, there cannot be degrees of being. There is just 'being'. If there are no degrees of being, there cannot be more 'being' in one direction than in another. The only geometrical shape that is identical in all directions is the sphere. So Parmenides concluded that the whole universe was one grand, homogeneous sphere of 'being', extending in all directions without limit.

Naturally, one might protest that this was not at all what the world looked like. Parmenides' answer was dismissive. He had reached his conclusions, he maintained, by sound reasoning. If one's senses told one that the universe was not one grand sphere of 'being', that simply showed that the senses were defective and incapable of revealing the true nature of the cosmos.

These philosophical speculations may not seem to have much to do with the question of ideas about the earth, but I mention them here, as they swiftly led on to the ideas of the Greek atomists, named above, which are of such importance in the history of Western thought and became of considerable significance many centuries later in thinking about the nature of the earth. Essentially what the atomists did was to try to find a middle way between the contradictory philosophical schemes of Heraclitus and Parmenides. The atomists imagined the great sphere of 'being' of Parmenides divided into innumerable small parts, each of which could move in an intervening space that was devoid of 'being': a vacuum. Each little piece of being was called an *atomos*, or that which could not be divided. Then all the change that Heraclitus might desire was made possible by the *atomoi* moving freely in the void, giving rise to an endless number of possible combinations or coalitions. Thus the atomists' theory was intended to account for both change and stability.

It will be seen that the atomism of the Greeks arose as a result of (almost) pure philosophical speculation.[17] For this reason, it should be distinguished from, say, the chemical atomic theory of John Dalton propounded at the beginning of the nineteenth century, where the atomic hypothesis was invoked to account for certain quantitative experimental results obtained by the interactions of gases in simple whole-number ratios by volumes.[18] In fact, Greek atomism was rejected for many centuries, the concept of a void (entailing the 'being of non-being') being too difficult to accept. As we shall see in the following chapter, atomism was eventually revived by Pierre Gassendi in the seventeenth century, and came to be deployed extensively as an explanatory model by the natural philosophers of that time, including Newton. But the theory, even then, was in large measure a specimen of speculative metaphysics.

Of highest importance in the early history of geology was the myth or legend of Noah's Flood, for as late as the nineteenth century it was widely taken to be a historical record of an actual event, which, moreover, was of special moral significance. Further, since there is definite evidence for the occurrence of great floods in the ancient Middle East, in Greece and in other parts of the world such as North America, and since this empirical evidence is supported in some measure by the folklore tradition of many parts of the globe,[19] geology and folklore studies could together hold out the hope of providing direct empirical confirmation of the veracity of the Bible. In fact, some of the geological evidence is rather persuasive, or at least highly plausible. So in northern Europe, what are today regarded as relics of glacial times were construed as vestiges of the biblical deluge. (Hence such materials were formerly called the 'Diluvium'.)

According to the results of twentieth-century scholarship, the Flood legend of Genesis is a compound of two sources (see note 8). These differ somewhat. *J*, for example, has a flood of forty days, while for *P* it took one hundred and fifty days. The idea of the ark arriving on Mount Ararat occurs only in the *J* version. But in any case, regardless of the differences, the legend can be traced back to a Babylonian version recorded in the epic of Gilgamesh, found in the library of Assurbanipal at Nineveh. And that version is thought to derive from a much earlier Sumerian story, perhaps going back as far as 3400 BC. It has important elements of the biblical account, such as a man – named Utnapishtim in the Babylonian account and Xisuthrus in the Sumerian version – being warned by a (sea) god of a flood impending because of humanity's wickedness; the building of a boat in which the man, his family and his animals were saved; and the sending out of birds after the storm was over. Water was said to have emerged from under the ground (from the 'fountains of the deep') as well as having fallen from the sky.

The whole question was examined in detail by the distinguished Austrian geologist, Eduard Suess (1904), who offered the following naturalistic explanations. It was, he suggested, the result of a tragic coincidence of a seismic event and a great typhoon blowing up the Persian Gulf. There was some preliminary seismic activity, which led a man to think of preparing (or at least caulking with asphalt) a large boat for himself, his family, and his animals. Then, when the main seismic event occurred – perhaps at the southern end of the Persian Gulf – he was fortunate to be prepared, and was washed northwards by a huge tsunami,[20] or great 'tidal wave', right across the plains of modern Iraq, to lodge somewhere on the mountains to the north – in that desolate region where, as I write, Kurdish refugees are endeavouring to eke out a harassed existence.

Whether the boat could have been washed right over the Kurdish mountains and on to Mount Ararat beyond, in remote north-east Turkey, one may doubt. But at least we do not need to invoke miraculous events to account for the main body of the legend. Suess (1904: 31) pointed out that the emission of water from out of the ground 'is a characteristic accompaniment of earthquakes in the alluvial districts of great rivers'. On this view, then, there is no difficulty about the question of the opening of the 'fountains of the deep'.[21]

As I say, the idea that some such catastrophic event occurred in antiquity is warranted by at least three sets of interconnected texts, of which the biblical one appears to be the youngest. And the whole may be given a naturalistic explanation, such as that suggested by Suess (although I do not claim that his theory is correct in every particular). In the Bible, of course, the whole becomes a grand myth, with a strong moral message; and it continues to be recounted and regarded in that way to the present day. What the peoples of Sumeria or Babylonia who survived the catastrophic events thought about them one can hardly say. But we may suppose that from their perspective too the catastrophe had a moral message. Why else would the normal course of their lives have been so disrupted by the earth?

Much of Greek mythology was eventually set down in Latin by the poet Ovid (43 BC – AD 17) in his famous work, *Metamorphoses*. As will be discussed in Chapter 3, an interesting attempt was made by Robert Hooke in the seventeenth century to construe parts of Ovid's text as having oblique reference to actual geological events or processes.

The most striking account of a natural phenomenon in mythopoeic terms, recorded by Ovid but clearly derived from the mythology of Homer and Hesiod, was the idea that the large Sicilian volcano, Etna, lay on top of the great giant, Typhoeus:

> Etna weighed heavily upon his head; as he lay stretched on his back beneath it, he spat forth ashes and flame from his cruel jaws. Often he strove to throw aside the weight of the earth, and roll off the towns and massive hills that secured him. At such times the earth trembled. (Ovid, 1955, p. 136).

Only when I looked into the orifice of Mount Yasur on Tanna did such a legend make sense to me. Surely, when one sees a volcano in action it does almost seem that some kind of living organism is caged up under the mountain. The legends of Tanna or Sicily seem just as likely – or unlikely – stories. The difference, of course, is that the Greek myth was eventually written down, becoming a classic text that has been preserved for centuries and which has given delight as a fable, if no more, to countless readers.

The last book XV of the *Metamorphoses* is different from the others in that it purports to give a direct account of the ideas of Pythagoras (*c*.580–*c*.500 BC) and is less overtly mythical in character than the other parts of the text. Also, it contains information that has directly to do with the ideas that the mathematician/philosopher had had about the earth.

Pythagoras, it was stated, believed (like Heraclitus) in an eternal flux. Earth changed into water, then into air and then into fire. Then the whole process reversed, with fire eventually ending up as earth.[22] For Pythagoras had, according to Ovid, seen (or at least envisaged) earth changed into sea and lands formed where once was ocean. Moreover, flowing water could cut valleys and mountains could be levelled by floods. 'Seashells lie far away from the ocean's waves, and ancient anchors have been found on mountain tops' (Ovid, 1955: 371). Just conceivably, the mention of the anchors provides corroborative evidence for the Noachian Flood legend, discussed above, although it may refer to nothing more than stones that looked like those used for anchors.[23] Be that as it may, there is evidence that Pythagoras, or the Pythagoreans, and presumably Ovid too, had a sense of geomorphological changes taking place over time, although we have no indication of how long the processes of change were thought to have taken. There was also mention of volcanic vents shifting their position. Etna was then inactive, but, it was thought, might become active again in the future.

As an example of naturalistic ways of thinking about geological processes in the ancient world, we may consider the ideas of the historian Herodotus (*c*.484–*c*.425 BC). He drew an analogy between the long gulf of the Red Sea and the valley of the River Nile, and suggested that if the former had a large river flowing into it like the Nile, constantly carrying loads of sediment, it might become filled up, like the valley of the Nile and its delta. Indeed, in what we might call a thought experiment, Herodotus (1954: 106) suggested that it might take about 20,000 years for the Red Sea to become silted up, if somehow the Nile were diverted into it (presumably at its northern end). He also reported shells found on the hills in Egypt, and referred to the quality of the black soil in Egypt. On such evidence, he concluded that the Nile Valley had slowly been filled up over thousands of years, and had built out its delta. Thus Herodotus evidently had the idea of large-scale geomorphological changes, brought about slowly over great periods of time.

Herodotus (ibid.: 456) also referred to the Plain of Thessaly in Greece, which he said had possibly once been a large lake, and was remembered as such in local tradition, being mostly surrounded by hills. The River Peneus cut straight through the hills to the east into the Aegean, rather than taking an apparently easier outlet to the south. Local belief had it that the gorge had been produced by the sea god, Poseidon, but Herodotus simply supposed that there had been a large earthquake that had opened up a cleft for the passage of the river.[24] It would appear that he was prepared simply to look at the earth before him, and ask himself how it might have come to be the way it appeared, by natural causes, perhaps over thousands of years.

In a less naturalistic vein, Plato (428–348 BC), in his dialogue *Phaedo* – which was chiefly concerned with the question of death – gave some consideration to the underworld, and hence the nature of the earth and its interior (Plato, 1977: 381–7). He supposed (ibid.: 375) that the earth was equipoised in the cosmos, and would appear as a huge ball covered with twelve patches of colour if viewed from above. Human beings lived in one of its hollows – a region that was injured and corroded – whereas other parts of the globe were much more perfect and beauteous. Some of the hollows, he surmised, led down into a system of subterranean channels and there was a large body of water at the centre of the earth called Tartarus. Surging around, it caused the flow of rivers, which run into lakes, the largest of which is the sea. And there was a constant circulation of the waters, some hot, others cold. Plato also envisaged great rivers of fire within the earth and streams of mud, 'like the rivers of mud that flow before the lava in Sicily, and the lava itself' (ibid.: 383). This model of the earth's interior very probably inspired the pictures drawn centuries later by Athanasius Kircher (see p. 36).

The idea of Tartarus was, however, discounted by Aristotle (384–322 BC) in his treatise on atmospheric and related phenomena, the *Meteorologica*. He understood the hydrological cycle, there being a balance between rainfall and evaporation (Aristotle, 1952: 145). However, he went further and, like Herodotus, envisaged interchanges between land and sea, by silting or by the encroachment of the sea. But, said Aristotle (ibid.: 109), the processes are so slow that they are not noticed in a lifetime. Also, he thought, different regions were wetter or drier in different epochs. Since he believed that time was infinite (ibid.: 119), there was no lack of time to perform great cycles of geomorphological or climatic change.

The *Meteorologica* is also of importance for the history of ideas about the earth in that it contained a hypothesis about the formation of different kinds of mineral substances according to their formation from either a dry (smoky) or a moist (vaporous) 'exhalation'. The former, said Aristotle (ibid.: 287), produced 'realgar,[25] ochre, ruddle,[26] sulphur and all other substances of this kind'. Metals were supposedly generated by the 'vaporous exhalation'. Thus was established a tradition which lasted in one form or another into the eighteenth century. But this was chiefly due to the overwhelming influence of Aristotle's philosophy, rather than the particular merits of the hypothesis. In fact, Aristotle had little to say about the mineral kingdom.

Mention should also be made of the geographer Strabo (*c*.63 BC–after AD21), whose famous *Geography* (Strabo, 1917) described the world as it was known in Roman times. It was chiefly based on his observations in the Mediterranean region, but included hazy information about regions as remote as India and China. Strabo had much to say about mining and quarrying and the various types of metalliferous deposits and gems. He also referred to the actions of earthquakes and volcanic phenomena and the sudden subsidences of land. He was probably familiar with tsunamis and attributed the presence of shells inland and distant from the sea to such phenomena.

Roughly contemporary with Ovid and Strabo was the Roman Stoic philosopher, Seneca (*c*.3 BC–AD 65). Like several of his predecessors, he was concerned with the problem of change, and in this light he proposed a grand cyclic theory of the cosmos, analogous in some ways to some modern suggestions that the universe has endless cycles of expansion and contraction, of which we only know the present expanding phase. For the Stoics, the dominant idea was the concept of 'tension' and its converse, 'relaxation'. The substance of the universe, then, passed through great cycles of 'tension', and as the 'tension' of matter increased, conversion of the elements occurred in the order earth, water, air and fire, and then the reverse. In his *Quaestiones naturales*, Seneca (1910) supposed that the world was in a phase of increasing tension, so that the oceans were gradually drying up. Eventually, all would end in a great conflagration, and then a new world would be formed out of the ashes of the old. Such ideas found their way into the Bible (2 Peter 3), which passage is thought to be of Stoic origin.

I have explained in outline above how the philosophy of atomism was arrived at. But we only have fragments of the works of the founders of atomism, or references to them in the works of other writers such as Aristotle. On the whole, the system was not satisfactory to the ancients because it suggested a universe where everything happened by chance. The atoms just formed their coalitions at random, and then might fall apart again. Yet the world did not seem to be like that. On the contrary, it appeared to have been designed by some designer or artificer; or in certain respects it had the characteristics of a living organism. Both Plato (1975: 55) and Aristotle (1939: 207) wrote of the heavens as being endowed with soul or life. In his *De natura deorum*, a vehement attack against atomist ideas, the Roman soldier, lawyer and philosopher, Marcus Tullius Cicero (106–43 BC), compared the universe to a water clock or a sundial – items of ancient technology that worked exactly 'by art not by chance' (Cicero, 1979: 207). Many other examples could be given. Cicero provided an early example of what came to be known as the 'argument from design', which has remained popular until the present, often with clocks or watches as analogues for a designed cosmos.

Despite the rejection of atomist doctrines, there was one major work by an atomist that must attract our attention, for it had some significant things to say about the earth (as well as matters of more human interest such as psychological phenomena or sexuality), and is preserved as a complete work. I refer to the celebrated Latin poem by Lucretius (*c*.99–*c*.55 BC), *De rerum natura*. In considering the phenomena of earthquakes, and specifically the Sicilian volcano Etna, Lucretius (1951: 238) suggested that they might be due to wind raging through caverns or passages, which might extend right through the body of the earth.[27]

Table 1.1 *Classical ideas about volcanic action and earthquakes*

Underground volcanic passages or pores within the earth	Air/wind the cause of fire earthquakes	Active wind within the earth the cause of fire	Air may be driven into the earth by the sea	Falling material may be the cause of movement of air or of earthquakes	Theory of lava as 'fuel'
Democritus	Anaximander	Aristotle	Aristotle	Anaximenes	Ovid
Metrodorus	Anaxagoras	Epicurus	Callisthenes	Epicurus	Strabo
Plato	Archelaus	Lucretius	Lucretius	Lucretius	Seneca
Aristotle	Democritus	Ovid	Seneca	Seneca	Pliny
Epicurus	Metrodorus	Strabo	Aetna	Aetna	Apollonius
Lucretius	Aristotle	Seneca			Aetna
Ovid	Callisthenes	Pseudo-Aristotle			
Diodorus	Epicurus	Apollonius			
Strabo	Lucretius	Aetna			
Seneca	Posidonius				
Pliny	Strabo				
Pseudo-Aristotle	Seneca				
Apollonius	Pliny				
Aetna*	Pseudo-Aristotle				
	Aetna				

* I.e. the anonymous poem

The idea was mechanical in character and was evidently inspired by the analogy of a forge, rather than some mythic being such as Hesiod's Typhoeus.

Lucretius was not alone in his suggestion. It appears most strikingly in an anonymous poem entitled *Aetna* (Anon., 1901, 1955, 1965). Various candidates have been suggested for authorship, including Seneca and Virgil, but on examining the evidence some years ago I came to the conclusion that the best candidate was one Lucilius Junior, Procurator of Sicily and correspondent of Seneca (Paisley and Oldroyd, 1979). This study led me to examine a large range of classical writings that had something to say about volcanoes and the earth, and the results shown in Table 1.1 were obtained for 'pre-*Aetna*' texts.[28] It will be noted that the author of *Aetna* (and others) supposed that the fire in a volcano fed on some kind of fuel. Lucilius (or whoever) took this to be lava. A few other ideas were also revealed in the texts I examined, as Table 1.2 indicates:

These tables indicate that there was in fact quite an extensive body of opinion concerning volcanic phenomena tucked away in the classical writings, although, being so scattered, it tends to get overlooked. It will be noticed that most of the ideas were not mythical in character, indicating that the effect of the presocratic revolution, if it may be called thus, was lasting. Moreover, since there are so many common factors in the various accounts of earthquakes and volcanoes in the ancient writings of Greece and Rome, we may conclude that there was some

Table 1.2 *Other classical ideas about the earth*

Mythical explanations of volcanoes (chiefly associated with myth of Typhoeus)	Analogies drawn with thunder and lightning	Employment of the theory of exhalations	Earthquakes ascribed to water	Earthquakes ascribed to fire
Hesiod Pindar Virgil Ovid Manilius Lucan	Anaxagoras Epicurus Seneca Pliny	Aristotle Diodorus Seneca Pliny	Thales Democritus Epicurus Lucretius Seneca	Anaxagoras

uniformity of thinking on the matter in the Mediterranean region of the ancient world and ideas were transmitted from one writer to the next, even if there was no scientific treatise dealing explicitly with the matter.[29]

In reference to volcanoes, there is one notable surviving account from antiquity of an actual volcanic eruption: that of Vesuvius in AD 79 Indeed, the famed Roman writer Gaius Plinius Secundus (Pliny the Elder) (AD *c*.23–79) himself perished in the eruption. He was at the time commander of a Roman fleet at Misenum, near Naples, and directing his ships in an attempted rescue operation without sufficient regard to his own safety he was choked to death by the fumes emitted by the erupting volcano. The details were graphically recorded in a letter by his nephew, Gaius Plinius Caecilius Secundus (Pliny the Younger) (AD *c*.62–*c*.113) to Cornelius Tacitus (Pliny the Younger, 1978: 135-37). Visitors to Herculaneum and Pompeii can, to this day, see the effects of the famous eruption of Vesuvius in AD 79 and meditate on the powers of the earth.

No doubt if Pliny the Elder had lived, he would have described the eruption in minute detail himself, for he was a man who liked to record all available knowledge. This he attempted to do in his celebrated *Natural History*, completed in AD 77 in thirty-seven books, the last five of which deal with the mineral kingdom. The work, which has been estimated to deal with about twenty thousand topics, is important to historians for its descriptions of practices of mining in the Roman era and its mention of the processes for making cement, plaster, glass, etc. and the preparation of building stones. There are also many descriptions of crystals and precious stones. For such matters, however, Pliny took many details on trust, especially with regard to the supposed magical powers of minerals and gems.

In many ways it is satisfactory to follow the suggestion of Thomas Kuhn (1962: 16) and to regard Pliny's work as an example 'pre-paradigm' science – a 'morass' of disparate and undigested facts. Information was collected without regard to theory or any other kind of guiding principles. Thus, in Kuhn's terms,

Pliny's uncritical empiricism did not constitute scientific research; nor was it strictly philosophical. Nevertheless, the *Natural History* offered a vast array of facts, albeit compiled uncritically, and it was transcribed repeatedly, presumably being thought to be a valuable storehouse of knowledge. It was one of the first books to appear in print in the fifteenth century, being published in Venice in 1469, and remains in print to the present. The term 'natural history', as used today, derives from Pliny.

We have seen that for the ancient Greeks the earth was indefinitely old. For Aristotle, it was regarded as stationary, at the centre of the cosmos, surrounded by a series of nested concentric 'crystalline' spheres, which carried round the sun, moon and planets in their paths across the sky. The stars were thought to be attached to, or in some manner part of, the outer sphere, which rotated about the earth daily. The dimensions of this sphere were not known, but it was not regarded as enormously great in size, perhaps because in very clear conditions, such as are rarely encountered today (but which may be found, for example, in central Australia), the brilliantly shining stars seem quite near and the whole star sphere seems to press down upon one. Anyway, although the overall dimensions of the cosmos were not known, the actual size of the earth, which was regarded as spherical as evidenced by the circular shadow cast during a lunar eclipse, was known with remarkable accuracy by an ingenious mixture of theory and measurement, due to Eratosthenes (*c.*275–*c.*194 BC), librarian at the great library at Alexandria. His actual description of his method has not survived and knowledge of it is known chiefly through a first-century BC author named Cleomedes.

Eratosthenes knew that at a place in the Nile Valley called Syênê[30] the sun was directly overhead at midday on the longest day of the year. (The sun shone right to the bottom of the wells of the district.) But to the north, at Alexandria, the sun appeared to be one-fiftieth of a circle from the vertical. Knowing the approximate distance along a north–south meridian, from Syênê to Alexandria, the size of the earth could be calculated. According to long-held tradition, the distance between Syênê and Alexandria was determined as follows. Syênê to Alexandria was a journey of fifty days, for camels that averaged a hundred stadia a day. Therefore, 5,000 stadia equalled one-fiftieth of the earth's circumference, which was thus equal to 250,000 stadia. Taking the travellers' stadion to be 157 metres (a matter about which there has been considerable debate in the literature), we obtain 39,250 kilometres, which is about 2% less than the modern value. Recent consideration of the matter (Dutka, 1993) has questioned the camel story, however, and the suggestion has been made that the distance between Syênê and Alexandria was available to Eratosthenes by previous direct survey. Be that as it may, it is the case that astronomers of the ancient world had a very fair idea of the size (and shape[31]) of the earth, if not the cosmos.

To explain the presence of sea shells inland round the Mediterranean, Eratosthenes supposed that the sea had once been a huge inland lake, with a water level higher than that of the Atlantic or the Black Sea. But with the breaking of the land barriers at the Pillars of Hercules (Strait of Gibraltar) and Pont-Euxin (the Bosphorus) the shells had been left stranded on the Mediterranean shore. Strabo disagreed, preferring the hypothesis of rapid changes of levels of land and sea, but at various localities and at different times.

The story of the Christianizing of the Western world and the transmission of Greek knowledge to posterity via the newly arisen Muslim cultures has been told many times and not much will be said about these matters here. The mysterious doctrine of the Trinity was formulated at the Council of Nicaea in AD 325 and Christianity absorbed ideas from Judaism, Stoicism, Platonism and Neoplatonism.[32] At this early stage, Aristotelian doctrine was not part of the theoretical structure of Christianity. So there was a tendency, exhibited for example by Lactantius Firmianus (AD *c*.240–320), tutor to the son of the Emperor Constantine, to prefer a flat-earth theory, as seemed to be sanctioned in the Old Testament. But this never became part of official Christian doctrine, and the matter was regarded as uncertain for many years. Indeed geography, and certainly geology were hardly matters of great concern to the early Fathers of the Church, and in the so-called Dark Ages that followed the collapse of the Roman Empire geographical knowledge deteriorated rather than advanced.[33]

The Fathers were, however, much concerned with time. Stemming from the Jewish tradition, the Christian concept differed radically from that of the Greeks and Romans, by virtue of the concern with a linear history leading from the first Creation of the world, as described in Genesis, to the hoped-for return of the Messiah and the coming of the Kingdom of Heaven. These events could supposedly be quantified by scholarly research and a little interpretation of certain biblical texts. Theophilus of Antioch (AD 115–81), the founder of Christian chronology, did what many were to try to do after him: work backwards through the historical information furnished in the Old Testament until the date of Adam and Eve could be determined. Then the very year of Creation might be known. Theophilus obtained the date of 5529 BC (Haber, 1959: 17). An important guiding principle was that of 2 Peter 3: 8: ‘One day is with the Lord as a thousand years, and a thousand years as one day.’ So it was supposed that the six days of Creation described in Genesis each corresponded to a thousand years of world history. For Lactantius, the first five millennia of the biblical chronology had already elapsed. So there was not too long to go before the Second Coming and the reign of Christ. The more influential St Augustine also envisaged a 6,000 year timetable for earth history.

So far as records show, the early Church Fathers had little interest in natural questions, except for issues of astronomy which had a bearing on the date of Easter. In consequence, for many centuries there were few changes in ideas about the earth, which were often very crude so far as geographical knowledge was concerned. Its age was 'thinkable' to human understanding. In the Middle Ages, when Christianity became interwoven with Aristotelian philosophy – most notably through the efforts of St Thomas Aquinas (*c.*1225–74) – one important aspect of Aristotelian thought that was not incorporated was the cyclic view of time and the great age of the earth. Otherwise, the cosmologies went well together. The earth was thought to be at the centre of the cosmos, with all the heavenly bodies circulating round it and beaming down their influence. It was a small, 'closed' universe, also intelligible with regard to size. The short timetable for earth history was, however, weakened to some degree in the Middle Ages by the incorporation of Aristotelian aspects into the Christian world view. Nicholas of Cusa (1401–64), for example, had ideas about a longer time span for the earth and cosmos. So too did Buridan, who will be mentioned below. But with the Reformation, with its emphasis on the Bible as the source of knowledge and moral authority, the shorter time frame was reaffirmed, and was sustained for most people in the West until at least the first half of the nineteenth century. Some Christian sects maintain this short time-scale even to the present.

Whereas the Church Fathers added little to knowledge of the natural world, the Arabs, who became the immediate heirs to the Greek intellectual inheritance, did make some significant contributions to matters pertaining to the earth. Scholars such as Ibn Sina (also called Avicenna) (980–1037) and Averroës wrote important commentaries on Aristotle's writings, and Avicenna wrote a text on the formation of stones (Avicenna, 1927), which added considerably to ideas about the earth from a 'proto-chemical' point of view.[34] Thus the classification shown in Figure 1.1 was offered. Remarkably, remnants of this classification (salts, sulphurs, metals and stones) were retained until the end of the eighteenth century.

Avicenna's book also had a section on 'The causes of mountains', which contains ideas of considerable interest. Aware of the ability of certain springs to

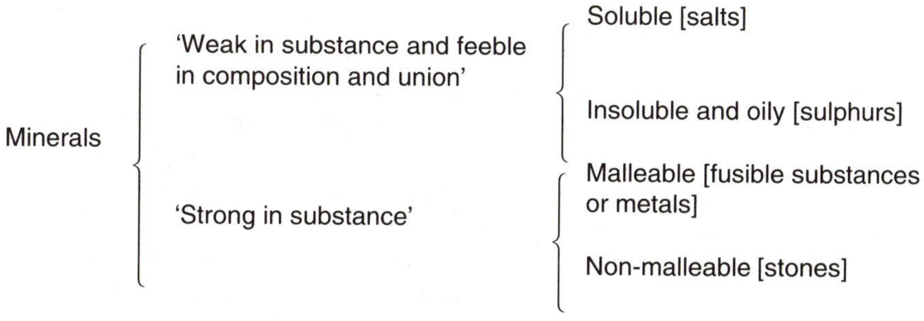

- Minerals
  - 'Weak in substance and feeble in composition and union'
    - Soluble [salts]
    - Insoluble and oily [sulphurs]
  - 'Strong in substance'
    - Malleable [fusible substances or metals]
    - Non-malleable [stones]

Figure 1.1 Avicenna's classification of mineral substances

petrify objects, he envisaged the idea of a 'mineralizing and petrifying virtue' in the earth. Valleys, he thought, had been excavated by the action of currents of water. The mountains themselves were formed by petrifaction of 'agglutinative clay' after land was exposed by the retreat of the sea (which was evidenced by the fossil shells found inland). Avicenna supposed that at present the mountains were chiefly decaying and disintegrating, but their layered appearance indicated that they were formerly produced by some kind of sedimentary process. Thus there was, perhaps, implicit the idea of a cycle in the formation and decomposition of mountains.

The idea of a kind of geological cycle was also to be found in the work of some tenth-century Muslim scholars called the 'Brothers of Purity'. They envisaged sediment slowly being deposited in the oceans, and that these would therefore eventually overflow, depositing fresh material on land. The whole idea was connected to that of the 'great year' envisaged by astronomers, who had discovered a slow but regular change in the alignment of the Milky Way and the positions of the 'equinoctial points' on the ecliptic.[35] One complete cycle occurred in a 'great year' (thought to be 36,000 years), and this was taken by the Brothers of Purity to coincide with their grand geological cycle (Duhem, 1958: 253).[36] Pierre Duhem, the nineteenth-century French physicist and historian of science who collected and collated so much material on ancient and medieval science, called the doctrine 'une théorie purement neptunienne' (ibid.).

Among Christian thinkers, ideas were not always so naturalistic. An illuminated manuscript from a thirteenth-century French bible, held at the State Library in Vienna and copies of which can be obtained at Norwich Cathedral (Figure 1.2), envisaged the universe being created directly by God (depicted, however, as Christ) – a kind of master craftsman. He appears to be at the stage of dividing (appropriately with a pair of dividers) the waters above the firmament (sky) from the dry land beneath. But the earth is not yet in its final form.[37] And, ignoring or forgetting Eratosthenes, the medieval writer Dante Alighieri (1265–1321) described the world as being so small that one might walk right through it – at least in one's imagination – over the course of an Easter weekend, visiting the several levels of Hell on the way.[38]

In a manuscript entitled '*Della composizione del mondo*' (1282), the Italian scholar Ristoro d'Arezzo (thirteenth century) imagined that the sky was responsible for keeping part of the earth free of water, and that there was a 'correspondence' between the pattern of stars in the sky and the pattern of mountain peaks and valleys in a landscape (Ristoro d' Arezzo, 1862: 145–66). (A comparable idea was also propounded by Dante, when he proposed that mountains were drawn up above the waters of the oceans by the fixed stars (Suess, 1906: 6).) Ristoro also envisaged the 'virtue' of the stars acting on the waters of the earth, drawing back the waters to uncover the dry land, as a magnet with its 'magnetic virtue' draws iron. However, he did examine rocks himself,

Figure 1.2 God/Christ, the 'Pantocrator', architect of the universe, dividing the land from the waters with a mason's compass, according to an illustration in a thirteenth-century French bible (by permission of the *Osterreichische Nationalbibliothek*, Vienna (Cod. 2554, fol. 1^{v}))

observed fossils therein and recognized the action of running water cutting into land to make valleys. (The Noachian Flood was also taken to be an important geological agent.) The difficulty with such a theory, or that of the Brothers of Purity, was that it lacked a mechanism for the formation of new mountains, and it might seem that the whole process would come to an end after one cycle, with all land levelled to a plane – either above or below the level of the sea. Avicenna's earthquakes, newly elevating land, would deal with this problem, but that would leave the question of the cause of the earthquakes to be determined. Ristoro was inclined to invoke the old idea of subterranean winds.

The issue was considered by the fourteenth-century scholar Jean Buridan (*c.*1328–58), Rector of the University of Paris, in a text entitled '*Questions sur le traité des météores*' (i.e. a commentary and elaboration of Aristotle's *Meteorologica*). Buridan boldly asserted that he was going to assume, like Aristotle, that the world was eternal, even if this idea was incompatible with 'our faith' (Duhem, 1958: 299). He further argued that there could not be a 36,000-year cycle. For in 18,000 years there would be a reversal of the land and sea, and a quarter of this change might have been expected to occur in about 6,000 years, since the time of the ancient Egyptians. But it was known from records or traditions that such a major change had not occurred in a time period of that order of magnitude (ibid.: 294). Buridan did, however, envisage a very slow cycle – 'a hundred thousand million years perhaps' (ibid.: 296) – such as might be required for all the planets, stars, sun and moon to return to their original configurations in the sky.

In response to these considerations, Buridan developed an ingenious model, which, following Duhem, has recently been described by Gohau (1991: 28–30). Buridan argued that, because the earth was heterogeneous, its centre of gravity did not necessarily coincide with its geometrical centre, although the centre of gravity must coincide with the centre of the cosmos. The sphere of the oceans had to be spherical, however, and its centre did coincide with the centre of the cosmos. So there were, so to speak, two globes: one of the earth and one of the oceans. One side of the earth tended to be exposed above the sea, while the other side was largely covered. (The idea may not appear so fantastic when it is considered that the Pacific Ocean is 'opposed' by a hemisphere where most of the world's land is concentrated.)

The exposed side, then, would be warmed by the sun and would be less dense in consequence. Erosion of the terrestrial side would occur, while sediments would be deposited in the waters of the oceanic hemisphere. But because of the very slow cycle of the heavens, there would (supposedly) be a very slow cycling of the oceans relative to the earth. In consequence, fresh land would be exposed, so that everything would not become levelled down and the seas totally filled with sediment. In fact, Buridan had the idea (Duhem, 1958: 304) that a town might slowly approach or retreat from the ocean. Thus, as the ocean slowly accomplished its displacement relative to the earth, the mean meridian of the habitable part of the globe would slowly change in relation to the sky, presumed fixed. In effect, he was contemplating something akin to pole wandering.

Buridan's theory, conceived within the womb of Aristotelian tradition, was quite extraordinary. It was opposed by his influential contemporary, Nicholas Oresme, but found support from other scholars such as Albert of Saxony, Marsilius of Inghen and Pierre d'Ailly. As we shall see, there was some distinct similarity between Buridan's ideas and those of Robert Hooke and Benoît de Maillet in the seventeenth and eighteenth centuries respectively, and perhaps even the Comte de Buffon. So far as I am aware, no direct historical connection has ever been established. Such a connection is not impossible, however. Hooke had a huge library and was an omnivorous reader.

As will appear in later chapters, one of the first problems that seventeenth-century scientific thinking about the earth had to deal with was that of shells embedded in rocks far from the sea, and perhaps quite different from those known alive today. But this problem had been considered much earlier by that great Renaissance polymath, Leonardo da Vinci (1452–1519). As is well known, his ideas were only recorded privately in his notebooks, in his own peculiar 'mirror handwriting', and because Leonardo's ideas remained unpublished in his lifetime they exerted negligible influence on his contemporaries. But they are significant in that they show that he regarded the shell problem as interesting and important. And they tell us something about what some people at least may have been thinking about the earth in the Renaissance.

Leonardo (1906: 106) was very matter-of-fact about the idea of the shells not having been emplaced by the Flood. Cockles and suchlike were slow walkers, so they could not have kept up with the rising waters. Also, the shells would not have been generated by some kind of astral influence, for if so why would they be so various, some old, some young, some broken, others whole?[39] Rather, the field evidence (which Leonardo described a little for the region round Florence and the lakes of northern Italy) suggested that the sea level had once been at the altitude of the shell beds. This would explain the presence of shingle beds (conglomerates) high above the present sea level.

Leonardo (ibid.: 109–10) also gave an account of the formation of fossils that we find entirely congenial:

> [I]n the course of time as the channels of the rivers become lower, these creatures [fish] being embedded and shut in the mud, and the flesh and organs being worn away and only the bones remaining, while even these have lost their natural order of arrangement, they have fallen down into the base of the mould which has been formed by their impress; and as the mud becomes lifted above the level of the stream, the water runs away so

> that it dries and becomes first a sticky paste and then changes into stone, enclosing whatsoever it finds within itself, and itself filling up every cavity.

Interestingly, there is one passage (ibid.: 104–5) which suggests that Leonardo had some knowledge of the ideas of Buridan. Leonardo hints at the idea of erosion acting so as to raise the 'light side of the earth' so that 'the antipodes draw near to the centre of the earth, and the ancient beds of the sea become chains of mountains'. Unfortunately, the idea was not pursued beyond this tantalizing hint. It is clear, however, that Leonardo, like the Greeks before him, placed no limitations of time on his thinking about the earth.

But Leonardo, as any Renaissance man, was always interested in life, as much as afterlife. And it was clear that the earth – mother earth – was the source of life in countless ways. So he employed a popular conceit, that there was a close analogy between man and earth – the 'macrocosm/microcosm' analogy:

> [W]e may say that the earth has a spirit of growth; that its flesh is the soil, its bones are the successive strata of the rocks which form the mountains, its muscles are the tufa stone, its blood the springs of its waters. The lake of blood that lies about the heart is the ocean; its breathing is by the increase and decrease of the blood in its pulses, and even so in the earth is the flow and ebb of the sea. And the heat of the spirit of the world is the fire which is spread throughout the earth; and the dwelling-place of its creative spirit is in the fires, which in diverse parts of the earth are breathed out in baths and sulphur mines, and in volcanoes, such as Mount Aetna in Sicily, and many other places (ibid.: 130–31).

In this fine passage, I suggest, we have a fair statement about the man in the street's way of thinking uncritically about the earth. It was a kind of organism, fecund, endowed with life; and it could be thought about by comparison with that entity with which man was most familiar – himself.

Apart from such highlights, most of the writings to do with the earth that have come down to us from the Middle Ages or Renaissance are 'encyclopaedic' in character, rather than theoretical, and had features somewhat similar to Pliny's *Natural History*.[40] Much detail has been given about such texts by Adams (1938), so much so that it is rather difficult to see the wood for the trees in his history. But in the well-known book by the French philosopher Michel Foucault, *The Order of Things* (1970),[41] there is a useful general picture of what the Renaissance scholar was about, and it is possible to extend Foucault's perceptive analysis specifically to Renaissance studies of the mineral kingdom (Albury and Oldroyd, 1977).

According to Foucault, the goal of the Renaissance scholar, when providing an account of some plant or animal, was to compile all possible information that was available about it. This would include not only descriptions of its appearance or behaviour but also its uses in medicine and magic, its symbolism in heraldry, its supposed sympathies and antipathies,[42] its 'correspondences',[43] and especially all the things that had been said or written about the plant or animal – which thus effectively became part of the very 'nature' of the thing. The Renaissance scholar's approach was by no means dissimilar to that of Pliny in the ancient world.

In this vein, then, we find the Swiss naturalist Conrad Gesner (1516–65), compiling his *De rerum fossilium, lapidum et gemmarum* (1565). It considered minerals according to their resemblances to geometric forms, celestial objects, animals, plants and artefacts. Gesner relied on information supplied by correspondents as well as his own observations, and this (as may often be the case) led him into some errors. For example, he depicted basalt columns as if they had pyramidal ends like quartz crystals (ibid.: 20).

Another Renaissance scholar who worked in a tradition similar to that of Gesner was Ulysse Aldrovandi (1522–1605). In a posthumously published collection of writings (1648) he treated minerals under the headings synonyms, definitions, origins, nature and properties, varieties, modes and places of occurrence, uses, historical references, sympathies and antipathies, temperaments, mysteries, miracles, moralia, adages, epithets, mythologies, dreams, symbols, and lapidati. The text of this mammoth treatise was copiously illustrated, but while some shells were carefully and accurately depicted, others indicate that Aldrovandi was making spurious analogies between stones and human artefacts.

In such writings, we find an interesting union of empirical work and supposed correspondences, rather than causal connections such as would interest a geologist today. But if the small lodestone could move iron, why could not a star 'move' a mountain? It is not useful to dismiss such ideas as quaint (as did Adams, 1938). Rather, one should recognize, with Foucault, that they were the product of a 'knowledge system' very different from that which we deploy today.

In addition to considering the character of Renaissance natural histories of minerals, we may think more precisely about some of the ideas from this period pertaining to the earth and minerals which were derived from what might be called 'proto-chemistry' and from the ancient philosophies of Aristotelianism, Platonism (and/or Neoplatonism), and Stoicism. A common feature was to think of the earth as in some sense alive. Thus there were ideas about the growth of minerals in the womb of the earth, which itself was a body with mineral veins. The miner might go down into the bowels of the earth. Another approach was to construct a hypothetical account about how the earth might first have been formed, which account would be consistent with the supposed ideas about the nature and interactions of matter.

A good sample of such eclectic thinking about the earth is to be found in the work of Paracelsus (1493–1541),[45] an itinerant scholar who wandered Europe criticizing ancient medical theory and practice and recommending the use of mineral substances (such as mercury) for medicinal purposes. He also sought to present a new chemical theory, in which the basic components of matter were claimed to be Salt, Sulphur and Mercury (the so-called *tria prima*),[46] although in some texts he also utilized the Aristotelian theory of four elements: earth, water, air and fire. Paracelsus was nothing if not eclectic. However, in the last analysis, all matter was for him but a corporified form of spirit.

Like many scientists and philosophers endeavouring to formulate new ideas, Paracelsus found it needful to coin new terms. Thus he wrote of a divine, immaterial spirit – the *iliaster* – which at the time of creation was distributed into four parts or regions (earth, water, air and fire) (Paracelsus, 1894, vol. 1: 203). Each of these served as a matrix or womb for the formation of different substances. Thus minerals might grow in the earth, but they did so, he supposed, from the matrix water, which was imagined to extend like a tree through the body of the earth, with visible external branches in the great rivers such as the Danube or the Rhine (ibid.: 92–3, 232, 241). All these branches terminated in the seas, which, he thought, formed a kind of 'canopy' over the whole aqueous tree.

The notion of the earth as a womb for the growth of metals and minerals stemmed as much from beliefs in the mining fraternity as from scholars or philosophers. I cannot provide a pictorial illustration of such a concept from Paracelsus's own writings, but we can get a fair idea of the nature of the beliefs that were current in the sixteenth century from illustrations that appeared in a later work by the Jesuit scholar Athanasius Kircher, discussed below.

The analogy of a growing seed (which can be traced back to the Stoic doctrine of the *logoi spermatikoi* or 'seminal reasons') was also deployed by Paracelsus. Water, he suggested (ibid.: 244), was like a 'bag filled with seeds of all kinds to be sown'. Somehow the Salt, Sulphur and Mercury were put together within the aqueous matrix to form the variety of mineral substances. This was easy of course, because there were many different kinds of salts, sulphurs and mercuries (though each of the same essence). All this work was somehow performed by a mysterious entity which Paracelsus called an *archeus*.[47] We might also call it an ordering principle.

As has been shown by A.G. Debus (1965, 1991) and several other writers, although Paracelsus did not achieve universal acceptance in his day he gradually acquired a considerable number of devotees, and we find a large collection of his works published in Geneva as late as 1658.[48] An important successor was the Flemish writer and medical man, Johann Baptista van Helmont (1579–1644). He is well known to historians of chemistry for his famous willow tree experiment, in which a willow stick was grown in water for several years, being kept well watered throughout. The stick was weighed beforehand and the sapling after, and (as we would expect, arising from the processes of photosynthesis) there was a considerable increase in weight of wood. Van Helmont concluded that there had been a process of transmutation of water into earthy matter.

Van Helmont's mineral theory was not dissimilar. 'Mettals,' he wrote, 'small stones, Rocky-stones, Sulphurs, Salts, and so the whole rank of minerals, do find their Seeds in the Matrix or Womb of the Waters, which contain the Reasons, Gifts, Knowledges, Progresses, Appointments, Offices and Durations of the same' (van Helmont, 1662: 693). 'Seeds . . . [issue] out of the dark Womb of the Water (which the voice of the Word hath here deposited as durable unto the end)' (ibid.). Here we discern the Neoplatonic concept of the *logos* (word[49]) and also the Stoic notion of 'seeds' ('seminal reasons').

The Stoic notion of 'seed' was, by van Helmont's day, becoming a theoretically interesting concept. It contained 'information' (as we might say) that determined how an object might develop. In some way, a 'seed' had an 'Idea' impressed on it that determined its subsequent development. In the case of human reproduction, 'regular Idea's . . . are implanted on the seed by the lust of the Generator' (ibid.: 548). Analogously, minerals developed in their aqueous matrices according to the Ideas there impressed. The Ideas themselves originated in the mind of God (a Neoplatonic doctrine), but the constructive work in the aqueous matrix was undertaken by His agent, an *archaeus*, which I take to be a kind of personified organizing principle. The whole process was likened to the organic activity of fermentation.

Van Helmont's concepts are never entirely clear from his published writings. Notably, I find it difficult to know whether he thought water was the matrix wherein the production of minerals occurred through a fermentatory process, or whether it was suggested as the material from which other substances were made, as Thales had long before supposed.

It is beyond the scope of this book to attempt to give an exegesis of all the late Renaissance writers who treated the question of the origin of stones, minerals and gems according to Neoplatonic or Stoic principles.[50] But as a further example we may consider the notions of the French Paracelsian Nicolas le Fèvre (*c.*1615–69), also called le Febure, who taught at the Jardin du Roi and later served as 'chymist' to Charles II in England. Le Fèvre (1664, Vol 1: 15) had the idea that there was a universal spirit on which certain characters or 'Ideas' could be imprinted by means of particular ferments. In this manner, the spirit became corporified, yielding a 'grosse body'. He postulated a continual flux of the spirit between the earth and the stars, and thus he advanced the idea of a general cyclic theory of mineral formation and decay. As the spirit emanated from the stars, it passed through the air (where some corporification might occur) and would then become corporeal in either a terrestrial or aqueous matrix. But the process was reversible: '[there are] as two Ladders, whereby the heavenly influences descend down, and re-ascend from the lower parts' (ibid.: 37).[51] Le Fèvre believed that when a substance vaporized it was becoming spiritual in character.

Engagingly, he wrote of the heating of an arsenical mineral. The garlic-like smell of the vapours indicated that some of the arsenical 'Idea' was still retained. Le Fèvre further suggested (ibid., vol. 2: 129) that it was light which by its 'ejaculation and irradiation did imprint and stamp the Idea and Character of its vertue in water as in a general and convenient Matrix'. Thus irradiated, water supposedly acquired a fermentative and generative principle such that it could give rise to metals, or whatever, when it percolated into appropriate matrices.

Another writer of the period, John Webster (1610–82), envisaged 'steams' rising with certain metallic seeds into suitable crevices within the body of the earth and there growing into mature metals (Webster, 1671: 71).[52] The process involved the impress of a 'seminal *Idea*', and this he equated (or conflated) with the Paracelsian *archeus*. Webster's text was one of the most influential seventeenth-century English writings on the mineral kingdom.

We see, then, that there was a gradual emergence of supposed causal connections between the world of the stars and the earth below. So, although such writings grew from the strange brew of Neoplatonism, Gnosticism, Stoicism, alchemy, Christian trinitarianism (which inspired Paracelsus's *tria prima*), and I know not what besides, they were, I suggest, becoming more modern in character in that they were concerned with supposed causal connections, rather than the simple 'correspondences' between earth and sky that had attracted the attention of one such as Ristoro d'Arezzo (thirteenth century).

There were also Aristotelian ideas at work that should not be overlooked, although to take them into account we must go back a little in time. The great German Aristotelian and Dominican priest, Albertus Magnus (*c.*1193–1280), teacher of Aquinas, wrote a major treatise on minerals entitled *Mineralia* (Albertus Magnus, 1967), which endeavoured to account for the nature and formation of mineral substances from an Aristotelian point of view. In this, Albertus advanced some way beyond Aristotle, who had only made incidental mention of the mineral kingdom in his *Meteorologica*.

What Albertus did was apply Aristotle's ideas about the generation of animals and the processes of digestion to the mineral kingdom. Just as in animal reproduction there is, according to Aristotle, the application of the active male 'form' (in the semen) to the passive female 'matter' (the menses),[53] so in the formation of minerals the active qualities, heat and cold, work on the passive qualities of matter, namely moisture and dryness. Albertus thought that the character of the stones depended on the places where they were formed. But as these seemed to be so various (including stones in the body and in clouds (meteorites?)), he found it difficult to make much progress with the problem according to Aristotelian principles.

Albertus did, however, deploy Aristotle's theory of the two kinds of exhalations, moist and dry (see p. 17). But even this was not always an advantage to understanding. For example, Albertus supposed that stones are found at the bottom of streams since there is heat from the river banks acting so as to congeal the vapours penetrating upwards and unable to escape because of the presence of the water in the river (ibid.: 31). Classification of minerals, gems, etc, was simply given according to alphabetical order.

Another writer displaying Aristotelian proclivities, but who had a much greater influence on the development of ideas about the earth was the German scholar and mining engineer, Georg Bauer (*c.*1494–1555), more usually known as Agricola. He is renowned for his magnificent posthumous treatise (*De re metallica*, 1556) on the practices of mining engineering, with its splendid woodcuts of the miners at work in their Saxon mines, and revealing much about the techniques of tunnelling, pumping, transporting ore and so forth. But, for Agricola's most significant ideas about the earth, we must turn to an earlier volume, *De ortu et causis subterraneorum* (1546).

In this work, the idea of a *succus lapidescens* was deployed, the term usually being translated as 'lapidifying juice'. Although it is somewhat anachronistic to do so, it is perhaps not unreasonable to regard the 'juice' as Agricola's equivalent of the modern notion of a mineral-bearing solution. But in saying this the Aristotelian context of Agricola's thinking should not be overlooked. Agricola (1546: 13) stated that his *succi* were produced by the 'coction' (or 'cooking'[54]) of a mixture of dry and humid material by heat, by waters abrading (literally 'licking') the earth, by moisture creeping round metallic matter and corroding it or by the expression of 'juice' from the earth by the action (*vis*) of heat. This last was likened to the exudation of resin from pine trees. The earth could produce *succi* endowed with various qualities, just as a living body could produce various 'humours'. Heat and cold were seen as the proximate efficient causes for the formation of mineral and stony matter, the changes being rung on these two variables in a somewhat desperate manner in order to try to account for the great variety of mineral substances discoverable in the body of the earth. Earth (the Aristotelian element of that name[55]) would become 'fat' or fertile when moistened and cooked by a gentle heat; it became 'lean' if the heat was excessive; or it might remain 'raw' due to excessive cold. Loose earths arose from dense ones (which had been 'thickened' by heat) when moistened. Cold could cause consolidation, while a moderate heat could break up lumps of earthy matter.

The *succus* was not, I think, simply a suspension of mineral matter. Although I have compared it to a mineral-bearing solution, it was perhaps more akin to the 'humours'[56] which (deriving from the theory of the ancient physician Galen) were supposed to fill the body, and which, according to the medical theory used in the sixteenth century, should be correctly balanced for a state of health. To form the *succus*, some kind of coction was required. An appropriate 'inspissation' (thickening) might be brought about by heat to produce solid mineral matter; or some kind of astringent matter had to be present. Thus the *succus* would deposit stony matter when heated, when cooled, on becoming cool or on exposure to the air. It was a versatile entity.

Agricola (1546: 46) spoke with considerable heat against the idea of identifying Sulphur and Mercury[57] with the two exhalations of Aristotle's *Meteorologica*: 'Albertus . . . debased the doctrine of Aristotle with the itch of the [al]chemists flowing with the bloody flux of quicksilver and the stench of sulphur'. The pure Aristotelian tradition had to be nurtured and sustained.

Classification was a fundamental problem for the Aristotelian natural philosopher. By an appropriate classificatory procedure one might hope to display the essential features of any object. So, looking for a classification in an Aristotelian manner, Agricola divided subterranean bodies into those which were 'exhaled' and flowed from the earth and those which had to be dug from the ground (fossils). Fossils could be composite (e.g. a conglomerate) or non-composite. The non-composite ones could be simple or mixed, in which latter case the character of the original constituents could not be discerned. The simple fossils were earths, congealed juices, stones or metals. Eighty-one kinds of earth were noted. The congealed juices could be 'unctuous' or 'harsh'. The metals might be liquid (mercury) or solid. Stones could be 'liquid plus earth' (e.g. halite or natrum); liquid plus metal (e.g. iron rust); or liquid plus mixed substances (e.g. alum). There were those that were hardened by heat, those that were congealed by cold, miscellaneous stones (e.g. magnetite, gypsum, pumice, etc.), gems, marbles and rocks. Thus was knowledge packaged as best it could be according to the principles of Aristotelian science/philosophy. One can hardly say that Agricola's classification displayed the essential features of minerals, rocks, stones or earths, although as an Aristotelian that was his intention.

In looking for 'essential definitions', with the help of a classificatory procedure, the Aristotelian had in mind the notion of deducing the properties of things from the defining essences. To use Aristotle's own example, if one knew that the essence of man was that he was a rational animal, then one could supposedly deduce that he was capable of learning grammar. This linguistic 'property' was supposedly deducible from, or 'flowed from', the defining essential character of man. Looking at the matter another way, one might say that the essence of man – perhaps his possession of a rational soul – conferred the property of being able to learn grammar, or perhaps just to speak.

For Aristotle himself, the link between essence and property was entirely logical. But later Aristotelians contemplated the idea that substances might have certain property-conferring essences or principles which made things the way they were, the connection being chemical rather than logical. This idea underpinned much chemical theory in the sixteenth, seventeenth and eighteenth centuries (Oldroyd, 1976–7) For example, Salt, Sulphur and Mercury were typical property-conferring principles, and as such there was an Aristotelian

Figure 1.3 Representation of Mount Aetna erupting, according to Athanasius Kircher (1678, vol. 1, between pp. 200 and 201) (by permission of the President and Fellows of the Royal Society)

Figure 1.4 The earth's interior, showing its supposed internal rivers and seas, according to Athanasius Kircher (1678, vol. 1, between pp. 186 and 187) (by permission of the President and Fellows of the Royal Society)

component to the Paracelsian theory, even though Paracelsus had no concern with supposed logical connections between his three chemical 'principles' and the properties of substances in which they were supposedly present. And his theory was an elaboration of Arab, rather than Greek, chemical theory.

To gain a picture of what the seventeenth-century man in the street may have thought about the nature of the earth's interior (apart from vague ideas about it being hellish and the abode of Satan), we may refer to the ideas of Athanasius Kircher (1602–80) in his *Mundus subterraneus* (1664–5/1678). A German Jesuit scholar, Kircher was a strange mystical figure, who concerned himself with cosmology, cosmogony, alchemy, magic, numerology, music theory, astrology and much else besides. He was willing to take a good deal of information on trust, for example figuring the legendary Atlantis on one of his maps. But in his

Figure 1.5 Matrices for the growth of metals and ores in the earth's interior, according to Athanasius Kircher (1678, vol. 2: 255) (by permission of the President and Fellows of the Royal Society)

eclecticism Kircher gave expression to the beliefs of his day, and in his illustrations he reveals clearly what many sixteenth- and seventeenth-century writers thought was going on in the earth's interior.

Kircher visited Mount Etna in 1637, and later depicted it (Figure 1.3) as a spiracle or chimney for the chambers and channels of fire which he supposed extended through the body of the earth. The earth itself was thought of as being stationary at the centre of the cosmos and subject to the influences of the sun and stars moving swiftly around it. There were two systems of channels within the earth, one for water and one for fire. The system for the internal rivers and seas is shown in Figure 1.4. If and when the water and fire interacted, earthquakes, volcanoes, springs, and storms resulted. Also, within the earth there were 'matrices', where metals and ores were generated (Figure 1.5).

Kircher called all available theory into action. There were sulphureo-salino-mercurial spirits penetrating the earth, which was also endowed with plastic and magnetic virtues. There were lapidifying juices, a panspermia and seminal principles. Earth, air, fire and water all had a part to play in fermentatory processes. And the earth was a kind of living organism which breathed and received nourishment. Mountain chains were like the skeleton of a body. The various processes of circulation within the earth maintained it in a healthy condition. All this was wonderfully speculative. But in Kircher's pictures we can see something of what Renaissance scholars such as Paracelsus thought about the earth's interior, and how it functioned. It was an organism as much as a physical body.

Another speculative writer who mingled cosmology with cosmogony, chemistry and a theory of the earth was Johann Joachim Becher (1635–82), a German alchemist, chemist, mining engineer and contributor to economic theory. In an early work entitled *Oedipus chymicus*, Becher (1664) adopted the idea (which had been previously found in the alchemical tradition, among miners and in Paracelsus) that minerals and metals were generated by a kind of copulatory process, with the principles of Sulphur and Mercury acting as the male and female entities respectively and with earth providing the matrix for the generative process. The two principles supposedly mingled with one another in the bowels of the earth, with the male Sulphur imparting form to its female counterpart, Mercury (ibid.: 26). However, in an eclectic fashion, Becher (ibid.: 63–4) also sought to reconcile his ideas about Sulphur and Mercury with Paracelsus's Salt, Sulphur and Mercury, but he was led thereby into obscurities rather than clarifications.

In his *Physica subterranea* (Becher 1669/1703/1738) (which provided the first enunciation of the famous phlogiston theory of combustion and other chemical reactions), Becher attempted to give an account of the creation of the globe, in accordance with the ancient Mosaic tradition of the Bible. Supposedly, there was an original chaos, which was designated '*panspermia*' (Becher, 1738: 11). From this, there separated a central globe with overlying aqueous and aerial layers. Two basic components for 'mixtion' (or compound formation, as we would say) were proposed: earth and water. Air was simply an agent ('instrument') for chemical change, but not itself a chemical constituent of things. Heaven (*coelum*) and fire were also proposed as instruments of mixtion. Heaven acted on air, air on water, and water on earth in a stepwise process which produced a gradual change from 'subtlety' to substantiality.[58]

Following the first stages of creation, Becher envisaged a stage of *productio*,[59] in which three possible combinations might be contemplated:

1. Earth + water → stars, animals, and plants, and certain minerals
2. Water + water → snow, frost, hail, ice, etc.
3. Earth + earth → earths, stones, metals.

But, suggested Becher, there were three possible types of earth: 'vitrescible', 'fatty' and 'mercurial'. So, eclectically, he was again linking his theory back to Paracelsus's *tria prima* of Salt, Sulphur and Mercury. In Georg Ernst Stahl's (1660–1734) subsequent version of the theory (see footnote 62), the fatty earth (*terra pinguis*) became known as phlogiston, and the mercurial earth was largely abandoned as a theoretical entity. The fatty and mercurial earths can also be regarded as descendants of the dry and moist exhalations of Aristotelian theory. This might explain why Becher thought that the first (vitrescible) earth (the 'mother' of stones) was found distributed throughout the globe, while the second (fatty) earth (the 'mother' of (ordinary, non-philosophical) earths) and the third (mercurial) earth (the 'mother' of metals) were found at the centre of the globe and in the sea.[60]

Becher (1703: 117–18) claimed that he could synthesize artificial stones and metals directly from his three 'earthy' principles. At some points (ibid.: 121, 123, 126), he referred to his first earth as quartz, so perhaps he was thinking of this as the closest available approximation to his first earth. Since beautiful and rare minerals and metals are often found associated with quartz, it is not surprising that he might also have spoken of quartz (the first earthy principle) as a womb or matrix for minerals, and also as a constituent of them. One might argue that he was not altogether wrong in this. Quartz seems to form a matrix for the growth of some kinds of minerals, and silica is an important constituent of rocks and minerals.

There were also some Neoplatonic and Stoic beliefs to be found in Becher's theory. He stated (ibid.: 128) that the 'idea' of his first earth was expressed in the form of a most subtle and penetrating spirit, which permeated the globe, nourishing ores and mineral veins. Moreover, the 'most special' principle for the formation of a body was deemed to be a specific semen or seed – which was itself supposedly made of earth and water (ibid.: 114). As was the fashion in his day, Becher then postulated a hierarchy of his principles (through *simpliciae* and *compositae* to *decompositae*), and thereby constructed a classificatory system for the mineral kingdom.

So far as the ontological status of Becher's principles was concerned, it is helpful to view them as Neoplatonic in character. For example, quartz was a close approximation to the first ('vitrescible') principle, but was not truly that principle. I suggest that Becher held in his mind's eye the notion of some principle that possessed the 'vitrescible' properties in a perfect or 'ideal' manner – so much so that it conferred vitrescible properties to substances of which it was a constituent. Quartz offered the closest empirical approximation to the 'ideal principle'; and the existence of the principle might have first been suggested by the tangible substance quartz – just as, for a Platonist, the 'ideal' triangle of the world of forms might be suggested by an actual triangle drawn on a philosopher's page.[61]

I have dealt with Becher's work in a little detail, partly because it was so important in the foundation of chemistry – specifically the transition from alchemy to a theory of matter which suggested the hunt for specific constituent substances in objects, and the idea of those constituents as having definite property-conferring powers.[62] But Becher is also interesting for our present purposes because of his speculative chemical cosmogony. His views may seem to us to be arcane ways of thinking about the earth and its constituents; and later writers such as Charles Lyell sought to show that such cosmogonies or theories of the earth were quite distinct from geology proper and that the conflation of geology and cosmogony had been 'the most common and serious source of confusion' (Lyell, 1830–3, vol. 1: 4).[63] Nevertheless, it was from the German (and Swedish) tradition of chemistry, metallurgy, and mining that much of the later work in mineralogy and petrology developed, and this tradition eventually gave rise to geological histories of strata too. In contrast, in the seventeenth and eighteenth centuries British and French authors tended to be more interested in 'natural histories' of fossils and astronomical ideas about the origin of the earth and in 'mechanical' rather than 'chemical' theories about the earth and all its earths.

In fact, Becher lay at the fountain-head of a tradition of chemical mineralogy that was eventually to lead to the work of Abraham Gottlob Werner (1749–1817), who was so very important in the history of ideas about the earth and the emergence of the science of geology at the end of the eighteenth century (see Chapter 4). Stahl republished Becher's works with added commentaries, popularized his ideas and developed them into the fully-flowered phlogiston theory of eighteenth-century chemistry. Among Stahl's pupils were Friedrich Hoffman (1660–1742), Johann Friedrich Henckel (1679–1744), Joachim Juncker (1679–1759), and Johann Heinrich Pott (1692–1777). Henckel spent twenty years at Meissen where he was associated with the celebrated porcelain industry. Later he became Director of the Freiberg Mining Academy and served as Councillor of Mines to the King of Poland and Elector of Saxony. He also became a member of the Berlin Academy. Pott worked for a time with Henckel at Meissen and was also a member of the Berlin Academy. Among Henckel's students at Freiberg was Andreas Siegmund Marggraf (1709–82), who made fundamental contributions to analytical chemistry. Werner himself studied at Freiberg, where he became familiar with the mineralogical tradition of Henckel, which is readily traceable to that of Becher. This tradition may also be carried forward into the nineteenth century, via the many pupils of Werner.

So these men formed an intellectual and practical tradition that gradually elaborated Becher's earlier crude and speculative ideas. The cosmogonical aspects of the theory were jettisoned, and as will be shown in Chapter 3 attention came to be focused on the existence of discrete kinds of earth, such as lime, silica, magnesia and alumina, which appeared to be constituents or components of minerals and rocks.

Before bringing this discussion to a close, however, it may be mentioned that the old 'organic' tradition of thinking about the earth lingered on into the eighteenth century, although by then most theories about the earth were either mechanical (Chapter 2), chemical (Chapter 3), or geological (Chapter 5). For example, the French botanist, J.-P. de Tournefort (1720), writing about a well-known cave in Crete, saw scratches made by visitors on the walls that seemed to have 'grown'. Whereas we might ascribe such a phenomenon to the percolation of lime-rich mineral waters, de Tournefort thought that he had found empirical evidence for the old idea of the organic growth of stone. Thus ancient ideas persisted well into the modern period.

Not only this, but some ideas discussed among geologists in the late twentieth century are in a certain sense returning to the venerable ideas about the earth as a kind of living organism (see Chapter 12). At the very least, it can be said that we are beginning to think in interesting ways about the role of life in the very chemistry of the planet, and the manner in which life may have played a fundamental part in the evolution of its 'body', and its atmosphere. This is not to say that an 'alchemical' picture of the earth is re-emerging. Recent discussions may, however, lead us to think of the seventeenth-century cosmogonists/(al)chemists/geologists in a new light, and perhaps more sympathetically than heretofore.

CHAPTER 2

'Mechanical' Theories of the Earth and Physico-theology

There are two marvellous passages by Owen Barfield describing what it felt like to be a man in the cosmos during the Middle Ages or Renaissance, and then in modern times after the seventeenth-century scientific revolution. Barfield tries to place himself inside the skin of the medieval man in the street and reconstructs the sorts of things that such a man might take for granted.

> [W]e will look at the sky. We do not see it as empty space, for we know very well that a vacuum is something that nature does not allow, any more than bodies fall upwards. If it is daytime, we see the air filled with light proceeding from a living sun, rather as our own flesh is filled with blood proceeding from a living heart. If it is night-time, we do not merely see a plain, homogeneous vault pricked with separate points of light, but a regional qualitative sky, from which first of all the different sections of the great zodiacal belt, and secondly the planets and the moon (each of which is embedded in its own revolving crystalline sphere) are raying down their complex influences upon the earth, its metals, its plants, its animals and its men and women, including ourselves. We take it for granted that those invisible spheres are giving forth an inaudible music . . . As to the planets themselves, without being specially interested in astrology, we know very well that growing things are specially beholden to the moon, that gold and silver draw their virtue from sun and moon respectively, copper from Venus, iron from Mars, lead from Saturn. And that our own health and temperament are joined by invisible threads to those heavenly bodies we are looking at. We probably do not spend any time thinking about these extra-sensory links between ourselves and the phenomena. We merely take them for granted (Barfield, 1957: 76–7).

In our own age – whether we believe our consciousness to be a soul ensconced in a body, like a ghost in a machine, or some inextricable

> psychosomatic mixture[1] – when we think *casually*, we think of that consciousness as situated at some point in space, which has no special relation to the universe as a whole, and is certainly nowhere near its centre. . . . Whatever it is that we . . . call our 'selves', our bones carry it about like porters. This was not the background picture before the scientific revolution. The background picture then was of man as a microcosm within the macrocosm. It is clear that he did not feel himself isolated by his skin from the world outside him to quite the same extent as we do. He was integrated or mortised into it, each different part of him being united to a different part of it by some invisible thread. In his relation to his environment, the man of the middle ages was rather less like an island, rather more like an embryo, than we are (ibid.: 78).

According to this viewpoint, some radical change in the way Europeans thought about themselves in relation to the cosmos occurred in the period of the 'scientific revolution': the time when, by common consent, the modern scientific movement began. Here historians customarily think of the new science ushered in by the likes of Galileo, Bacon, Harvey, Descartes, Boyle, Newton and so forth. Certainly something new occurred in the seventeenth century. There was a quest for clarity, both in the models used to try to comprehend the phenomena of nature and in the language in which the models were to be expressed. There was a changed practical emphasis to knowledge (as Bacon in particular stressed).[2] Experimental and mathematical ways of proceeding were deployed more regularly, and observations seem to have become more exact, as revealed by art and book illustrations. A new social system was developed for the production of knowledge – legitimated by the establishment of well-patronized academies, which published their proceedings through the filter of edited (and later refereed) journals. In particular, the 'new science' sought to construe the complexities of the world with the help of mechanical models, which could often be conveniently expressed in mathematical terms. The old system of virtues, sympathies and correspondences, occult forces and powers, declined. Or so the older histories of science tell us.

Unfortunately, this gratifying story is only partly correct. Even Newton's notion of gravity was regarded by some critics as having the attributes of the supposedly outmoded 'occult' powers. As we saw in Chapter 1, the old 'organic' idea of the cosmos continued to manifest itself in the chemistry of Becher; and his cosmogony was based on a speculative tradition involving such factors as trinitarianism and matter theories derived from Aristotelian, Neoplatonic or Stoic traditions. Kircher's thoughts about the earth were an eclectic mixture of earlier beliefs. Indeed, the chemical theory of property-conferring principles that developed in Germany, and later in France and other parts of Europe, was by no means congruent with the 'new science' as traditionally understood. Yet it was from the German tradition of Becher and Stahl, associated with the long-established crafts of mining and metallurgy, rather than the 'mechanical philosophy' of Descartes and the Cartesians, that the science of chemistry first established itself. This was not necessarily surprising. In its early phases at least, chemistry was inevitably a more 'obscure' subject than (say) observational astronomy. And the German chemists certainly did not stand back from empirical enquiry. Indeed, some of them may be regarded as typical 'sooty empirics'.

Yet there was a theoretical chemistry of a kind – or perhaps I should say a speculative matter theory – that flourished in the seventeenth century which was quite consonant with some of the leading ideas of the mechanical philosophy. And, as will be shown, this 'mechanical chemistry' found a number of applications in seventeenth-century theories of the earth.

By the term 'mechanical philosophy' I mean the following. For the mechanical philosopher, explanations of phenomena – the cosmos and all things therein – were given by means of mechanical analogies. The supreme example of this way of thinking was to regard the cosmos as a huge piece of clockwork. This became very acceptable theologically in the seventeenth century.[3] For clocks have makers or designers, and if the universe was like a clock then it too must have had a designer. One could also think of matter as being made up of minute corpuscles, which might interact with one another mechanically like mini- billiard balls.

There were two versions of the mechanical theory of matter: the first spoke of corpuscles, which might be of various sizes and shapes and might be divided or joined; the second likewise envisaged the ultimate particles of matter as tiny corpuscles – but they could never be divided or 'wear out'. This second version was a kind of atomic theory. It was a direct intellectual descendant of the atomic theory of the ancient world and was reintroduced into European intellectual life by the French philosopher/theologian Pierre Gassendi (1592–1655). However, so far as ideas about the mineral kingdom were concerned, Gassendi was a kind of transitional figure. He subscribed to the doctrine of mineral and metallic 'seeds', but supposed that the seeds were to be thought of as somehow atomic in nature (Gassendi, 1678, vol. 8: 321). And when it came to explaining the shapes of crystals he eschewed the organic model and simply suggested that the external shapes arose from the shapes of the constituent atoms. For example, diamonds had octahedral particles; amethysts had hexahedral units (ibid.: 268).

The virtue of the mechanical theory of matter was that it seemed intelligible.[4] One could easily imagine corpuscles (or atoms) joining together in various ways; and it was plausible that their micro-shapes could give rise to the particular shapes or properties of macro-objects. Or the pressures of gases, and the gas laws, could be understood in terms of the properties of the constituent particles behaving like mechanical systems.[5] Moreover, the mechanical philosophy held out the promise of a successful mathematization of nature, even though one might be puzzled how action at a distance might occur.[6] Of course, as our extract from Barfield quoted above shows, action at a distance seemed simple enough to the medieval man in the street. But, once propounded, the mechanical philosophy was received with enthusiasm, even though, when pressed, some of its concepts were obscure.[7]

In France, the chief spokesman for the mechanical philosophy was the philosopher René Descartes (1596–1650), and in the major 'textbook' exposition of his scientific views, his *Principia philosophiae* (1644), he detailed an interesting and influential theory of the earth, along with an exposition of his astronomical theory. Some details of this are given here.

As is well known (e.g. Aiton, 1972), Descartes supposed that matter – or substance – and volume were one and the same: the greater the volume the greater the quantity of matter. This proposition Descartes regarded as self-evidently clear and distinct, and thereby true; and since volume could be divided (mathematically) without limit, so too could matter. There could, of course, be no vacuum, according to Cartesian principles. In Descartes's cosmogony, it was supposed that God imparted an initial motion to the matter of the universe. The basic type of motion was circular or in vortices. This avoided the problem that linear motion would leave spaces or vacua 'at the ends', so to speak, which to Descartes was a contradiction in terms. As a result of the motion of the corpuscles and their mechanical interaction, three basic types of particle began to be generated. There were (1) small spherical corpuscles formed by the rubbing of original larger pieces; (2) the very fine 'rubbings' that filled the interstices between the spherical corpuscles; and (3) larger particles that might be formed by the coalescence of the smallest kinds of particles, or were perhaps residues from the larger pieces arising from the original division of matter. These three types of corpuscle corresponded roughly with the air, fire and earth of earlier theories, although Descartes supposed that the first kind of matter was that by means of which light was transmitted (as a 'pression') through the sky.

In a somewhat naïve manner (though certainly one that maintained the Cartesian desideratum of being clear and distinct), Descartes supposed that the different objects of the macro-world might be made up of appropriate corpuscles in his hypothetical micro-world. For example, quicksilver was supposedly constituted of rather large, round and smooth corpuscles; oils might have slithery spaghetti-like corpuscles; and so on. In his cosmogonical 'just-so story' Descartes imagined fiery particles (second element) collecting at the centres of celestial vortices to form suns. But 'blisters' might form at the cooler surfaces of these bodies, rather like slag collecting on the surface of the metal in a furnace. These 'blisters' were sunspots. If they accumulated to a sufficient extent, the whole surface of a sun might be covered and its light extinguished, and it would become a body with a solid crust. In such circumstances, it might be susceptible to capture by another vortex (Descartes was not very clear – to me at any rate – why this would be so), and then the 'dead' sun would form a planet in a different solar system/vortex, the material of the collapsed vortex being absorbed into that of one of its neighbours.[8]

Meshing with his cosmological theory, Descartes was also able to propose a theory for the gradual evolution of the earth, and he conveniently represented for us four stages of the process. These are shown in Figure 2.1. In this figure we have the first stage represented by the top right quadrant of the figure (at A), with the next three stages in the succeeding quadrants, clockwise. I represents a region of fiery particles. M is a compact and opaque region (formed by the coalescence of sunspot material). C is a layer of irregular branched corpuscles. D is a liquid layer (later to form the waters of the oceans) consisting of flexible water corpuscles. E is another layer of solid material. And A and B represent the earth's fluid atmosphere.

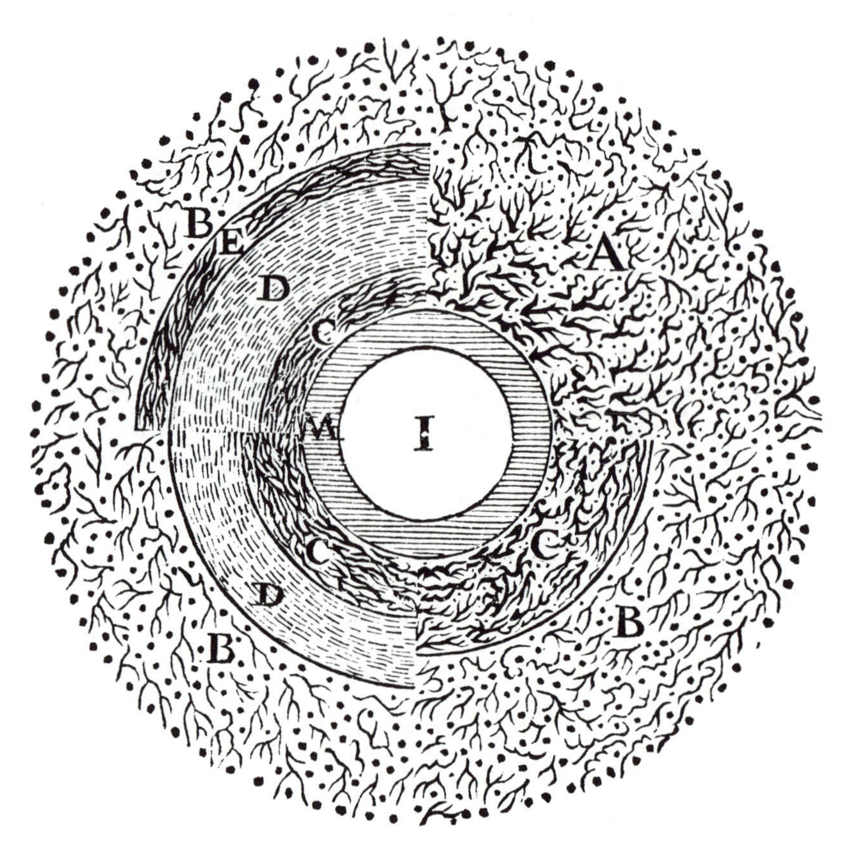

Figure 2.1 Four stages in the development of the earth, according to René Descartes (1656: 151) (by permission of the President and Fellows of the Royal Society)

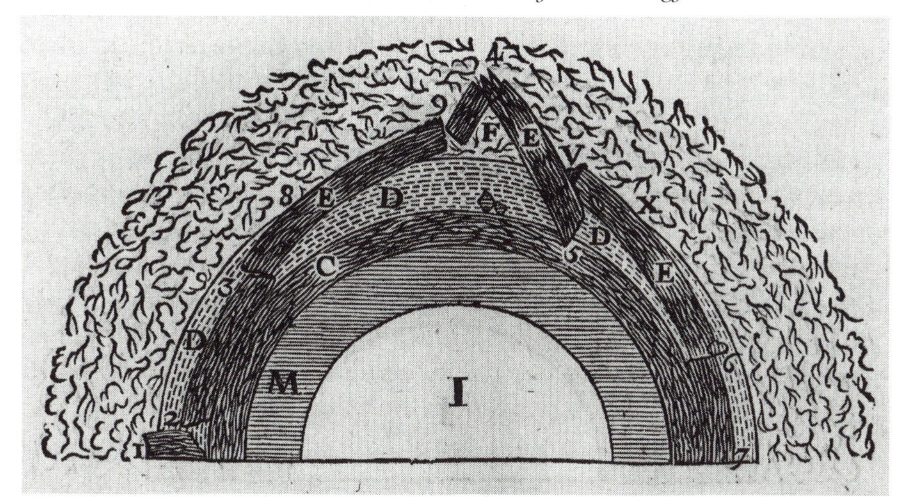

Figure 2.2 The further development of the earth, according to René Descartes (1656: 155) (by permission of the President and Fellows of the Royal Society)

Then, Descartes imaginatively suggested, in hot weather corpuscles of the liquid layer D might be squeezed up through pores in E and, on combining with aerial particles of region B, might form further material for layer E, and consequently be unable to return in cooler weather to the region D whence they came. The process being repeated over a number of years, 'spaces'[9] might develop below E, until eventually rupture of the crust could occur, yielding collapse structures, as shown in Figure 2.2. In this manner, Descartes endeavoured to develop a theory of the earth that accounted for the existence of mountains (as at 4 in Figure 2.2), with the occurrence of waters both above and below the main crust (E) as at 2 and D, in accordance with the commonplace observation of the existence of oceans and also the biblical tradition of the existence of subterranean waters.[10]

The main point to note here, I suggest, is the development of a speculative 'theory of the earth' on mechanical principles, which was designed to give an adequate 'story' as to how the earth, known to us by observation, might have come into being. The story was intermeshed with Descartes's general cosmology and cosmogony. One might suppose, therefore, that something rather similar might have occurred elsewhere in the cosmos: it was compatible with the conception of a 'plurality of worlds'. But the envisaged time-scale, though not specified precisely, seems to have been quite short – given that Descartes supposed that major geological changes were induced over a period of (seemingly) just a few years.

This brings me to an important point. In the seventeenth century there was, as said, the emergence of the modern scientific movement. This had side-effects in many aspects of cultural life, not least that of the world of religion.[11] Moreover, with the developments of the Reformation, there was increasing emphasis on the strict and literal interpretation of the Bible. The reason for this is not hard to find. In the older Catholic tradition, the source of religious authority was the Church itself, but for Protestants this source had been discredited. So one was enjoined to read the Bible in order to understand how to behave, what the point and purpose of life were about and even some broad outline of cosmology and earth history. However, if the information about the origin of the earth vouchsafed in Genesis were untrue, perhaps the whole Bible might be suspect – and then what would become of the Protestant faith?

So in the seventeenth century, and more particularly in the Protestant parts of Europe, we get a serious effort to show that there was a satisfactory harmony between the words of the Bible and the emerging ideas of the new science. Thus arose the movement known as physico-theology. It involved a kind of interweaving of science and religion: the theology supported the new science, and reciprocally the science supported the theology – at least, that was the intention. It might do this most effectively by showing that the cosmos, the earth and its living inhabitants manifested design. This line of reasoning was actively pursued from the seventeenth until well into the nineteenth century, and still has some devotees to this day, although the term 'physico-theology' was not much used after the eighteenth century. Natural theology came to be the preferred term.

Thus, whereas in earlier years the ark of Noah had been seen in allegorical terms as representing the body of the Church and was figured thus in iconography (Allen, 1963), in the seventeenth and eighteenth centuries a more literal and 'scientific' approach was taken, and quite elaborate calculations were undertaken to try to estimate the size of the ark, the number of animals it might have been able to convey, the amount of food that would have to have been stored to feed them all and the amount of waste matter that they might have produced. There was, I believe, a mad consistency in the whole approach – and this can even be said for the modern descendants of seventeenth-century physico-theologists, the so-called creation scientists. They recognize that, if some parts of the Bible are shaky, then the divine origin of the whole is in question and in principle all may be doubted.

An interesting corollary of all this was the gradually shortening time scale that affected Western ideas about the earth in the seventeenth century.[12] Indeed, the Bible, if taken literally, seemed to offer a viable way of calculating the age of the earth. One had to work backwards through the history of Israel described in the Old Testament, using the family trees therein, until one reached back to Adam himself. Such a task was attempted by no less than Sir Walter Raleigh, the well-known Elizabethan sailor (and buccaneer?) (Raleigh, 1614[13]), and many others,[14] but the most famous effort of the kind, in Britain at least, was undertaken by James Us(s)her (1581–1656), Archbishop of Armagh and Primate of Ireland, who calculated a very precise date of 23 October 4004 BC (Usher, 1658: 1):

> In the beginning God created Heaven and Earth, *Gen.* 1. *v.* 1. which beginning of time, according to our chronologie, fell upon the entrance of the night preceding the twenty third day of *Octob.* in the year of the Julian [Roman] Calendar, 710 [= 4004 BC].
>
> Upon the first day therefore of the word, or *Octob.* 23. being our sunday, God, together with the highest Heaven, created the Angels. Then Having finished, as it were, the roofe of the building, he fell in hand with the foundation of this wonderfull fabrick of the world, he fashioned this lowermost Globe, consisting of the Deepe, and of the Earth; all the Quire of Angels singing together, and magnifying his name therefore. . . . And when the Earth was void and without forme, and darknesse covered the face of the Deepe, on the very middle of the first day, the light was created; which God severing from the darknesse, called the one day, and the other night.

As has been pointed out by Gould (1993), drawing on the investigations of Barr (1985), Usher's work in fact involved solving an exceedingly intricate problem. The biblical chronicle of the history of Israel record is not complete, notably from the time of Ezra and Nehemiah to the birth of Christ, so one has to try and cross-link the biblical record with that provided by other cultures such as those of the Chaldeans, Persians and Romans. Further, the Jewish chronology was based on lunar months. There was the problem of leap years, which had led to confusion in the old Julian (Roman) calendar and its replacement by the Gregorian calendar of Pope Gregory XIII in 1582. But the Anglican Usher wanted to have nothing to do with a 'popish' calendar. So to produce any sort of dating that was satisfactory to Protestant scholars was in fact an intellectual *tour de force* on Usher's part. Yet history has allotted him the part of some benighted obscurantist.

Be that as it may, since the Usher dates were sometimes printed in the margins of the King James Authorized Version of the Bible (for example the 'Bishop Lloyd' reprint published by Oxford University Press, 1701), they were seen by many on Sundays or at family prayers.[15] It was Barfield's man in the street's view being given written sanction by the 'experts'. Usher sought to give the precision to his scholarship that was expected in the age of the new science – the coming age of reason. Also, perhaps unconsciously, he deployed the metaphorical language characteristic of the mechanical philosophy. The world was a 'fabrick' with a roof that had to be constructed. Light was 'severed' from darkness.

Such a brief time-scale had a remarkable effect on thinking about the earth.

The whole of earth history had to be collapsed into a few thousand years. Yet the history was perhaps plausible on empirical grounds, for the following reason. Much of northern Europe is smothered with 'boulder-clay', and there are also many 'erratics' left by the (geologically) recently departed glaciers of the last glacial epoch – or so modern geology tells us. This glacial detritus could easily be construed as the residuum of the Noachian Flood. So, well into the nineteenth century, the biblical history of the earth had many supporters and seemed warranted by empirical evidence.

Descartes's theory, outlined above, also shows just the beginnings of the emergence of the idea of a short time-scale with a scientific warrant. In France, there were numerous disciples of the Cartesian system in the latter years of the seventeenth century and into the eighteenth century. Even among those who might have preferred to regard themselves as chemists, rather than physicists, the typical Cartesian approach to matter theory manifested itself, appearing in a number of writings that had to do with the earth or with rocks and minerals. Always, the claim for the 'mechanical' view of things was that it was (to use Descartes's words) 'clear and distinct'. The point was well made by the Cartesian J. Rohault (1620–75):

> From the Nature of Stones both precious and common which we have now laid down, I don't see how we can deduce certain Properties which are mentioned by the Writers of Natural History: As, for Instance, that a Blood-Stone, worn by the Person who has the *Bloody-Flux*, will stop the Distemper, and that other Stones will cure other Distempers. And therefore we find by repeated Experiments, that these sorts of Properties are falsely ascribed to the greatest part of these Stones (Rohault, 1723, vol. 2: 162).

This was the negative side of the argument. The old ideas of sympathies and correspondences were repudiated as the scientific revolution took hold. The positive side was in the direct application of mechanical ideas to explain the nature of rocks, crystals, minerals, etc. For example (ibid.: 159–60), the production of sand grains was ascribed to the mechanical interaction of (Aristotelian) 'exhalations' with 'fine terrestrial particles'. Or the formation of six-sided crystals from an initially fluid condition was ascribed to their having being consolidated from plastic spheres pressed together, 'every Piece of Crystal . . . [being] surrounded and compressed between six others'.

Another example of 'mechanical mineralogy' is provided by the French academician, Etienne François Geoffroy (1672–1731) (Geoffroy, 1716). He envisaged two kinds of corpuscle for the formation of stones: uniform fine thin plates and varied irregular particles. Crystals were formed from the first kind; stones resulted from an admixture of the second. Thus, there was a kind of scale, from purest rock crystal to coarse and irregular stones. The presence of the first type supposedly aided fusion by heat. Material constituted of the second kind underwent calcination on heating.[16]

Such ideas were taken up and developed by Geoffroy's fellow-academician René Antoine de Réaumur (1683–1757), in a series of papers presented to the *Académie* in 1721, 1723 and 1730. He envisaged the 'lapidifying juice' as being formed by the mechanical attrition of stony matter. A hooking together of corpuscles in appropriate ways could give rise to crystalline growths in cavities in the earth. Or there might be some rocks with fine pores such that only aqueous particles could squeeze through. So deposition of mineral crystals or metallic ores from the juice might occur. Réaumur was also interested in the properties of earth. He supposed that sand had solid angular corpuscles, whereas those of soil were porous. The addition of water to dry earth might bring about an interlocking of corpuscles. But the corpuscles must be somewhat flexible to account for the properties of moist clay. Other ideas were proposed, and even a little experimental work was described, which to my knowledge represents the first experimental investigations in sedimentary petrology.

The Swiss naturalist, Louis Bourguet (1678–1742), went so far as to offer a speculative quasi-mathematical account of the processes of lapidification. In an essay that sought to give an account of the crystalline structure of belemnites,[17] he pointed out that the mass of a particle decreased according to the third power with decrease in diameter, while the surface area decreased according to the second power (Bourguet, 1729: 41). So fine particles could be held in suspension or solution by a moving fluid, but, if the movement slowed, coalescence or crystallization might be expected. Again, different kinds of crystals might be attributed to the different kinds of constituent corpuscles or '*particules intégrantes*' (ibid.: 42). But speculative mechanical mineralogy did not advance much beyond such proposals. And from about 1730 onwards it was largely superseded by the chemical approach of the Stahlians, although this, as was observed in Chapter 1, had more ancient roots.

It may occur to the reader that the foregoing remarks may seem to be at odds with the account of eighteenth-century natural history that was given by Foucault, which I also mentioned in Chapter 1. This is true. But the works of the mechanical mineralogists and the chemists were not one and the same as natural history.[18] Nor indeed were they very successful. While the principles of the mechanical philosophy were beautifully 'clear and distinct', they were also highly speculative so far as the inner constitution of matter was concerned. For the natural historian, it was generally better to eschew such speculations and attend to the external, visible, parts and to classify objects accordingly. This is what the majority of eighteenth-century naturalists endeavoured to do, for all three kingdoms of nature. The difficulty, however, was to find the appropriate natural system of classification, a problem that remains right up to the present, particularly with respect to biological classification.

Returning to the question of physico-theology, there was a major difficulty for naturalists in relation to the question of fossils. The word 'fossil' originally meant no more than something dug from the ground. But, for the present purposes, I shall simply consider the term as having the same meaning as it does today: fossils are the remains or traces of animals, plants or other living organisms that have been preserved by some means in the earth's crust.

Suppose, then, that we wish – as physico-theologists – to effect an interweaving of our empirical knowledge of the earth's crust and its contents and the biblical history of the earth. Fossils present a problem in a way that would have been quite absent for ancient writers such as Herodotus. According to the 'biblical' view of the earth, ratified by Usher's calculations, it is quite a youthful object – perhaps 6,000 years old. It was formed by a divine artificer and was in a pristine condition at the time of Adam and Eve. Since then, the only 'known' significant event that might have produced 'geological' changes was the Noachian Flood.

But fossils presented a major problem for such a way of thinking. They are often found embedded deep in rocks, far inland or near the tops of mountains. They may also be very different from any organisms known today.

Several possible solutions might be offered to these enigmas. As to the difference between modern and fossilized organisms, this might be ascribed to the fact that the organisms were still extant, but had not yet been discovered. After all, only a fairly small part of the earth's surface had been explored by Europeans in the seventeenth century; and forms unknown to the moderns might one day be found in the hidden depths of the ocean (as, in fact, did sometimes subsequently occur). But that did not deal satisfactorily with the inland fossils or those found high in the mountains.

So one solution was simply to deny that the fossils were organic in origin. For example, the York physician and Fellow of the Royal Society Martin Lister (1639–1712), reacting to ideas published by Nicolaus Steno (which will be discussed in the next chapter) and basing his argument on the fact that the material of fossils was generally similar to that of the surrounding rock, proposed that fossils were just peculiar stones (*lapides sui generis*) formed in the rocks without living creatures ever having been involved in the process (Lister, 1671: 2282). The Oxford naturalist Robert Plot (1640–96) suggested that fossils were generated within the earth by some 'plastick virtue' (Plot, 1677: 111–12). But how that 'virtue' operated was not clear; and it was hardly in keeping with the principles of the mechanical philosophy. Edward Lhwyd (1660–1709), who prepared a fine illustrated catalogue of the fossils in the Ashmolean Museum, Oxford (Lhwyd, 1699),[19] found himself quite unable to believe that the Noachian Flood was responsible for the entombment of fossils and was driven to offer the unlikely suggestion that shells 'might be partly owing to fish-spawn, receiv'd into the chincks and other meatus's of ye earth in the water of the Deluge', which spawn had then developed in the rocks (Lhwyd, 1945: 389).[20] So, although Lhwyd could not understand how the Flood might have emplaced full-grown shells physically, he nevertheless felt pressed to deploy it as a kind of geological agent, for want of better.

Lhwyd's correspondent, the celebrated Essex parson-naturalist John Ray (1627–1705), also obviously troubled by the problem, considered it at some length in his *Miscellaneous Discourses Concerning the Dissolution and Changes of the World* (Ray, 1692a). He weighed up all the views and arguments on both sides of the question of the nature of fossils and their implications for the biblical history of the earth, and seemed to reach the conclusion that there was at the time no satisfactory solution to the problem. At least, he offered none (ibid.: 104–32). In effect, he put the problem in the 'too-hard basket'. He did, however, hold that the Noachian Flood had played a major role in shaping the earth's surface.

Ray played a prominent part in advancing the 'design argument' in Britain. And his texts give a clear picture of the way in which the people of his day saw the earth as an object of beneficent intention:

> That the earth should be made thus . . . with so great variety of parts, as mountains, plains, vallies, sand, gravel, lime, stone, clay, marble, argilla, &c. which are so delectable and pleasant and likewise so useful and convenient for the breeding, and living of various plants and animals; some affecting mountains, some plains, some vallies, some watery places, some shade, some sun, some clay, some sand, some gravel, &c. That the earth should be so figured as to have mountains in the mid-land parts, abounding with springs of water pouring down streams and rivers for the necessities and conveniences of the inhabitants of the lower countries; and that the levels and plains should be formed with so easie a declivity as to cast off the water, and yet not render travelling or tillage very difficult or laborious. These things, I say, must needs be the result of counsel, wisdom and design (Ray, 1692b: 190–1).

Thus did the devout parson think about his earth – although fossils unquestionably puzzled and worried him.

Of the several writers in the seventeenth-century British physico-theological school, the case of the physician John Woodward (1665–1728) is perhaps the most interesting. Woodward was an avid collector of fossils, and the materials that he gathered together formed the nucleus of the great palaeontological collection named after him and held at the University of Cambridge. Unlike Lister and the proponents of the inorganic origin of fossils, Woodward had no difficulty in accepting the idea of their being the remains of former living creatures. But how did they get into the rocks?

Writing in the period of the great impact of Newton's theory of gravitation and wishing to account for the presence of fossils deep within rocks, he proposed (Woodward, 1695: 29, 74) that, at the time of the Flood, all stones and minerals 'lost their solidity'. It was as if God had 'switched off' gravity, so to speak – which would, I suppose, be perfectly feasible for an omnipotent being. Then all was swept up into one gigantic promiscuous sludge of water and earth; and after the Deluge was over (when, I suppose, God switched gravity back on again) layers of sediment settled out, in order of decreasing density, with their fossil contents entombed therein. How Noah and his charges were supposed to survive all this I do not pretend to know – it was, perhaps, achieved by miraculous means.

As to the fossils, Woodward was questioned by Elias Camerarius as to why they did not fall to pieces when the force of gravity was suspended. His answer was ingenious and in keeping with the principles of a Cartesian-style mechanical philosophy. The organic corpuscles, suggested Woodward (1726: 94–5), were of intricate shape, interlocked with one another. So the organisms would retain their shapes and structures even if the cohesive forces were not acting.

Years ago, when I first became acquainted through secondary sources with the writings of these British physico-theologists, I thought they must have been slightly mad or singularly stupid. For historians such as Adams (1938/1954), the problem was that the naturalists of the seventeenth century were insufficiently careful observers. But this was not, I think, the problem, as may be seen from the most beautifully drawn illustrations of fossils prepared for Robert Plot's *Natural History of Oxfordshire* (1677) (Figure 2.3). Such careful attention to detail shows that the physico-theologists did not fail in geological theory by reason of inadequate observation. On the contrary, the detail and beauty of their observations were inspired by their desire to know God's works. Indeed, according to a view among historians of science that has some currency, the source of modern science is to be found in religious beliefs, as much as in (for example) economics, geographical exploration or advances in technology.

So to think of the seventeenth-century physico-theologists as stupid or unobservant is anachronistic. Their basic assumptions (protopostulates) were such that they were inevitably faced with intellectual difficulties when it came to explaining the problem of fossils. For all that Woodward's speculative hypotheses may strike us as bizarre, they were conformable to his theology and his mechanical view of the nature of matter – also to the idea of Newtonian forces. But his science was different from ours. It was trying to work within a very limited time-scale; and it was prepared to entertain *ad hoc* hypotheses for which there was very little scope for independent testing. But, in his day, Woodward was a man whose opinion counted for something. He developed elaborate systems for the classification of the mineral kingdom (e.g. Woodward, 1728), and, as I say, his fossil collection was one of considerable extent and importance.

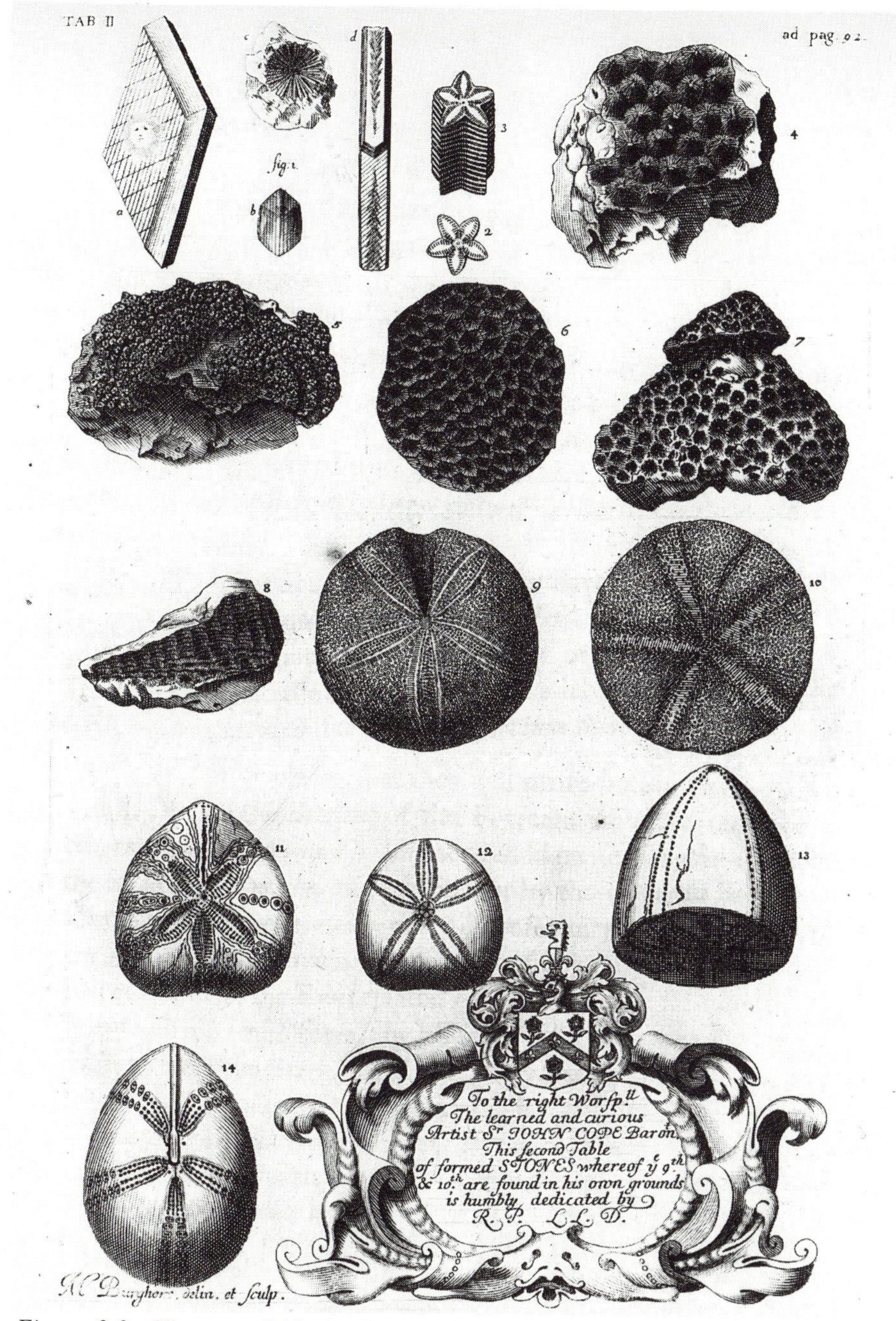

Figure 2.3 Figures of 'formed stones' (fossilized corals, sea urchins, etc.), prepared for Robert Plot (1677, facing p. 92) (by permission of the President and Fellows of the Royal Society)

I would also emphasize (with Herries Davies, 1989) that the physico-theologists that I have been discussing worked within a tradition of antiquarian (or archaeological) and philological research rather than geology as we understand the term. Lhwyd was chiefly a philologist. Woodward was a collector of antiquities. Lister studied Roman inscriptions. Given the belief in the small age of the earth, structures such as Stonehenge or the burial chambers of southern Britain were thought to have been made not very long after the Creation itself. So a study of such archaeological sites naturally linked up with a study of the history of the earth. Early human history and early earth history were intimately connected. It might be as appropriate to talk about physico-archaeology (where ancient records or traditions are corroborated by digging in the ground) as physico-theology. The Anglican physico-theologists were astute and assiduous scholars, with a programme of research that was beginning to enter the field, rather than being restricted to the study or cloister, as had been Bishop Usher.

Mention of Newton in connection with Woodward brings to our attention Newton's contribution to an understanding of the shape and size of the earth, which led to an elaborate controversy with the French Cartesians, which was eventually settled in favour of the Newtonian cosmology. Some points are worth mentioning, as they certainly have much to do with thoughts about the earth, although they are perhaps more strictly concerned with histories of cosmology and geodesy than with geology. Newton was a physico-theologist *par excellence*, his gigantic intellectual effort constantly inspired by an attempt to interweave his religion with his science.

Apart from a few remarks in his correspondence, Newton wrote nothing of significance on geological matters. His great goal was a mathematical description of the cosmos and its understanding in terms of his celebrated laws of motion and the idea of a universal force of gravity, operating according to the inverse square law. Newton did not have much to say about the origin of the solar system, other than the fact that it was the product of divine will. (He was, however, interested in the age of the earth, for which he adopted an essentially biblical time-scale. And he did some work – utilizing his knowledge of the rate of precession of the equinoxes – which endeavoured to offer a reconciliation of biblical chronology with what was known of the histories of ancient Greece, Rome and Egypt.[21])

Newton's cosmos was an orderly place, heliocentric or Copernican in character. There were the innumerable distant stars, of which the nearest was our sun. The planets moved round the sun in stable elliptical orbits, and moons 'ellipticated' some of the planets. All were interacting with one another because of the mysterious force of gravity, which operated according to the inverse square law. The earth itself rotated on its axis, and this explained the apparent diurnal motions of the sun and stars. The whole system was a grand perpetual motion machine, for there were no frictional forces to impede the cosmic motions, outer space being presumed to be a void. Also, in 1687 Newton (1934) showed that his three laws of motion, Galileo's law of free fall, the inverse square law and Kepler's laws of planetary motion[22] were all mutually compatible.[23]

With the earth rotating on its axis, and having presumably first formed out of some chaos of matter (as the Bible stated), one could assume that it was at first more fluid than at present. From this it would follow that the shape of the earth would be that of an oblate spheroid. Descartes's theory offered a quite different picture. There was no void anywhere. (By Cartesian protopostulate, 'anywhere' was constituted of matter.) The great system of vortices – infinite in extent – was such that matter was transmitted from one vortex to another, from the equatorial region of each vortex into the poles of adjacent vortices. In keeping with this view, Cartesian theory supposed (although it didn't really have a proof) that each planet was the shape of a prolate spheroid. Thus there was a difference between the Cartesian and the Newtonian theories which might offer the possibility of an empirical test being made to differentiate between the two.

For Newton, such a test would probably have seemed superfluous. His theory was capable of mathematical expression and could be tested precisely by means of astronomical observations. In contrast, the theory of vortices was, as Newton (1934: 543) put it, 'pressed with difficulties'. It was a 'qualitative' theory – albeit admirably clear and distinct. It had no hope of accounting for Kepler's laws, and in particular the precisely formulated mathematical expression of his third law.

But the Cartesians were not to be talked down so easily. Considering Kepler's third law, while it was relatively simple to determine a planet's orbital period, determination of the size of its orbit with precision was another matter altogether. So, well into the eighteenth century, the debate between the Newtonians and the Cartesians had not finally been settled. Eventually, it was brought to a close by the physical survey of the earth to determine whether it was an oblate (Newtonian) or a prolate (Cartesian) spheroid.

In order to accomplish this task, one had to measure the number of miles in a degree at different latitudes, i.e. near the equator and also towards the pole.[24] (A degree was determined astronomically and also by chaining.) This task was undertaken by a number of French expeditions in the eighteenth century. Of course, they had other objectives than settling the old Cartesian–Newtonian dispute. Nevertheless, it was still an important issue. And eventually, through the efforts of and controversies between men like Jacques Cassini (1677–1756), Pierre Moreau de Maupertuis (1698–1759) and Charles-Marie La Condamine (1701–74), the figures came out[25] showing that the earth was indeed an oblate spheroid, although it was only slightly flattened. So, by the middle of the eighteenth century, it was fully established that it was a spheroidal object, spinning round the sun in empty space, and its actual dimensions were known, by refinements of the method pioneered by Eratosthenes back in antiquity.

As Georges Cuvier was later to put it: 'Genius and science ha[d] burst the limits of space; and observations, explained by just reasoning, ha[d] unveiled the mechanism of the universe' (Cuvier, 1813: 3–4).[26] But the question of time was another matter altogether. The earth was boxed into a time frame of a few thousand years. There were rocks, minerals and fossils. These were collected, named and classified. But that they might be used to read a history of the earth was scarcely realized by the persons mentioned in this chapter. Indeed, the study of earth history, other than for antiquarian purposes, might seem a superfluous undertaking. For the essential history of the earth was already known, by courtesy of sacred writ.

The manner in which the history of the globe came to be written on the basis of empirical studies of its strata and their fossil contents, rather than mythopoeic principles, studies of human history or abstract 'mechanical' (or other) theory, is the subject of the next few chapters. In broaching this subject, we approach the beginnings of geology as an independent scientific discipline, and something significantly different from the study of the mineral kingdom, alongside the animal and vegetable kingdoms, as the source of objects to be collected, named and classified.

CHAPTER 3

The Beginnings of Geological Science: Detachment from Cosmogony and Mineralogy

What is it about geology that makes it a science distinct from a branch of natural history or theories of the earth? The term derives from the Greek words *Gaia* and *logos*. *Logos* means 'word', 'discourse' or 'reason', and in modern English usage gives us a word-ending for some science or branch of study, as in biology, astrology or whatever. Geology, then, means the science that treats of the earth; or rather less generally it is the science that investigates the earth's crust, the layers thereof (strata), their mutual relations and the changes that have brought them into the situation and condition that we see them to possess today. Defined in this way, geology is a historical science. It endeavours to work out the history of the earth (or at least its crust) by examination of the strata and the objects contained therein. It utilizes knowledge of minerals, fossils, *etc*., but the work of the mineralogist or palaeontologist is different in that it does not necessarily have a temporal or historical dimension. The mineralogist is sometimes content to collect, classify and display his or her materials, or discover their chemical constituents, without concern as to how or when they were formed.[1] The palaeontologist is sometimes only concerned with the classification of organisms, or their reconstruction from fragmentary remains, although some concern with age or time is nearly always present today.

Geology is also distinct from cosmology or cosmogony. We have seen examples of the latter in the work of Becher and Descartes. These were speculative schemes. Descartes looked at the earth in a general way, with its land masses, mountains, plains, springs, rivers, *etc*., and then devised a 'just-so story' – compatible with his general matter theory and his ideas about cosmogony and astronomy – that could give an account of how such things on the earth came to be. But essentially, Descartes started with his 'clear and distinct' notions of matter and motion and gave an account of the earth on this aprioristic basis. In contrast, the geologist (say of the nineteenth century) typically examines rocks, minerals, fossils and strata, enters information about these entities on a map and, with the help of available theories (or new ones that may be devised as required), endeavours to work out a plausible sequence of events that could have given rise to the disposition of the strata as observed at the present time. He (or more recently s/he) seeks to provide a history of a portion of the globe. Eventually, all the surface of the globe would be mapped, and a composite history could be drawn up. By this time, one might be in a position to theorize more generally about the origin of the globe, its internal constitution and the grand processes by which it has been formed and by which it changes.

In fact, geology has departed significantly from this nineteenth-century programme. As will be described in subsequent chapters, modern geological theory is much concerned with the inner, invisible parts of the globe and needs to know something about such matters in order to make sense of the history of the external parts. Also, some geologists interest themselves in theories of the origin of the earth. Even so, cosmogony falls more in the domain of the astronomer than that of the geologist.

At the time of writing, a grand theory about the structure and behaviour of the earth over time has emerged, and has attracted widespread acceptance. This is the theory of plate tectonics, which will be discussed in Chapter 11. The establishment of this theory involved, in a sense, returning to the grander goals and aspirations of the men who wrote 'theories of the earth' in the seventeenth and eighteenth centuries, but which were mostly eschewed in the first half of the nineteenth century as being too ambitious and generally speculative.

So in this chapter, and the two that follow, we shall be interested examining the emergence of geology as a science – not geogony, 'theories of the earth', mineralogy or earth science (a term which, since the development of plate-tectonic theory, the rise of geophysics and geochemistry and the interest in the earth as a planetary object, has to some extent displaced the older term geology).

As might be expected, the process of the emergence of geology from natural history, 'theories of the earth', etc., began before the term geology was actually coined.[2] And it was a gradual rather than instantaneous process. We can see early intimations of the process of 'detachment' in England with the work of Robert Hooke (1635–1703), which was quickly followed up by Nicolaus Steno (1638–86) in Italy, and then a number of other Continental writers such as Lehmann, Füchsel and Arduino.

The polymathic Hooke, well known as an important figure in the early Royal Society of London, opponent of Newton, inventor, experimentalist, astronomer, architect, microscopist and much else besides, is interesting as a transitional figure in that while offering a 'theory of the earth' in the older sense, he also saw the possibility of reading a 'history' of the globe from its strata and their contents. Moreover, he thought it possible to corroborate his ideas from the mythological and historical information that had been transmitted from antiquity. But being so taken up with many other concerns, his ideas were not followed through successfully and remained largely programmatic in character.

Like his contemporaries such as Robert Plot (see Figure 2.3), Hooke was keenly interested in the occurrence of fossils in rocks. Coming from the south of England, he was familiar with the remarkable giant ammonites that may be found in several places along the south coast. He observed such objects minutely, as may be seen from the figures of fossils published in his *Posthumous Works*, (Hooke, 1705), which were delineated with even more care than those of Robert Plot.

Hooke never doubted, as did some of his previously mentioned contemporaries like Martin Lister or Johann Joachim Becher, that such objects were indeed the remains of former living creatures. And their presence inland indicated to Hooke that there had been alterations in the relative levels and positions of land and sea. In discussing the enigmatic fossil remains in his 'Lectures and discourse on earthquakes' before the Royal Society in 1668 and later in 1687, Hooke proposed that the earth was an irregular sphere,[3] surrounded by an oblately spheroidal envelope of water. The earth had a polar axis that was approximately fixed with respect to the heavens (although it changed its direction slowly and steadily, giving rise to the phenomenon of the precession of the equinoxes). But, Hooke proposed, the earth had a slow motion additional to that of its daily spin on its axis, its annual passage round the sun and its precessional motion. There might, he suggested, be a further motion, such that the equatorial regions gradually moved towards the poles and the polar regions moved towards the equator. In other words, he hypothesized the idea of polar wandering on a grand scale.[4]

Two major effects would be expected to follow from such a motion: (1) the forces acting on any given parts of the earth's crust would gradually alter (because the centrifugal effects are greater at the equator than at the poles); (2) because of the spheroidal shape of the earth's envelope of water, parts of the surface that were once land would pass under water and become layered with sediments, while other parts would gradually become exposed and subject to erosion. Earthquakes might result from the gradual alteration of the magnitude and direction of the forces acting on different parts of the crust, and cause collapses and slippages of strata. The second effect (slow successive exposures or submergences of land) would give rise to cyclic episodes of erosion and deposition. Thus the occurrences of fossils in places distant from the sea might be explained. Hooke went further and speculated that while certain forms might have perished by exposure or drowning, others might have come into being (he didn't suggest exactly how) because of the changes in environmental circumstances. Then he went even further, suggesting that one might 'raise a *Chronology* out of them [i.e., the fossils], and . . . state the intervalls of the Times wherein such . . .

Catastrophies and Mutations have happened' (Hooke, 1705: 411). In this passage, we have the first intimation that a history of the earth might be compiled by examining its fossil contents. This would, I suggest, have been the true beginnings of historical geology had the programme been carried through.

But it was not carried through. Hooke was a desperately busy man, ensconced in London and without the leisure to go out into the field to undertake such an enterprise. In any case, as can be seen from the remarks above, his contribution was essentially another speculative theory of the earth. Yet there was something special about Hooke's theory, for he thought of a way of testing it empirically. It occurred to him that, if the positions of the poles were moving as he envisaged, then the direction of the meridians would also be slowly shifting. To test this idea, he devised an accurate method for directing a long telescope at the pole of the heavens. Then, by hanging two plumb-lines from the telescope, one could draw a line on the ground corresponding to the direction of a meridian. Come back ten or twenty years later and repeat the determination and perhaps one might find that the direction of the meridian had changed, thus supporting the hypothesis of pole wandering.

However this suggestion was but a plan of research, rather than something that Hooke actually accomplished. He did try the meridional determination with the members of the Royal Society in attendance one night (Birch, 1756–7, vol. 4: 527), but the sky was cloudy, and the project was forgotten so far as an astronomical test was concerned. But Hooke did not quite give up. It is clear from the above that he was, like his contemporaries, thinking of a relatively short time-scale for earth history, otherwise there would have been no possibility of detecting changes in the direction of the meridian in a human lifetime.

For such reasons, no doubt, his contemporaries rejected his ideas out of hand. Indeed, it is well documented (Turner, 1974; Oldroyd, 1989) that the natural philosophers in Oxford repudiated Hooke's whole theory since, they said, it was well known, from biblical history, that events such as Hooke envisaged had *not* taken place. Hooke was more or less forced to agree, but in his later years he spent an inordinate amount of time combing through ancient texts such as Ovid's *Metamorphoses* and Plato's account of the legend of Atlantis, to see whether there might be evidence hidden in the texts that the earth had indeed had a turbulent history, as implicit in Hooke's theory of earthquakes.

That is, Hooke went back to ancient texts (including some in the Bible) that we might regard as chiefly mythological in character to see whether they might furnish evidence in favour of a modern scientific theory (in this case Hooke's theory of pole shifting). This procedure may be called Euhemerism,[5] but I prefer the term 'physico-mythology' (Birkett and Oldroyd, 1991), which is suggestive of an intimate interweaving of science and studies of myth. It chimes well with the physico-theology of Hooke's day, discussed in Chapter 2.

Again, Hooke's project was unsuccessful. Though willing to make liberal interpretations of ancient texts, in the long run his efforts led nowhere, and to my knowledge they had but little influence on the history of geology, interesting though they were as specimens of seventeenth-century ways of thinking about the earth. Myths were too weak a reed to lean on for writing a history of the globe. And the very idea that human texts, rather than strata, could provide the empirical information needed to write the geological history of the globe or proof of a theory of the earth, indicates that Hooke was not a geologist as we understand the term today. As was said, he did have the idea of looking at rocks and fossils and 'raising a chronologie' out of them. But in the event nothing came of the suggestion so far as Hooke himself was concerned. He turned to his library and his physico-mythology as a way of thinking about the earth. This was not geology.

It is a nice question whether Hooke's contemporary, Nicolaus Steno (1638–86)[6] should be regarded as a geologist, for though best remembered today for his ideas about fossils and the earth's strata, he was trained in medicine and ultimately devoted his life to the Catholic Church. Steno was born in Denmark, where he received his initial medical training. After further work in Holland, Paris and Montpelier and after obtaining important results in his anatomical investigations, he travelled to Italy, where he obtained a position as court physician to the Grand Duke of Tuscany and converted to Catholicism. Steno did not, however, remain many years in Tuscany and he returned to northern Europe, where he undertook the thankless task of trying to spread the Catholic gospel in the Protestant world. However, while in Tuscany, he had the opportunity to travel extensively in some of the mountainous areas of the region, and was able to prepare a small introductory treatise, dedicated to the 'Most Serene Grand Duke' (Ferdinand II), which set forth the results of Steno's observations and his theorizing about the earth. The small Italian dukedom was taken to serve as an epitome for understanding the geology of the whole globe.

As a skilled anatomist, Steno was requested to undertake the dissection of a huge shark captured near Leghorn in 1666. In carrying out his dissection, Steno naturally made a close examination of the animal's teeth, and came to the conclusion that they were so similar to objects that had been dug from the ground in Malta and elsewhere that the latter could *only* be sharks' teeth. This was an important conclusion. Earlier writers on these objects from Malta had thought that they might be serpents' tongues or birds' tongues, or they might just have grown in the earth. A very ancient tradition recorded by Pliny held that they just fell from the sky on moonless nights. Another legend had it that when St Paul had been shipwrecked on Malta he had been bitten by a viper and had cursed the snakes of the island, making them harmless; the 'tongue stones' (*glossopetrae*) were the petrified teeth of the snakes. It was not a trivial thing to assert that such objects were the remains of once living fish. As mentioned in Chapters 1 and 2, many thinkers were troubled by fossils, and some came up with extravagant explanatory theories. (Leonardo's ideas were both exceptional and unknown.)

But, if the 'tongue stones' were indubitably petrified sharks' teeth, the question as to how they might be found within sediments had to be answered. Steno undertook to answer this question, but in a completely general and rather abstract manner: '*[G]iven a substance endowed with a certain shape, and produced according to the laws of nature, to find in the substance itself clues disclosing the place and manner of its production*' (Steno, 1968: 141). He undertook precisely this task in a short treatise, dedicated to the Duke of Tuscany, and curiously entitled *The Prodromus [introductory work] to a Dissertation on Solids Naturally Contained Within Solids*, published in Latin in 1669.[7]

Both form and content of the *Prodromus* deserve comment. As can be seen from the italicized words quoted above, Steno was endeavouring to solve a problem in general terms from first principles. Thus the treatise had some of the features of a mathematical text. Indeed, had the subject been more susceptible to mathematical treatment, the text could well have emerged looking a bit like Euclid's *Geometry*. But, needless to say, studies of the earth were not then amenable to formal mathematical exposition.

A second point that the reader may notice is that the *Prodromus* was beholden to Descartes's theory of the earth and to his matter theory. Thus Steno tried to account for the phenomena of crystallization in terms of the accretion of Cartesian-type corpuscles. This point need not delay us particularly here, but his method of depicting crystals is interesting. It will be noticed from Figure 3.1 (1–7) that different specimens of the same kind of crystal are shown as maintaining a constancy of angle between homologous faces. And Figure 3.1 (14 and 17) shows 'cut-out' representations of the faces of two specimens of the mineral haematite in a manner that clearly foreshadows the methods used by eighteenth-century naturalists such as Linnaeus (see p. 193). It has been suggested by Ellenberger (1988: 282) that Steno may have found how to generate a solid of the form required by Figure 3.1 (17) by making suitable bevellings of a cube. If this is so, then Steno's method of crystal analysis and depiction was essentially the same as that developed by Romé de l'Isle over a hundred years later (see p. 195). But Steno was not looking forward to the eighteenth century. *His* concern was the fact that geometrically structured crystals were *not* akin to organic fossils.

Even more important for present purposes were Steno's generalizations about strata. These included the idea that sediments were deposited from fluids and would originally have been deposited with their layers horizontal; that if all the particles of a stony stratum are of the same nature and fine size they were 'produced at the time of Creation from a fluid that then covered all things' (ibid.: 163); that if there are fragments of other strata or remains of animals or plants in a stratum, then the stratum was *not* produced at the first creation; that if a

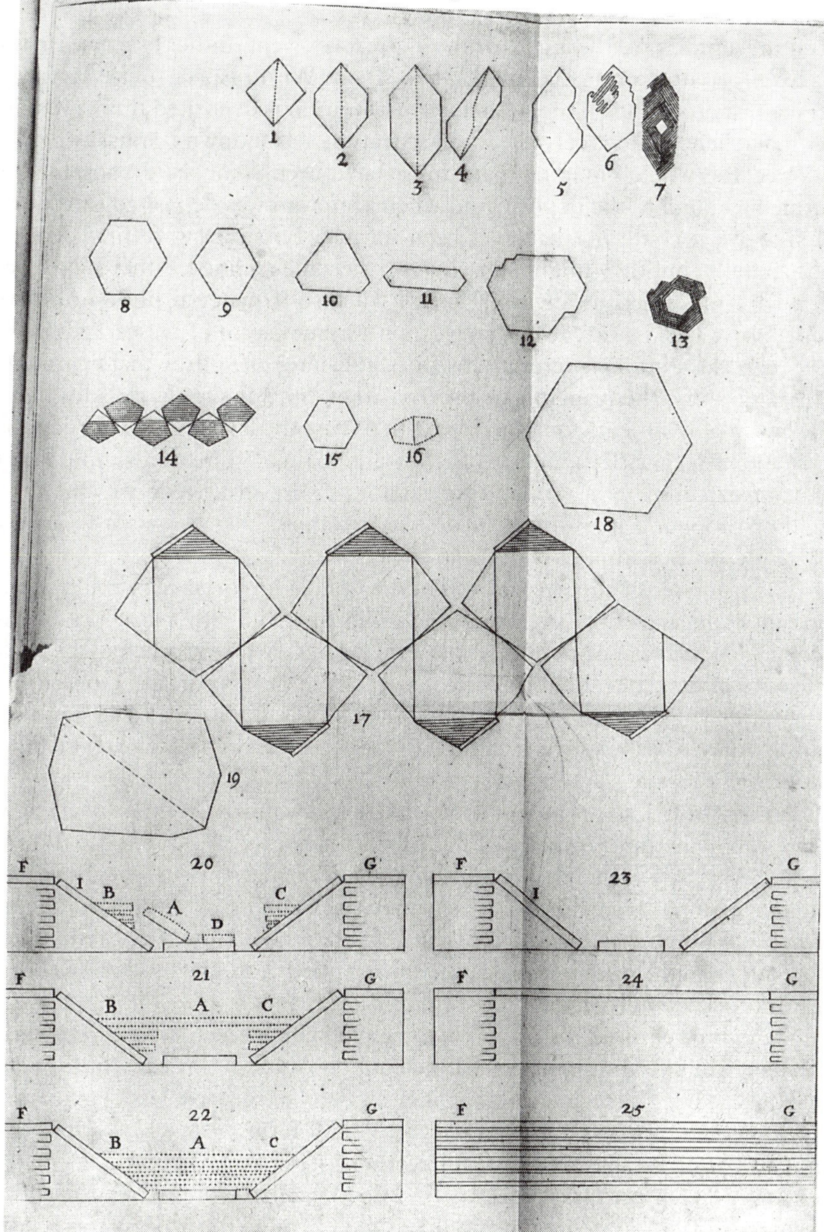

Figure 3.1 Examples of Nicolaus Steno's procedure for the geometrical analysis of crystals, and his schema for representing the geological history of the valley of the Arno River, near Florence (Steno, 1669, unnumbered folded page, preceding p. 1) (by permission of the President and Fellows of the Royal Society)

stratum contains tree branches or the like it was 'swept thither by a river in flood or by a torrential outbreak' (ibid.); and so on. With respect to the layering of strata, it was obvious to Steno that when a stratum is deposited there must have been one beneath to receive it; when a stratum is laid down it must either have covered the whole globe or there must have been some solid substance that formed a bound to the deposit; and when a stratum was deposited the one now seen above it could not have yet been formed. So strata were first deposited horizontally, but they might subsequently become inclined, either by the 'sudden flare of subterranean gases' or by collapse if material underneath were somehow washed away, which suggestion is reminiscent of Descartes's theory.

Steno is traditionally credited with establishing one of the major principles of stratigraphy: the principle of superposition. This is simply the idea that for the layers of sediments found in the earth's crust the lower ones were deposited first and are older than the ones that lie above them. This is a simple idea (or may appear to us to be so), and we cannot be sure that Steno was the first to hold or propound it. But the *Prodromus* is the oldest text known to state the principle, at least implicitly if not in so many words.

The principle of superposition is of the utmost importance in geology, for it provides a means of linking what one sees in the field with a discoverable time sequence. Hence, in principle, we have the necessary theoretical tool for writing a history of the globe, or at least some part of the earth's surface. One can work backwards from what one sees of the earth's present surface to an account of how it came to be the way it is, getting the sequence of events in correct temporal order.

In point of fact, the principle sometimes lets one down. As is now well known, there can sometimes be large-scale overturnings of strata due to earth movements. Also, the principle says nothing about changes due to the intrusion of molten material from below, which may, in some cases mean that the geometrically lower rock is younger than the one above. However, Steno himself only intended the principle to be used for stratified sediments and in this context it works successfully much more often than not.

Steno himself used his principle to help elucidate the temporal sequence of events leading to the situation that he could observe in Tuscany in the seventeenth century. His reconstruction of events was summarized in a famous set of six sections, reproduced in Figure 3.1 (20–25). Here we see (25) the layers of the earth as originally laid down at Creation – or so Steno supposed. Then (24) the lower layers were washed away by a flood. Collapse of the uppermost layer occurred (23). More sediment was deposited (22), and the process was repeated (21), giving the rocks as we see them today (20). (Note that the number order is the reverse of the temporal order envisaged by Steno.)

Although this sequence appears to be highly schematic, and hardly related to any specific geological site, it is possible that it does in fact correspond to an actual locality in Tuscany.[8] If this is so, Steno's model was not a pure abstraction, but was devised to help him understand some very solid rocks in a particular part of Italy. Notably, the strata such as I and A of Figure 3.1 were lithologically distinct. So at least two stages of sedimentation could be distinguished.

It appears that Steno's patron, the Grand Duke of Tuscany, was well pleased that his small dukedom was an exemplar for the geology of the whole globe. But it was perhaps this feature which made Steno's geology unhistorical as we would understand the term today. His principle of superposition provided a 'tool' for the historian of the earth. But although the tool was of general application, the result of its use in Tuscany was not. When historical geology was fully launched in the late eighteenth/early nineteenth century, it became clear that the task was to piece together all the little 'subhistories' such as that which Steno proposed for the Arno Valley. And for this purpose some techniques for cross-correlation were required. These Steno did not provide, and leaving Tuscany for northern Europe he soon lost touch with geology, entered the priesthood and became wholly engaged in the affairs of his Church.

It should be noted also that Steno's time-scale, like that of Hooke and others in the seventeenth century, was extremely short. His events in Tuscany were squeezed into the biblical time frame; and it seems that he believed that there were but two major periods of sedimentation: one at the time of Creation, and the other at the time of the Noachian Flood. A geology that had to submit to such time restraints was (I suggest) impoverished and limited in application. This said, it must again be emphasized that Steno did establish the all-important principle of superposition; and he looked at rocks directly and wondered how they might have been formed – in ways that were compatible both with his religious beliefs and his understanding of the (Cartesian) mechanical philosophy. His *Prodromus* was, then, a further example of physico-theology; and in so far as it was interwoven with the myths of Genesis I might also call it a specimen of physico-mythology, although Steno did not, like Hooke, specifically turn to myths in order to find support for his ideas about the earth.

The persons I have mentioned so far in this chapter may properly be referred to as scholars, although in the seventeenth century, as one of the concomitants of the scientific revolution,[9] scholars began to work in a more practical fashion and take note of the skills and practices of artisans – as Francis Bacon recommended. This is well seen in the work of Newton, with his alchemical experiments and his laborious efforts at grinding and polishing mirrors and lenses. Robert Hooke was an experimentalist *par excellence*. So too was his aristocratic patron, the Honourable Robert Boyle (although perhaps he planned and oversaw experiments rather than conducting them himself).

There was, however, a much older empirical tradition, so far as the earth was concerned, namely the technical skills and practices of the mining trade. Much mining was carried on in the ancient world, as is evidenced, for example, by the reports of Pliny (1962). But for written accounts of mining practices in Europe in the Middle Ages and the Renaissance, nothing survives from before the beginning of the sixteenth century. The earliest known publications are some anonymous booklets (some undated) on mining and assaying (the *Bergbüchlein* and the *Probierbüchlein*[10], which originated in the mining districts of Saxony and were perhaps first published in 1505 (Sisco and Smith, 1949: 50). These booklets, illustrated by crude woodcuts, gave simple accounts of the procedures used in the mining and assaying of metals and ores. The series of booklets on assaying represent some of the earliest printed indications of the emancipation of practical chemistry from alchemy, although the theory of ore formation in the *Bergbüchlein* was 'organic',[11] revealing the ground in which Becher's theories, discussed in Chapter 1, were nourished. More detailed writings of higher quality in the way of printing soon followed, such as Vannoccio Biringuccio's *De la pirotechnia* (1540), and Lazarus Ercker's *Treatise on Ores and Assaying* (1574/1951). But by far the most famous work in this tradition was Georgius Agricola's *De re metallica* (1556;1912/1950), which by its charming woodcuts (frequently reproduced) gives intimate glimpses into the practices of mining in the sixteenth century.

Thus within the mining communities there was considerable practical knowledge of ways of identifying the ores of the different metals, such as gold, silver, lead, copper, tin, bismuth, mercury, and iron, and the metals themselves. Methods for assaying gold and silver were of special importance, the technique of cupellation being used and 'touchstones' deployed. The assayers (and the bullion merchants) fully understood the use of the balance in their assay work, long before chemists acknowledged its use and formally introduced the concept of the principle of the conservation of mass. Moreover, to all practical purposes, the assayers deployed the idea of elements as simple substances, occurring as the last terms of analysis, long before Antoine-Laurent Lavoisier (1743–94) defined elements thus in his *Traité élémentaire de chimie* (Lavoisier, 1789a). So while the chemists of the seventeenth and eighteenth century, up to the time of Lavoisier's 'chemical revolution' when 'modern' chemical theory was initiated, wrote copiously on notions of Aristotelian elements, of abstract chemical principles, or on mechanical explanations of chemical phenomena – and made rather little headway with the matter – the practical men carried out their work successfully, apparently unimpeded by the chemical theories of the time.

In the period between, say, Agricola and the time of the chemical revolution (i.e. the 1770s and 1780s) there was an uncomfortable ambiguity about the concept of 'earth'. From one perspective, the term denoted an abstract theoretical entity, ultimately traceable to the Aristotelian earth (with its elementary colleagues, water, air and fire). But at the same time, we see in the eighteenth century the emerging recognition that there were several *different kinds* of simple earths – notably silica, lime, alumina, magnesia and baryta – which might appear as the end products of analytical work.

One approach towards discovering the constituents of earthen bodies was to try to synthesize desired products (such as porcelain) by heating together various kinds of earths. This was tried by Johann Heinrich Pott (see p. 40) in a vain attempt to discover the secret of Meissen porcelain, but although he tried thousands of different combinations, his efforts led nowhere. Pott (1753, vol. 1: 7) did, however, recognize four broad categories of earthy substances: calcareous, gypseous, argillaceous and vitrifiable (or siliceous). These behaved differently under the action of heat and acids. Pott didn't really extract simple earths from raw materials; but his recognition of siliceous earth as a special kind was important. It was, of course, what had long been known of in crystalline form as quartz.

Knowledge of lime as a particular kind of earth goes back to antiquity, as it is readily formed by heating limestone or marble and is the essential ingredient of mortar. But it was only in the eighteenth century that it became recognized as one of a family of earthy substances with somewhat similar chemical properties.

The first recognition of 'argillaceous earth' (alumina) as an individual chemical entity – as opposed to the muds and clays of common experience – seems to have been due to the German chemist Andreas Siegmund Marggraf (see p. 40), his ideas being presented to the Berlin Academy in 1753 (Marggraf, 1762). He dissolved alum in water, filtered the solution, and precipitated it with alkali, giving 'earth of alum'. He was also able to extract much the same substance from clay by treatment with vitriol (sulphuric acid), followed by precipitation with alkali.

Marggraf (1762, vol. 2: 411) also confirmed a suggestion made at the beginning of the century by Friedrich Hoffmann (1660–1742) that there was a specific kind of earth, magnesia. Marggraf showed that there was an earthy component in the salt contained in the mother liquor of brine that was different from the alkaline component of the salt itself.

Baryta was discovered by the Swedish chemist Carl Wilhelm Scheele in 1774, in the course of his investigations of the manganese ore, pyrolusite (Scheele, 1901: 52–95). The presence of baryta in 'heavy spar' (barite: barium sulphate) was demonstrated by Scheele's countryman, J.G. Gahn (1745–1818).

In fact, the main developments in mineral chemistry in the latter part of the eighteenth century were as much in Sweden as Germany. Axel Cronstedt (1722–90) systematized and popularized the techniques of blowpipe analysis (Oldroyd, 1974a). This involved very simple, portable apparatus – a charcoal block, a candle, a mouth blowpipe, and a few chemical reagents – which enabled the prospector to raise a piece of ore or whatever to quite a high temperature and see what visible changes occurred. This wasn't strictly chemical analysis – rather, a simple kind of diagnostic test. Even so, it greatly facilitated the recognition of similar mineral bodies that had otherwise been identified as distinct kinds.

The great breakthrough in mineral analysis occurred when the Swedish chemist T.O. Bergman (1735–84) published a *general* scheme of mineral analysis in the wet way (Bergman, 1777). A mineral or rock (or, so far as Bergman's initial publication was concerned, a gem) was first brought into solution by fusion with alkali. Bergman knew the general properties of the different 'simple earths', mentioned above, and, by a skilfully devised sequence of precipitations, filtrations, and solutions (using water and different alkalis and acids), he was able to effect a separation of different earths from the original material. Bergman's first effort was, we now know, hopelessly inaccurate. But the general approach proved viable, and soon gifted chemists like Nicholas Vauquelin (1763–1829) and Martin Klaproth (1743–1817) took up the principles of Bergman's method and greatly improved it. Hence the chemical compositions of many minerals and rocks were gradually determined, particularly in the early nineteenth century. The technique was time-consuming and, with the clumsy apparatus of the time, often using reagents of doubtful purity, the task must have been extremely difficult. But with the guide offered by the principle of the conservation of mass (now enunciated by Lavoisier), there could be a kind of bookkeeping check on the procedures; for the sum of the masses of the earths extracted in the analytic procedures was expected to equal the mass of the mineral or rock with which the chemist started.

As said, such procedures grew from the tradition of mining, which became conflated with assaying procedures and ultimately with analytical chemistry. The miners themselves had their own tacit knowledge and their own vocabulary, some of it derived from the old organic traditions about the earth – mineral *veins* and *matrices*, ore *bodies*, *mother*-lodes and so on. The miners spent their working hours in the *bowels* of the earth. A comprehensive list of vernacular rock terms has been compiled for the English language by Arkell and Tomkeieff (1953), and doubtless similar glossaries, unknown to me, are available for other languages.

But, while miners knew how to excavate pits and wells, drive shafts, pump out mine waters and recognize signs (in the form of rocks, minerals and fossils) that might indicate profitable or unprofitable digging, and while such practical knowledge increased and improved considerably in the eighteenth and nineteenth centuries concomitantly with the early Industrial Revolution, mining did not, in itself, furnish new ways of thinking about the earth. The miners dug in the earth and had rules of thumb as to how to proceed if a coal seam, for example, were cut off suddenly at a fault. Eventually, in Germany, there arose also a new kind of 'philosophical mining', which gave rise to the first establishment of the basic outline of the stratigraphical column by Abraham Gottlob Werner, lecturing at the Mining Academy in Freiberg. Consideration of this very practical form of geology will, however, be deferred to Chapter 4.

Before discussing Werner and his school, we should refer to some more general eighteenth-century activities that served as foundations for the emergence of geological theory, and which grew from the knowledge of miners and mining engineering. It is my view that the emergence of geology as a science that we might recognize as such today – as opposed to the study of natural histories of the mineral kingdom, or the collection and classification of fossils – occurred when it came to be realized that the earth had a history that might be deciphered by examination of the rocks and fossils of the earth's crust. We have already noted intimations of this in the ideas of Hooke and Steno; but, as we have seen, these seventeenth-century writers, partly due to their particular personal circumstances, did not carry through the ideas that may seem (to us) to be implicit in what they wrote. In both cases, a fundamental problem, limiting the elaboration of their ideas, was that of time (or lack of it) in theorizing about the earth, because of constraints flowing from literal readings of the Bible. So the rise of historical geology did not occur until towards the end of the eighteenth century, after more fieldwork had been done on the surface of the earth, not necessarily in mines within it.

Nevertheless, important developments occurred in the earlier years of the eighteenth century, providing a wider empirical base on which subsequent changes in ways of thinking about the earth could develop. In part, the increased knowledge consisted of information about rocks, minerals and fossils, and their classification by a variety of means. To an important extent, the wider empirical base was furnished by interest in the earth's economic aspects, and so the mining industry again played a significant part in the developments.

In the eighteenth century, the French were the first to establish a scientifically based topographic survey of their nation, this being useful for military and commercial purposes. To solve the problem of the shape of the earth (see p. 57), the French government established the Paris Observatory, employed eminent astronomers such as the several members of the Cassini family, and sent expeditions to various parts of the globe to measure the length of a degree at different latitudes. In France itself, the younger Cassinis (Jacques II and his son César François) worked assiduously preparing a detailed map of the kingdom, which was completed just in time for the French Revolution.[12] Thus during the course of the eighteenth century, good topographical maps of France became available. Intended for commercial and military purposes, they also assisted the development of geology in France.

Primarily for commercial reasons, but also out of scientific interest and as part of the eighteenth-century French map-making movement, the naturalist Jean-Etienne Guettard (1715–86) undertook the massive task of preparing a mineral map of the kingdom. His aims were limited, and his project was not the production of what we would call a geological map. Rather, the map, or at least its first version, which was presented to the Académie des Sciences in 1746 (Guettard, 1751), simply showed the localities of different kinds of mineral

Figure 3.2 Mineralogical map of north-west France and southern Britain, prepared by Phillipe Buache to illustrate Jean-Etienne Guettard's Académie memoir (1746/1751) (by permission of the President and Fellows of the Royal Society)

substances in France (and parts of England). The character of the work is displayed in Figure 3.2. It should be noted, however, that Guettard and Buache linked up strata of similar rock type on opposite sides of the English Channel. Such extrapolation, or induction, is of the utmost importance in geological map work.

In 1766, with financial support from the French government and with the assistance of the young Lavoisier, Guettard began a more detailed survey of France, with the intended production of no less than 214 sheets. But the work, when eventually published in 1780 (Guettard and Monnet, 1780), had only sixty sheets, although a small-scale single sheet for the whole of France appeared in 1784 (Guettard, 1784). Guettard himself did not complete the full map, and thirty-one of the sheets were produced by the mineralogist Antoine-Grimoald Monnet (1734–1817). Lavoisier worked on the project for a relatively short time in its early stages. The final outcome of Guettard's *Atlas* was, therefore, a rather miscellaneous production (Rappaport, 1969).

Although intended chiefly to indicate the occurrences of ores and economically important materials such as limestone or coal, Guettard's map did show three main *bandes*: sandy, marly, and schistose or metalliferous, the last of which included schists, slates, granites, marbles and metallic ores. However, Guettard did not seem to have any view of the stratigraphical relationships between the three units. This is not surprising, given that he did not think of the deposits as having been formed historically (ibid.: 279). Indeed, he supposed that large sedimentary deposits were laid down at once, and that the appearances that we attribute to bedding were due to drying and cracking that was for some reason in a horizontal direction. (He did, however, recognize a difference between 'primitive' and sedimentary rocks, the former supposedly having been formed at the earth's creation and the latter subsequently.[13])

In contrast, the young Lavoisier, drawing on ideas that he had learned from his chemistry teacher G.-F. Rouelle (1703–70), recognized a difference between pelagic deposits (*e.g.* some mudstones) and littoral deposits (*e.g.* some sandstones or conglomerates), and found that these might occur one on top of the other. This suggested the idea of marine incursions and regressions, and so Lavoisier began to think on the lines of a historical sequence of events leading to the appearance of the rocks that he was surveying and entering on the maps. He also drew up general sections (*coupes générales*) to illustrate his ideas,[14] and the maps. Guettard seems to have accepted these as being useful. But, wishing to eschew theory, he did not move toward the historical view of the strata adumbrated by Lavoisier (1789b). And the younger man soon moved off to devote himself chiefly to chemistry, marriage, and money making (as a tax official). So his novel geological ideas had rather little impact in their day, being overshadowed by his own contributions in chemistry. One may say, however, that Lavoisier displayed tremendous potential as a geologist, and would almost certainly have made as great a contribution to geology as he did to chemistry if he had sustained his thinking about the earth as a major preoccupation. But already specialization was beginning to get a grip on science.

One of the most important regions of France, geologically speaking, that attracted Guettard's attention was the Massif Central of the Auvergne district.

Here he recognized that there were lava flows that might be traced back to the cone-shaped hills, chiefly made of heaps of ash and cinder, near the town of Clermont-Ferrand; and hence he realized that the hills were themselves formerly active volcanoes (Guettard, 1752/1756). But it was some years before Guettard (1779, vol. 1: 128–54) came to think that basalts were of igneous origin, for with their sometimes columnar structure (due to cracks occurring on cooling) it was usual to regard them as having crystallized out of some fluid – perhaps even a primeval ocean.

The basalts of the Auvergne were interpreted as being of igneous origin by Nicholas Desmarest (1725–1815), Inspector of Manufactures at Limoges, in a paper presented to the Académie in 1765 (Desmarest, 1774, 1777). The previous year he began a detailed mapping of the numerous lava flows of the region near Mont-Dore (a popular hillside spa resort in the Auvergne), eventually publishing his work in full as late as 1806. The great merit of this work was that Desmarest constructed a *history* of the various lava flows, recognizing that they could be regarded as of different ages according to the degree of weathering and erosion that they had undergone and whether they were or were not still connected to the volcanic orifices from which the lava must once have poured. Desmarest also noticed that a lake (Lac Aidat) had been formed by the damming of a stream by a lava flow (Figure 3.3[15]). Thus, for a small region at least, he was able to look at the rocks and reconstruct hypothetically the sequence of events that had given rise to what he was able to observe. He was, then, 'doing geology', as we would understand the process. However, by the time Desmarest finally published his work on the Auvergne (*i.e.* Desmarest, 1806), the notion of a geological history of the globe was close to being well established among 'earth scientists'.

In Germany, Johann Gottlob Lehmann (1719–67), who taught mining and mineralogy at Berlin and who later joined the German scientific community in St Petersburg, first published a treatise on ore bodies and their 'matrices' which lay well within the tradition of Becher and the German miners (Lehmann, 1753). Shortly thereafter, he published an interesting volume *Versuch einer Geschichte von Flötz-Gebürgen* in 1756 (French edition, Lehmann, 1759), in which the idea was developed that there were three types of rocks. The first of these, making up the *Gange-Gebürgen* or veined mountains, were supposed to have been formed at the original creation of the earth and so might be regarded as 'primitive' (or primeval). They were generally high, sometimes in long ranges, and were made of crystalline rocks, often with mineral veins. After the devastating Noachian Flood, which supposedly affected the whole globe, much material was deposited in basins in sedimentary layers, forming the *Flötz-Gebürgen* (layered mountains) (*C* to *K* in the top left-hand picture of Figure 3.4 – *C* being at the bottom of the layered sequence and *K* at the top). The third type, chiefly volcanoes, were supposedly due to 'local accidents or revolutions'. There were also materials such as gravel, or earth and soil,

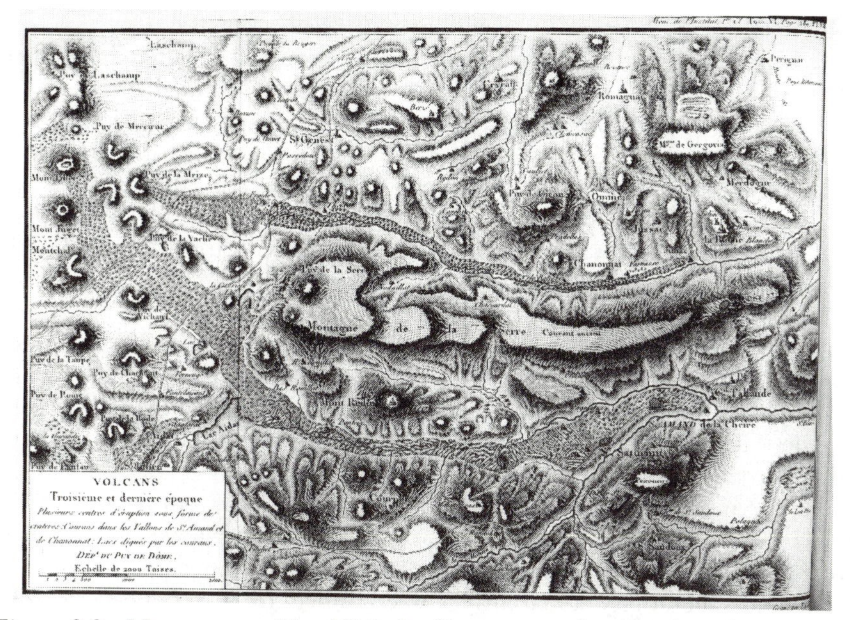

Figure 3.3 Map prepared by Nicholas Desmarest, showing lava flows in the vicinity of Clermont-Ferrand, Auvergne (Desmarest, 1806, Plate 7) (by permission of the President and Council of the Royal Society)

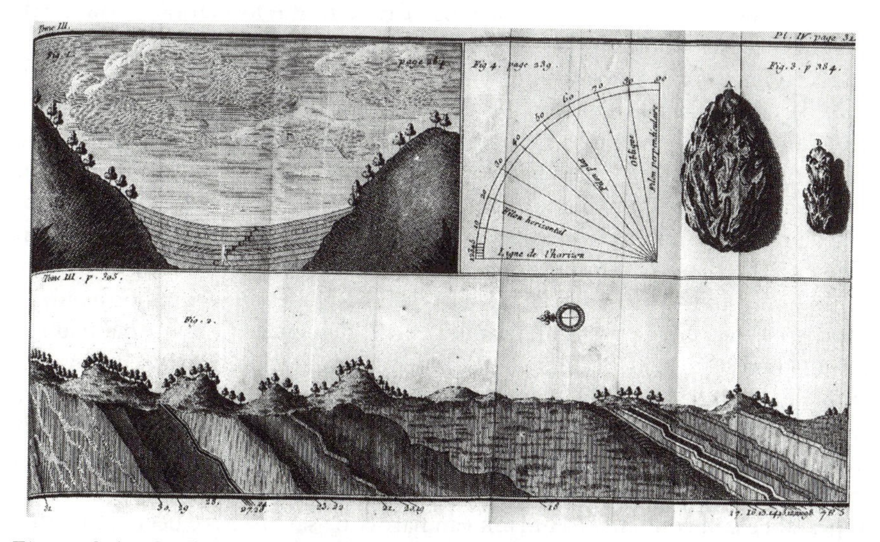

Figure 3.4 Geological sections of strata in Thuringia, according to Johann Gottlob Lehmann (Lehmann, 1759, Plate IV) (by permission of the President and Council of the Royal Society)

imperfectly consolidated. Lehmann hypothesized that after the Flood the Noachian waters either gave rise to the oceans, evaporated or disappeared into subterranean abysses.

All this may seem to betoken a rather primitive, time-constrained, theory – as indeed it was. But the significance of Lehmann's work was that he had a clear idea of a sequence of stratified rocks, and he figured such a sequence in section, the area of investigation being between Ilfeld and Mansfeld in Thuringia (Figure 3.4, bottom). It was this aspect of his work that marked a fresh approach to the study of the earth. As may be seen from Lehmann's diagram, thirty distinct bands of rock were recognized and figured,[16] many being designated by the local miners' terms. Some of these have passed thereby from the vocabulary of miners into the geological literature: for example, *Zechstein* or 'mine-stone'. (This corresponds to the Magnesian Limestone of British geology.) Each rock unit was described and named, sometimes colourfully. For example, the second unit was referred to as a malodorous earth – a calcareous stone that emitted the smell of cat's urine when rubbed. It is interesting that the bottom of unit 18 was referred to as a 'base', and the boundary with unit 19 is shown as being very irregular. Since unit 18 is what we now call the 'New Red Sandstone' (Triassic),[17] Lehmann's boundary apparently corresponded to that between today's Mesozoic and Palaeozoic rocks. So this major stratigraphic boundary was vaguely foreshadowed in Lehmann's diagram.

In Lehmann's work, there was undoubtedly an intimation of the idea of stratigraphical sequence as an indicator of time in the formation of the earth, and the representation of this concept by means of a section was a major advance. His sections were obviously much less abstract than those of Steno, previously considered. In fact, Lehmann (1759: 304) specifically referred to himself as a *historian* of the earth, and he claimed – correctly so far as I am aware for the Western world – that no one before him had treated the layered mountains in this fashion. He regarded his stratigraphic sequence, elucidated in Germany and based upon differences in lithologies not fossils, as being of universal application. This had the virtue of offering a general programme for geological research: one could check the sequence in other parts of the globe. On the other hand, by supposing that there was a unique and general sequence of events that had laid down strata worldwide Lehmann offered a generalization that was not arrived at by the historical method of enquiry appropriate to geology. The earth historian needs to work out sequences in different parts of the world and try to 'knit' these together. Also, geologists (as we understand their enterprise) try to envisage the circumstances under which the rocks were formed, on the basis of the their appearance and contents. Lehmann can hardly be said to have been doing this. A geologist's history of the globe comes from the strata, perhaps supplemented by some general physical theory or theories, not from a supernatural sequence of events described in a religious text. Lehmann's 'history', though sequential and related to observations of strata, was only partly historical in character, and was not a fully-fledged specimen of historical geology.

Nevertheless, work such as that of Lehmann was soon developed further. Also working in Thuringia, a physician at the court of the princes of Schwarzburg-Rudolfstadt, Georg Christian Füchsel (1722–73), published two important geological works (Füchsel, 1761a, b, 1773). The first of these, written in two parts entitled '*Historia terrae et maris*' and '*Eiusdem usus historia suae terrae et maris*', is important in that it offered what were arguably the first geological maps (as opposed to mineralogical maps such as those of Guettard). As in Lehmann's work (to which Füchsel referred), a definite sequence of strata was recognized for Thuringia, but with more subdivisions than had been proposed by his predecessor. Certain fossils were found to be associated with each particular stratum, although strata were actually distinguished by their

Figure 3.5 Georg Christian Füchsel's geological map of part of Thuringia (fold-out plate bound into back of *Actorum Academiae Electoralis Moguntinae* . . . , 1761, Vol. 2) (by permision of British Library)

Figure 3.6 Geological sections of strata in Thuringia, according to Georg Christian Füchsel (fold-out plate bound into back of *Actorum Academiae Electoralis Moguntinai* . . . , 1761, Vol. 2) (by permission of British Library)

lithologies and colour. Füchsel's map of Thuringia (Figure 3.5) showed a kind of 'perspective plan' of the region to the south-west of Jena, with the outcrops of different strata crudely delineated on an approximate topographical map of the area. Another diagram (Figure 3.6) showed combinations of cross-sections and plans.[18] And in another diagram, not reproduced here, Füchsel showed how inclined layered strata might have different outcrops according to different surface topographies – the very essence of geological mapping.[19] Indeed, the fundamentals of geological mapping may be recognized in Füchsel's work.

However, Füchsel's identification of strata on his maps by means of numbers or letters did not make for easy understanding, and it was not until colour became extensively used in the early nineteenth century that the potentialities of geological maps for the understanding of earth history were realized. In any case, Füchsel's work, published in Latin in a minor journal, did not attract much attention at the time, although it may be noted that the stratigraphic principles and some of the terminology that were later popularized so effectively by Abraham Werner were in fact found in Füchsel (and some in Lehmann too). All three were drawing on the discourse of the German mining fraternity, so that terms such as *Muschelkalch* and *rothe tode Lager* (later to become transmogrified into the Old Red Sandstone, see p. 119) entered the geologist's regular vocabulary. The connection between Füchsel and Werner may have been effected by a review of the former's work by J.S. Schröter (1775). Füchsel's work was most favourably reviewed long after his death by Christian Keferstein (1830), who opined that Werner adopted the ideas of Füchsel but without following their logic sufficiently.

Füchsel utilized the terms *stratum* (bed or stratum), *situs* (bed or stratum), *positus* (position) and *series montana* (chain of mountains or formation). The concept of *situs* differed from that of *stratum* in that it had the connotation of being adjacent to particular neighbours in a stratigraphic series. A mountain might have the same stratigraphical status, or *situs*, as some other mountain, and yet occupy a different geographical location (*positus*). He emphasized that the strata had settled out from a fluid and hardened after initially being soft (Füchsel, 1761a: 97), but did not simply appear in the order of their specific gravities, as Woodward, for example, had supposed. Rather, the order of the layers corresponded to the times of their formation (as was implicit in the writing of Steno).

Having established his general stratigraphical principles and his terminology, Füchsel began to unravel the stratigraphical history of Thuringia: 'One may combine the preceding observations both with each other and with others, and deduce events or a history from them' (ibid.: 81). In addition, a particular age might conveniently be named by the strata to which it gave rise.

Füchsel discussed the possible conditions of deposition for the formation of strata – for example, in terms of streams carrying sediments in particular directions at particular times, or wind storms producing particular sedimentary materials in oceans at given periods. He speculated on marine transgressions and regressions as having been due to 'shakings' of the globe – possibly due to disturbances produced by gradual accumulation of sediments. His time-scale was restricted to the order of hundreds of thousand of years, and he made an attempt to reconcile his geological findings with the Mosaic account of the history of the globe (Füchsel, 1773: 264); but his work was more strictly historical than that of Lehmann, according to the considerations that I have suggested above.

Turning now to Italy, we should note the somewhat earlier work of Antonio-Lazzaro Moro (1687–1764), who was for many years Professor of Philosophy and Rhetoric at the seminary at Feltre near Padua, and subsequently chapel master at the cathedral of Portogruaro, in Friuli, east of Venice. Moro was particularly interested in accounts of the emergence of new islands from time to time in the Mediterranean. This led him to develop a strongly 'Plutonist' view of geological history. He thought of the earth as having a vast hot interior, and the pressure arising from this subterranean heat might, from time to time, elevate land and form new islands or even mountain ranges. The mountains might themselves be volcanic in character (Primary) or stratified (Secondary). He noted that certain rocks, pushed up from beneath the Aegean, were covered with oysters (as depicted in the frontispiece of Moro's book); and then these rocks became covered by volcanic ash. This allowed Moro to account for the presence of shells high in mountain ranges.

Moro's theory was specifically proposed so as to be compatible with the biblical account of Creation. He prepared diagrams showing hypothetical cross-sections of the globe, with great subterranean cavities which contained the materials spewed out by volcanic action. These diagrams were speculative, and suggest the influence of earlier theorists such as Kircher and Descartes. But there was also a plate showing folded strata with structures that appear to be what later came to be called unconformities (see p. 94); and these seem to have been based on direct observation, although the vertical scale is greatly exaggerated, and the theoretical importance of such structures was not recognized. As an exponent of a 'fiery' theory of the earth, Moro (1740: 427) seems to have supposed that all sediments originated as material blown out from volcanoes (as volcanic ashes or tuffs, we would say). These may seem speculative ideas of a physico-theological character, as indeed they were. But he did think of the lower layers as containing 'plant remains and fruits corresponding to each season' (ibid.: 431); and he believed that there would be a recurrence of such events, so that one day our own civilizations would be overwhelmed by new strata, and people of the future might one day dig into the ground to uncover our remains. Evidently, Moro's whole theory was shaped by the environment of Italy and the Aegean. But, as an early and quite influential Plutonist, he is important to the history of ideas about the earth. His 'history' was as much biblical as geological, but had a strong empirical component.

Moro's countryman, Giovanni Arduino (1735–95), Director of Mines in the Vincente Province and in Tuscany, and subsequently Professor of Mineralogy at Padua, is more important than Moro for our present purposes. For professional reasons, as metallurgist, mining engineer and surveyor, he was required to have an intimate knowledge of the landscape, rocks, minerals, strata, mountains and mines of northern Italy, and there is clear evidence, in the form of sketches, geological sections and mine plans held at the Verona Library, that he undertook extensive fieldwork. Also, as Vaccari (1993) has shown, Arduino spent time collecting books on mineralogy and studying their systems of classification, which were numerous.[20]

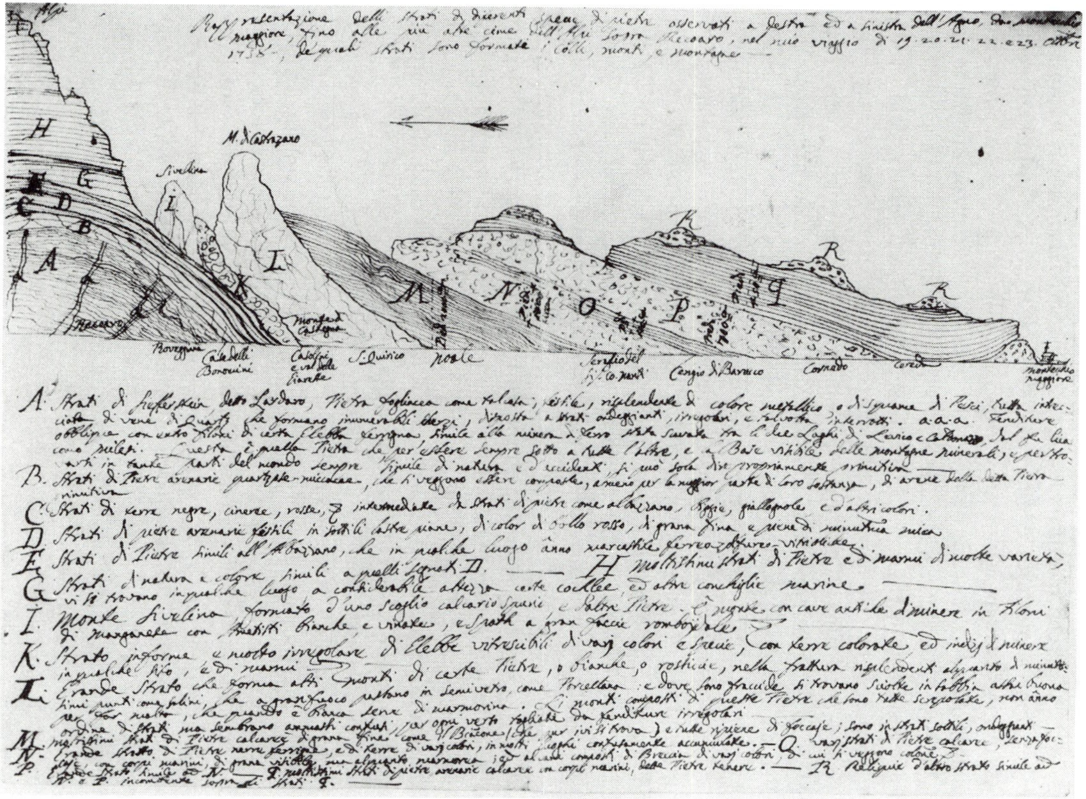

Figure 3.7 Giovanni Arduino's manuscript sketch section of the geology of the Agno Valley, near Vicenza (*Carteggio 'Arduino', busta IV, Fascicolo 'Schizzi topografici'*) (by permission of the Verona Public Library)

Particularly interesting is a record of a journey undertaken by Arduino in October 1758 up the valley of the Agno from Montecchio Maggiore to the high alpine peaks above Recoaro to the north-west of Vicenza. As a result of this journey, Arduino produced a remarkable geological section, shown in Figure 3.7.[21, 22]

While, in the nineteenth century, such sections, arrived at simply by walking along a valley floor, sometimes led to serious geological errors, for the eighteenth century Arduino's sketch was, in my view, an indication of a valuable new empirical approach to the study of strata. The results were made known to the world in the form of two letters (Arduino, 1760), which did not, however, include the geological section of Figure 3.7. In the second of these letters (ibid.: 158), Arduino proposed the division of the earth's crust into 'four general and successive orders': Primary, Secondary, Tertiary and Quaternary, a classification that has left its mark on the stratigraphical column to this day.

The Primary order, said to be first formed, contained a basement of schists (*roccia primigenia*), and mineral-bearing crystalline rocks (*vetrescibili*), sandstones and conglomerates (*monti primari*), which are today classified as Palaeozoics. These appeared to Arduino to be unfossiliferous. They correspond to units *A*, *B*, and *C* in Figure 3.7. The Secondary order (*monti secondari*) (*D* to *M* in the figure) consisted of marbles and stratified limestones with fossils, and corresponds (approximately) to our Mesozoics.[23] The Tertiary mountains (*monti terziari*) (*N* to *R* in the figure) were made up of gravels, clays, fossiliferous sands and some volcanic matter. They correspond approximately to the strata that we still classify as Tertiary. The fourth unit made up the plains (*pianure*) and were made up of alluvial materials. Vaccari (1993) has shown that the various divisions shown in Figure 3.7, made on the basis of lithological rather than palaeontological criteria, correspond remarkably closely to those used today for this area of Italy.

Arduino gave a particularly good description of the line of demarcation between Palaeozoics and Mesozoics in northern Italy, as understood to this day.[24] He also noticed that the rocks higher in the stratigraphic column contained fossils more like creatures still extant than did those in the deeper and older rocks. Not only this, Arduino in effect stated the principle of uniformitarianism ('the present is the key to the past') that is usually associated with the names of Prévost and Lyell[25] (Stegagno, 1929: 24). On the other hand, Arduino also spoke of great 'catastrophes and changes' that the earth had undergone (ibid.: 29). So his uniformitarianism was more methodological in character than an expression of the notion of 'steady-statism'. In the later part of his life, after he moved to Venice and was doing less fieldwork, he began to develop the notion of there having been four 'epochs' in the earth's past. But his thoughts on this remained unpublished.

Arduino did not produce geological maps other than one (unpublished) rough sketch diagram of the area traversed in the Agno Valley, which Vaccari has shown can be conceptually linked to the section reproduced in Figure 3.7. But from the point of view of thinking in terms of earth history Arduino's sections were probably of more immediate value; and as mentioned his fourfold division of strata has been of lasting significance and influence.

Continuing this sketch of some of the eighteenth-century writers who laid the basis of the modern science of geology, without establishing what one might call a full-blown paradigm, we should consider the work of the intrepid German traveller, Peter Simon Pallas (1741–1811), who occupied a chair of natural history at the Imperial Academy of St Petersburg and led extensive expeditions into Siberia (1768–74) on behalf of Empress Catherine II, in the course of which he examined the Urals and the Altai Mountains.

According to Pallas (1777–8), one finds granite at the core of all great mountain systems. Typically, it is found surrounded or lapped by unfossiliferous schists, serpentines and porphyries, resting against the granite in highly inclined layers. Then follow argillaceous schists and shales and thick beds of fossiliferous limestones, generally less tilted. The lowland regions are usually made up of sandstones, marls and clays, rich in fossils, often including plants. Thus Pallas envisaged three main kinds of mountains or strata: Primitive,[26] Secondary and Tertiary. He commented that the Secondary and Tertiary mountains served as 'Nature's archives, prior to even the most remote records and traditions that have been preserved for our observant century to investigate, comment on and bring to the light of day' (Pallas, 1778: 46).

That Pallas was in a certain sense offering a theory of the earth that had a distinctive historical dimension has been well shown in a recent re-evaluation of his work by A.V. and M. Carozzi (1991). Besides translating Pallas's work from the original German, rather than using the incomplete French version as have some other commentators, they have made a careful examination of the 'mineralogical' map that Pallas prepared for the region of the Ural Mountains, and have compared it with the geology of the area as understood today. Further, they have given a fascinating reconstruction of Pallas's theory, illustrating it with a series of sections of the Urals in temporal sequence, from which it can be seen how Pallas was indeed proposing a theory that was intended to provide a history of the development of the Urals which would account for the field evidence collected in the mineral-rich Sverdlovsk region.

There was, Pallas suggested, initially a long granite ridge – existing as an island in an original primeval ocean. Exposed to weathering and erosion, materials were washed off the island and deposited in the adjacent seas, being more sandy to the west and more argillaceous to the east. Then combustible organic and ferruginous materials were floated towards the ridge and penetrated the sediments, forming flammable pyritous deposits, which in time provided material for volcanoes. These acted so as to lift up the sediments, converting them into schistose materials and allowing the injection of metallic ore bodies.

Further sediments were then formed, including the deposition this time of extensive massive limestones. The process of accumulation of combustible volcanic material was repeated, with the eruption of a second set of volcanoes. Some of the limestones were tilted into a vertical posture and converted to marble. On the western side, there was yet a third sequence of volcanic activity, with the sedimentary rocks all being raised above sea level. The Ural range was thus asymmetrical. A huge inundation from the east completed the geological history of the area, bringing with it coarse gravels and the like, together with the remains of animals such as rhinoceros and mammoth.

It is an interesting and important question as to whether such reconstructions of the earth's past should or should not be construed as histories. Pallas's scheme was certainly speculative. It was, however, based on extensive fieldwork and a primitive form of mapping. And, even though there was only the simple designation of mineral localities and the like, as in Guettard's map, rather than the designation of different areas of ground by some suitable signifier (such as the colours of nineteenth-century geological maps), the Carozzis have been able to collate Pallas's empirical data those found in modern maps. The work was certainly different from that which we shall be considering for nineteenth-century map-makers (see Chapter 5), who were concerned to establish stratigraphical boundaries on the basis of differences in lithologies and according to the distinctive fossil contents of different strata. The later map-makers sought to build up a layer-by-layer history, each layer being like a page in an archive. Such work could be pursued effectively without too much intrusion of theory, and the resulting maps and sections were the preferred means of presentation of the results of historical studies of the earth. Thus, in a previous study, I have suggested a clear distinction between eighteenth-century theories of the earth and nineteenth-century historical geology (Oldroyd, 1979).

I believe that this distinction is still useful, but it is clear from the Carozzis' close examination of Pallas's work that it cannot serve as a universal discriminator for differentiating eighteenth- and nineteenth-century studies of the earth. Pallas was reconstructing the history of the Ural region in that he tried to tell a plausible stage-by-stage account of the area, even though it was based on mineralogical evidence and an implausible theory of volcanic action.

In Sweden also we find the development of a threefold division of the rocks of the earth's crust. The author was Torbern Bergman (1766), previously mentioned in connection with his work in mineral analysis, and it is likely that Pallas's ideas were developed in the knowledge of, and with reference to, Bergman's writings.[27] Bergman referred to four major classes or layers of rocks: *Uråldrige*, *Flolägrige*, *Hopvräkte* and *Vulkaner*. The first of these were the primitive rocks, supposedly formed soon after the earth itself first formed. They often contained veins and were devoid of fossils. The second division was made up of the bedded rocks – limestones, clays, coal, sandstone, etc. The third was made up of what today would be called alluvial deposits; and fourth there were volcanoes. Bergman was familiar with many earlier writings, including those of Lehmann. (He sought to apply Lehmann's divisions of the bedded rocks, established in Thuringia, in Sweden.) But Bergman does not seem to have been familiar with Arduino's work in Italy. Besides influencing Pallas, Bergman also influenced Abraham Gottlob Werner, previously mentioned, whose 'geognosy' will be discussed in the next chapter.

We find, then, in work in several parts of Europe in the second half of the eighteenth century the beginnings of field mapping, the drawing of geological sections and the first steps towards the establishment of the divisions of the stratigraphical column. The earth was studied as an object in its own right, and not as a branch of cosmology. Biblical considerations were not excluded, but declined in importance for practical purposes. The enquiries were empirical and showed intimations of a historical approach. High theory was not always to the fore, and one might have looked to a programme of research based on the careful empirical study of strata, with an unravelling of their history. But theory was needed to stimulate empirical enquiry, which otherwise might have become parochial and mining-based. Also, institutional developments were needed to make possible the coordination of individuals' observations and theories.

By the end of the eighteenth century, the old idea of the growth of minerals in the body of the earth was outmoded so far as educated persons' thinking about the earth was concerned, as also were the simplistic 'mechanical' theories of the seventeenth century. Alternative ideas were developed in the nineteenth century, but these depended on new chemical and physical knowledge rather than grand metaphysical doctrines. From the early nineteenth century, there developed an elaborate science of historical geology that grew from studies in the open air rather than the bowels of the earth. Such work extended well beyond the areas that were of specific economic importance, as the founders of geology, through mapping, sought to gain a total picture of the rocks forming the earth's crust. But the mining industry also played an important part, for new institutions were established for the professional training of miners and mining engineers, and the teachers in these institutions were willing to develop views that transcended strictly pragmatic considerations. So new ideas of interest and importance were developed, both from the field and from the mines.

From the figures we have considered in this chapter, we may think of Arduino and Lehmann as contributors to the emerging science of geology from the mining industry. Bergman was chiefly a chemist and mineralogist. Pallas was an explorer or traveller. Füchsel we might call a gentleman-scholar. It was by a synthesis of such disparate activities that the science of geology and a 'geo-historical' way of thinking about the earth emerged in the nineteenth century. The mining industry was already institutionalized in Germany by the end of the eighteenth century. Institutionalization of field geology came a little later.

CHAPTER 4

Thoughts on Heat, Fire and Water

In the previous chapter, we looked at some of the practical work concerned with the earth carried out in the eighteenth century, particularly to do with mining, the beginnings of chemical analysis, early fieldwork and the drawing of maps and sections. But, in 'thinking about the earth', we must concern ourselves with theory as well as practical matters, for in the emergence of all sciences theory and practice have been linked, rather than standing independently. As mentioned earlier, speculations about the origin of the earth (such as those of Becher) did not lead very far towards the establishment of geological science – although Becher's work provided a route towards the establishment of chemistry. So far as geology was concerned, two grand (and sometimes opposed) theories about the major agencies of geological change – fire and water – were widely discussed in the eighteenth century, and indeed well into the nineteenth century. It was debates about the relative significance of these two agencies that provided the major source of theory as geology emerged as a science at the end of the eighteenth century. The theorists that we discuss in this chapter did (in some cases) do much fieldwork, but they were not, by and large, notable for their map work. This chapter is, then, concerned with the grander eighteenth-century thoughts about the earth – often so grand that they were sometimes vehemently repudiated in the nineteenth century, as 'geology proper' was established.

Obviously, both fire and water were credible causes of geological change, and there was no reason in principle why one might not develop a view of the earth that invoked both agencies. The action of water in rain, rivers, oceans and floods seemed evident, both from everyday observations and from the testimony of sacred writings and myths. The power of volcanoes – especially evident to those living in the Mediterranean region, but also reinforced in mythography – was the second obvious candidate as a major force for change in or on the earth. But, this said, there was ample scope for debate as to which agency – fire or water – should be granted precedence.[1] Or were they to be accorded equal weight?

For a theory that focused chiefly on a 'fiery' origin for the earth, but which also had an important role for water, we may mention first a somewhat curious work by the great German polymath Gottfried Wilhelm Leibniz (1646–1716). Leibniz's theory was initially developed in the seventeenth century, and came to be written for rather curious reasons, first in the form of a synopsis of only three hundred and sixty words (Leibniz, 1693; Oldroyd and Howes, 1978), which was later extended to a substantial book, published posthumously in Latin under the title *Protogaea* (Leibniz, 1749).[2] So its impact came rather late, although the manuscript is believed to have circulated quite widely before its eventual publication.

Among Leibniz's innumerable duties and along with his multitudinous interests, he served as court historiographer to the House of Brunswick, and was employed in that capacity to write a history of this petty dukedom in order to try to establish certain territorial claims on its behalf. To perform a thorough job, Leibniz was eventually led into a consideration of the origin of the earth as a whole.[3]

Leibniz's theory for the origin of the earth was essentially Cartesian in character. He supposed, as with Descartes's vortex theory, that the earth had originated as a star, the surface of which had become obscured by slaggy sunspots. (Observations of chemical furnaces were extrapolated to the whole earth.) The original surface would have been glassy in character, but this material attracted water[4] unto itself, as does quicklime. The separation of the land from waters supposedly occurred as described in Genesis. As in Descartes's doctrine, there were collapses of parts of the earth's crust into subterranean cavities, formed as the earth cooled and solidified; and hence the waters drained away and dry land was exposed. The previous cover of water over what is now dry land was indicated by the occurrence of marine shells in the layered sediments. Sands were produced from the original glassy matter, being broken up by the action of water. Thus Leibniz's theory invoked the agencies of both fire and water. He was perhaps the first to make a distinction between 'primitive' and 'secondary' rocks: i.e. between those produced when the earth first formed and those formed subsequently by weathering and erosion from the first 'crust'.

Leibniz was adamant that fossils were the remains of once living organisms, and in this he was (it is believed) influenced by Steno, who was with Leibniz for a period in Hanover. Becher's idea that fossils were produced by some subtle force in the earth that mimicked animals[5] was rejected. Leibniz suggested that fused mineral matter might have come to occupy the spaces left by what were once living animals.

Although drawing on the mechanical ideas of Descartes, Leibniz believed that the biblical Flood was a definite historical event,[6] and (according to his editor, Christian Scheidt) he invoked it to account for the presence of fossils buried in strata. To explain the occurrence of the Deluge and its subsequent dispersal, Leibniz toyed with several ideas, including the old suggestion of Buridan (see Chapter 1) that there might have been a movement of the earth with respect to its envelope of waters; the idea of the action of a passing comet; and that of the moon approaching the earth for some reason. Eventually, Leibniz settled for what was probably the favourite idea of his day, namely that the waters had been disturbed by the collapse of portions of the earth's crust into subterranean cavities, followed by the draining of water into the cavities. He proposed that there were initially two great sets of cavities within the earth. When the waters first retreated, they filled the upper chambers while air occupied the lower. At the time of the Flood, the walls of the upper reservoirs fractured and water poured out. As the crust over the upper cavity subsided in ruins, the overlying sedimentary strata broke and became tilted, as in mountainous districts. In this respect, his theory was similar to that of Descartes. Cracks and fractures supposedly provided sites for the formation of mineral veins (in which Leibniz had a practical interest through his association with a mining venture).[7]

For a notable 'aqueous' theorist, we may mention the work of the well-travelled French diplomat, Benoît de Maillet (1656–1738). His ideas were presented in a somewhat unusual literary form: a dialogue between a hypothetical Indian scholar named Telliamed (obviously de Maillet's name reversed) and a French missionary. The book itself was itself entitled *Telliamed.* Written some time between 1692 and 1708, when the author was French consul in Egypt, it circulated in manuscript for a number of years, eventually being published posthumously by the Abbé Jean Baptiste le Mascrier in 1748 (de Maillet, 1748; Carozzi, 1968). Its notable features were the detailed evidence assembled for gradual diminution of sea levels, a proposal for a hydrographic station that would carefully monitor such changes, a suggestion for exploration of the sea floor by diving and by means of a primitive submarine (a design for which was offered), the suggestion of strong marine currents which moved sediments into submarine mountains and valleys (which would be exposed as further desiccation of the planet occurred), the origin of terrestrial life from seeds derived from outer space, the transformation of organisms by some kind of slow evolutionary process as conditions changed, and an age of the earth of the order of billions of years.

It is hardly surprising that such iconoclastic ideas evoked strong opposition (Voltaire thought de Maillet was a charlatan), and the book, when eventually published, was bowdlerized, particularly so far as the question of the age of the earth was concerned. Nevertheless, it proved popular, passing through several editions. And, since some of the early manuscript versions have survived (though not de Maillet's original text), there is sufficient extant material to form a clear picture of the author's original intentions. This scholarly task has been carried out most faithfully by A.V. Carozzi (1968).

De Maillet suggested that mountain ranges of different forms might be attributed to the different circumstances of oceanic currents that had given rise to their initial formation under the sea. He envisaged a distinction between primitive and secondary mountains. The former were supposedly produced by the initial heaping up of debris on the ocean floor when the sea was exceedingly deep – too deep for the existence of living organisms. Then, as the primitive mountains were exposed, they provided their own contributions to sedimentation, and the secondary mountains, formed on their flanks in shallower water, were able to support life and were richly fossiliferous.

De Maillet's theory had a number of sources. Particularly important was Descartes. The theory envisaged that the earth originated as one of Descartes's extinguished stars, which had gathered unto itself an envelope of water, dust and ashes from the vortex in which it found itself. But with the desiccation of the earth, it might eventually ignite in a grand conflagration with the help of combustion of material produced from fossils (oils). And then a new star would be regenerated.[8] But de Maillet had also picked up ideas from Arab sources as a result of his sojourn in Egypt. (He was fluent in Arabic.) There is, for example, direct reference in *Telliamed* to Omar-al-Khayyam (Carozzi, 1968: 152–3), who, by considering the evidence of sea shells in hills well above sea level and by examining old maps for evidence of changes in coastlines, had come to the conclusion that there was strong evidence for a decrease in the level of the sea.

In sum, de Maillet had ideas that were a blend of observation, speculative cosmology, ancient knowledge and daring and original hypotheses. He did not enter directly into the eighteenth-century debates about the roles of fire and water in the history of the globe. But, in France at least, his ideas were not without influence.

Whereas de Maillet was an outsider in the French scientific establishment in the eighteenth century, the next person we shall consider was anything but. Georges Louis Leclerc, Comte de Buffon (1707–88), was born into a wealthy Burgundian family. He increased his wealth by judicious investments in forestry and arms manufactures; but he was also an energetic and distinguished natural historian, with interests in most fields of natural knowledge, including cosmology and Newtonian physics. He was elected Associé to the Académie Royale des Sciences in 1739, and the same year was appointed Director of the important Jardin du Roi. Ten years later, Buffon began the publication of his celebrated Histoire naturelle, which eventually reached forty-four volumes.[9] The work had three introductory volumes (the first of which had to do with geological matters), twelve volumes on mammals, nine on birds, and five on minerals. There were also six supplementary volumes, the fifth of which, entitled *Epoques de la nature* (Buffon, 1778, 1962; Roger, 1962), contained a major statement of a theory of the earth. With all this effort, Buffon was regarded as one of the major writers of his day, being especially renowned for his style.[10] Though an eighteenth-century natural historian *par excellence*, Buffon can also be regarded as a transitional figure in that his ideas involved a combination of cosmological speculations about the origin of the earth, its behaviour as a whole according to physical laws, the formation of its strata, and details of particular mineral kinds.

Buffon's first volume, published, be it noted, just a year after the printing of *Telliamed* and in the year that the full text of Leibniz's *Protogaea* appeared, offered a quite elaborate '*Théorie de la terre*' and was intended as a basis for the information in the other parts of the *Histoire naturelle* on the animal kingdom.[11] Buffon took up an idea previously suggested by the English Newtonian astronomer William Whiston (1667–1752), who had supposed that the earth had originated from a fragment of the sun, knocked off by a comet (Whiston, 1696). Thereafter, the formation of the earth's strata occurred rather as de Maillet had envisaged, with submarine currents constructing submarine hills and valleys. But Buffon, developing ideas of Jean Jacques Dortous de Marain (1678–1771), also had the idea that the interior of the earth was still hot, and, starting in 1767 (Roger, 1989: 467) and using a hint from Newton himself (*Principia*, Book III, Prop. 41, prob. 21), Buffon began some remarkable experiments to attempt to determine the age of the earth, assuming that it had cooled from an initially molten body.

In the *Epoques de la nature*, we read of seven stages of earth history. As the globe cooled, it first solidified on its surface, and developed giant blisters, cracks and cavities in so doing. Thus the 'primitive' mountains were formed right at the beginning. Further cooling led to the consolidation of metallic bodies, which collected in the hollows of the primitive crust or vaporized and deposited along fissures in the crust. Thus ore bodies were formed.

Yet further cooling allowed the precipitation of water, and thus a new phase began, with vigorous interactions between earth, water, air and fire. Sand and clay were formed, but clay was later indurated to form shale, slate and schist. Life began in the hot fetid waters of this epoch, and limestones were formed from the remains of shellfish. Under the oceans, and also under the influence of the rotation of the earth, sediments were heaped up into large ridges, or even submarine ranges, as had been supposed by de Maillet and in the '*Théorie de la terre*'. Then the sea level fell to expose the mountains, as a result of the drainage of the waters into great caverns in the earth's interior – the roofs of the giant original blisters supposedly having collapsed.

Conditions were now regarded as being suitable for the development of great forests, and the vegetable matter that they produced, mixed with pyritous matter, accumulated to form the great coal deposits of the world. The combustion of this material then supposedly gave rise to volcanoes as the pyrite became heated by contact with water.[12] This aspect of the theory was similar to that of Pallas published the same year, discussed in Chapter 3.[13]

Buffon also supposed that life could form spontaneously, it being an inherent property of matter, which would naturally manifest itself under appropriate environmental circumstances. Under the conditions of a cooling earth, conditions propitious for life would first appear near the polar regions; and then, with cooling, animals and plants would gradually migrate towards the equator. Further, secular cooling would lead to the migration of different kinds of organisms towards the equator. Buffon thought that the first life-forms were commonly larger than those existing today, since in the hot conditions of the primeval earth their constituent 'organic molecules' were thought to be particularly active. Thus, according to Buffon, organisms were 'degenerating' rather than 'evolving', as the earth supposedly cooled. (The molecules of the deceased organisms were supposedly recycled to form new organisms, the forms changing as conditions deteriorated.)

Noting some similarities between the ancient animals of North America and Eurasia, Buffon reckoned that the two continents must formerly have been connected. Hence he proposed as one of the last events of earth history a foundering of the strata of the area of the present Atlantic Ocean into one of the giant cavities in the earth's interior that his theory envisaged.[14] The final 'epoch of nature' was marked by the emergence of man.

Obviously, this was all highly speculative, and as much beholden to earlier theorists, such as Descartes, Newton, Leibniz and de Maillet, as to field observations. However, what is especially interesting for present purposes was Buffon's effort towards an experimental determination of the age of the earth, in terms of the theory outlined above and using Newton's law of cooling.[15] Working at his iron foundry at his country estate at Montbard (near Dijon), Buffon prepared a set of iron spheres of different sizes, heating them until they were almost molten. Letting them cool in his cellars, he measured the times taken for the different spheres to reach the temperatures of the air in the cellars. With data for the times required for a range of spheres of different sizes to cool to cellar temperature, he then estimated the time that might be required for a sphere the size of the earth to cool to its present temperature. He made a few additional adjustments to his calculations, using mixtures of metallic and non-metallic materials, and allowing for the heat received by the earth from the sun in the course of cooling. The figure eventually published for the age of the earth, since it was a molten globe, was about seventy-five thousand years. But Buffon speculated privately that it might be some three million years old.[16] Eventually, with continued cooling, the planet would become uninhabitable and dead; or so Buffon thought. Buffon's thinking, it may be noted, allowed for both fire (or heat) and water.

It is inappropriate to think of Buffon as a geologist – indeed the term did not exist in his day. He was a theoretician and man of the cabinet, rather than a worker in the field. He did not build up a history of the globe on the basis of observations of rocks, minerals and fossils and the due recording of information about such objects on maps, followed by efforts to synthesize the information thus gathered into theories as to how the earth reached its present condition. He did not observe (or at least he did not report upon) processes of geological change as they may be observed occurring today. Buffon was the grand theorist, and the path to modern geology did not run directly through him, influential though he undoubtedly was.

There was a distinct difference in the mode of thinking about the earth in '*Théorie de la terre*' (Buffon, 1749) and *Epoques de la nature* (1778). The first contemplated cycles of change and the maintenance of an equilibrium in natural processes (Taylor, 1992.) The second offered a 'directionalist' interpretation. The changes envisaged by the later Buffon were irreversible. Indeed, he even predicted the future extinction of life as the earth gradually cooled. He was interested in the supposed laws that governed the earth's changes, rather than establishing exactly what the earth's changes had been as contingent historical events.

Yet theorizing in the late eighteenth century was soon to lead to important developments in ideas about the earth and to the establishment of a discipline of geology in the nineteenth century, institutionalized and fairly rapidly professionalized. In Germany, the leading theorist was Abraham Gottlob Werner (1749–1817), who taught at the Mining Academy at Freiberg in Saxony and has been mentioned several times already in previous chapters. He gave chief emphasis to the role of water in the history of the earth.

In Britain, the major theorist was a Scottish farmer and businessman with philosophical and scientific tastes, James Hutton (1726–97). While not unconcerned with the effects of water as an agent of erosion and deposition, as a theorist Hutton was chiefly concerned with the idea of the central heat of the earth and the manner in which the agency of heat might operate in the changes that the earth had undergone in the past and would undergo in the future. We shall deal with the ideas of these men in turn. They both had immense influence on the history of ideas about the earth.

Although originally trained as a medical man, Hutton spent much of the early part of his adult life as a farmer, working on his two properties in Berwickshire to the east of Edinburgh.[17] Here he spent much time and energy endeavouring to improve his farms according to the latest scientific principles of agriculture, and he was eventually successful in this, to the point at which he was able to retire to Edinburgh at a fairly early age, engaging in some business, but devoting himself chiefly to science and philosophy. Though reared a Christian, Hutton seems to have lost his faith in a personal god at quite an early age, putting in its place very strong deistic principles. That is, he held to a belief in a wise and beneficent deity who had specially designed the earth and its inhabitants for the well-being of mankind. Thus – as was commonly believed during the period of the Enlightenment – Hutton thought that one could argue for the existence of God on the basis of the evidence furnished by the seeming design evidenced in nature. The supposed evidence for the existence of God revealed through the life of Jesus was discounted by deists.

Specifically for a farmer such as Hutton, one might contemplate the rich soils[18] produced by the weathering of the subsurface rocks, animals and plants together playing a part in the process. But then it occurred to him – perhaps as he toiled away on his farm – that the good soil was constantly being washed into the seas by erosion, thus apparently spoiling the fine design of nature, so suitable for mankind. For, having adopted deism, Hutton, unlike so many of his contemporaries, did not feel obliged to keep to the biblical time-scale in his thinking. Indeed, his imagination taking wings, he contemplated an indefinitely long period since the earth was first formed and he disregarded how it might have been formed in the first instance.

But, if the soil was slowly but inexorably being washed away and if the beneficent design was to be sustained, some kind of cyclic 'geological' process had to be envisaged, for otherwise all the land surface would be reduced to a flat plain by erosion, the farming land would no longer be generated, and life as Hutton knew it would come to an end. Thus he began to think deeply as to how there could be a cyclic process for the renewal of the agricultural land on which mankind lived. Several factors may have assisted Hutton in the development of his theory.

During his travels, Hutton visited the new industrial regions of the English Midlands, and he corresponded with James Watt. In England, he saw and was much impressed by the new steam engines, in which a huge complicated mass of machinery was kept in motion by a single source of heat. It was also known in Hutton's day that when a gas was compressed it became heated.[19] And he may have been disposed to think in terms of a cyclic theory from the fact that as a young student of medicine he had written a thesis on the circulation of the blood – 'in the microcosm', as he put it (Hutton, 1749; Donovan and Prentiss, 1980). Putting these pieces together in a kind of intellectual jigsaw puzzle, Hutton came up with the following remarkable theory of the earth.

Hutton postulated (as others had done before, of course) that the earth's interior was hot and consisted of a great store of subterranean fuel, produced, over long ages, by the action of plants receiving light from the sun (Allchin, 1994). Further, Hutton proposed that, from time to time and in different places, the internal material of the earth would begin to expand, forcing its way into the crust above, bending the layers of the crust upwards in the process. Sometimes the molten rock would break through to the surface in the process, giving rise to volcanoes. Sometimes it would merely raise hills or mountains, with the phase of expansion ceasing before the magma could break through. This material would then cool and solidify, giving crystalline rocks such as granite,[20] which long after would be revealed – stripped bare by erosion and found at the central parts of the high country, but yielding rock which could in time produce new soils for the well-being of mankind. On this view, the high-altitude, crystalline rocks would be younger than the surrounding lower-level sediments, and this, in a certain sense, contravened Steno's principle of superposition. (But this principle was, of course, intended to apply to sedimentary rocks.)

The foregoing describes the 'constructive' side of Hutton's theory of the earth. It was balanced by the processes of weathering and erosion, which would give rise to the accumulation of layers of sediment in the seas. In time, these would be covered by other layers of sediment, and the lower ones would become strongly squeezed and hence consolidated as hard sediments. Then, with the ever increasing pressure, the temperature of the sediments would rise and fusion might occur, leading to the expansion of the 'magma' and its emplacement as described above. Hutton referred to the emergence of the Isle of Santorini in the Mediterranean as an instance where new land was being formed from beneath the sea (Hutton, 1795, vol. 1: 147). He insisted (Hutton, 1899: 194–5) that no observed rocks were primeval or 'primitive'.[21]

The weak point in the theory was, I suggest, the change-over from a compressive to an expansive phase in the working of the 'earth-machine'. But if this were accepted and if, above all, one could contemplate the prospect of virtually unlimited vistas of time – which Hutton did, for to all intents and purposes his earth was a perpetual motion machine running to eternity – then the whole process might seem reasonably intelligible. But Hutton went further and asked himself what particular field observations might be anticipated if his theory were correct.

The rising granitic matter (magma) might be expected to tilt the superincumbent strata. But then, an expansive phase being finished, the rocks would all be subjected to erosion, and in time the worn-down edges of the strata might be exposed to the elements. If these exposed edges were at some later stage, due to unspecified earth movements, lying under the ocean, then they might in time be covered with layers of sediment, to be consolidated in their turn. Then, if the whole were uplifted, one might see revealed to view some approximately horizontal strata lying across the edges of inclined or even vertical strata.

That such configurations or structures (today called unconformities) might be found was a prediction of Hutton's theory. He went looking for them and was successful in several parts of Scotland: on the northern tip of the island of Arran, on the cliffs at the side of the River Jed at Jedburgh, just north of the border with England; and at Siccar Point, on the coast a few miles to the north-east of Hutton's (see Figure 4.1). In each case, he located an unconformable contact between the famous Old Red Sandstone and a hard grey grit (greywacke), which is today classified as Silurian. Hutton must have been familiar with these rocks, for one of his farms was located on the greywacke and the other on the sandstone (Oldroyd, 1994).

Figure 4.1 'Hutton's unconformity' at Siccar Point, Berwickshire, showing sandstone lying unconformably on vertical greywacke ('schistus') (by permission of the Director, British Geological Survey; NERC copyright reserved)

The case of Siccar Point is particularly interesting. Hutton knew that the Old Red Sandstone cropped out on the coast, north of his farms near Cockburnspath. The greywacke was exposed on the cliffs a few miles to the east. All he had to do was to sail along the coast until he found the contact. This he did at Siccar Point, where the exposure shown in Figure 4.1 may be observed. It was a remarkable prediction, made with the help of theory and grounded in Hutton's metaphysics of the wisdom and goodness of the Creator, sustaining the earth as a place suitable for mankind's life and well-being. Moreover, it provided empirical evidence, so far as Hutton was concerned, for the enormous age of the earth.

To my knowledge, Hutton's prediction was perhaps the first carefully reasoned and successful piece of geological argument of the form: 'If *p* then *q*; *q*; therefore *p*'.[22] This form of reasoning is not strictly deductive. Indeed, it is fallacious, involving the so-called fallacy of affirming the consequent. Nevertheless, Hutton made a successful prediction, and such success generally tends to attract assent to one's ideas. It did so in this case among Hutton's friends in Edinburgh, even though his theory as a whole had many speculative and uncertain features.

Figure 4.2 Granitic veins, Glen Tilt, photographed 1987. Author *in situ*

Hutton made another important prediction. It seemed to him that if large quantities of molten rock were from time to time intruded into the crust of the earth from below one might hope to find veins of the crystalline rock extending from the main mass of granite into the surrounding material.[23] During a journey into the southern part of the Highlands of Scotland in 1785, Hutton visited Blair Atholl and travelled up Glen Tilt for a few miles. The common rock of this region is a schist (Dalradian Schist), which Hutton called a 'schistus'. (There is also limestone in the valley.) He presumably had some prior information from the local landlord that a mass of granite was to be found some distance up the glen, for he travelled up the valley with an apparent expectation of finding the contact between the granite and the schistus. In this he was not disappointed, for in the river valley one can find magnificent granitic veins, such as are shown in Figure 4.2. The significance of this finding in relation to Hutton's debates with the disciples of Abraham Werner will be discussed below.

Hutton's ideas were first published in the form of a small pamphlet (Hutton, [1785] 1987); then in the *Transactions of the Royal Society of Edinburgh* (Hutton, 1788);[24] and finally in book form as the celebrated *Theory of the Earth, with Proofs and Illustrations* (Hutton, 1795–1899).[25] Four parts were planned for this, but the third was only printed in 1899, more than a hundred years after Hutton's death, so it had little influence; and the fourth was never published. The first two volumes have a few illustrations, the most interesting one being a representation

Figure 4.3 James Hutton's theoretical section of the northern part of Arran, drawn by Clerk of Eldin (by permission of Sir John Clerk of Penicuik)

of the unconformity at Jedburgh, but they do not give the full flavour of Hutton's theory. Fortunately, some years ago a magnificent set of drawings pertaining to the *Theory of the Earth* was discovered (Craig et al., 1978). Among the pictures, there is a remarkable section of the island of Arran (which Hutton visited with friends in 1786), and this is reproduced in Figure 4.3.[26]

From this drawing, the nature of Hutton's theory is easily understood: a huge mass of granite had forced its way upwards from the bowels of the earth, heaving up the overlying strata (which are older than the granite) into a dome-like structure. Then, after erosion had got to work, the granite was revealed as a large mass of crystalline rock. Today, this forms Goat Fell in the northern part of Arran, and is rightly regarded as one of the classic areas in the history of geology. The hypothetical 'magma chamber' located below the granite mountain should be noticed particularly.

But there is another quite different way of 'viewing' or thinking about such a structure. One might suppose that the granite was formed first in some manner, and then the surrounding sloping sedimentary strata were formed around the granite 'island' as the sea for some reason receded. This was the way the rocks of Arran would have been interpreted according to the leading geological theory of the last quarter of the eighteenth century and for some time into the nineteenth century. I refer to the 'aqueous' (or 'Neptunist'[27]) theory of Werner.

Several mining academies were established in eastern Europe in the second half of the eighteenth century, to provide technicians and technologists for the important mining regions of Saxony, Slovakia, etc. They were the first institutions in the world to provide something akin to a formal training in practical geology, although it might be said that the practical emphasis was so strong that they were more like technical colleges than universities. However, at the most famous of these institutions, Freiberg, Werner developed a very significant theory about the earth's past, and the training was quite 'philosophical' in character, to the extent that several notable 'humanist' scholars such as von Humboldt, Novalis and Goethe thought it worth their while to attend Werner's lectures. Many important nineteenth-century geologists received their initial training under Werner.

Werner, whose family background was in the mining and metallurgical industry, was himself a student at Freiberg. From there he went to Leipzig to study law, and while at Leipzig he wrote a useful handbook on the identification of minerals by means of their external characteristics (Werner, 1774, 1962). This led to his appointment as lecturer in mineralogy at his Alma Mater, and he spent the whole of the rest of his career teaching at Freiberg.

Students of Michel Foucault's influential *Order of Things* (1966, 1970) will find that Werner's text on the external characters of minerals fits the book's thesis very well.[28] Foucault suggested that during the so-called 'Classical Period' (*c*. 1660–1800) it was typical for natural historians to examine objects and attempt to sort them according to the similarities and differences of their external features, thus yielding a hierarchical taxonomy.[29] One did not enquire into the inward nature of things or try to elucidate their histories (in the sense of their past circumstances), but simply looked at the external features of objects and classified them accordingly. At the time Werner was preparing his mineral handbook in Leipzig, this was certainly the accepted procedure. However, there are great difficulties in classifying the objects of the mineral kingdom in this way. Substances of the same chemical composition may occur in various forms. Conversely, substances of different chemical composition may have similar appearances or crystalline forms. Moreover, it is difficult to say what a mineral individual is, whereas there is usually little difficulty in deciding this matter for plants or animals. Again, what is a mineral species? The question of the definition of a species still troubles biologists to this day, but the problem is far worse for the mineralist: one cannot put a male and female mineral together to see whether they will breed successfully. Given such considerations, the fact that chemical procedures for the analysis of minerals were in a rudimentary state in the 1770s, and that mineralogical nomenclature was in a highly confused state with many synonyms, it was entirely appropriate on Werner's part to attempt to devise a mineral classification that dealt only with external features, with descriptions directly related to those features.

Even so, after his appointment at Freiberg, Werner gradually moved away from this approach and began to study minerals in the light of their supposed histories as they had been produced (or deposited) since the formation of the globe. Thus his mineral studies took a definite theoretical turn, away from the 'positivistic' or phenomenological stance of his earlier work. The result of his labours was an important pamphlet: his *Kurze Klassifikation und Beschreibung verschiedenen Gebirgsarten* (Werner, 1786, 1971). Briefly, the theory developed was as follows.[30]

Werner envisaged the earth forming in some fashion by the aggregation of matter, perhaps from some primeval chaos. And, just as in the theories of de Maillet, Buffon, etc., it was supposed that there was a large initial envelope of water – a universal ocean – formed round a primeval core of solid matter. Werner further hypothesized that this universal ocean contained – in solution – great masses of material, which crystallized onto the irregular surface of the primeval core. Granite was the first substance to be deposited, and then other rocks in regular order: (1) granite; (2) gneiss; (3) micaceous schist; (4) argillaceous schist; (5) primitive limestone; (6) trap; (7) porphyry; (8) syenite; (9) serpentine; (10) topaz rock; (11) quartz rock; (12) siliceous schist.[31] Following the completion of this sequence, it was supposed that the water level of the ocean had declined to such an extent that the higher parts of the core – now covered with a mantle of crystalline material such as granite – were exposed. And thus a process of weathering and erosion could begin, depositing sediments in the ocean basins. These sedimentary rocks too had their own definite layers and order. Werner called these layered, fossil-bearing rocks of mechanical origin the *floetz* rocks, and it is convenient to continue to use his term. At a later stage in the development of his theory, Werner introduced the notion of 'transition rocks', which were partly mechanical and partly crystalline in origin and character. Finally, he had volcanic and alluvial materials as minor contributions to the whole. He supposed that volcanoes were produced by the combustion of flammable materials such as coal.

To my knowledge, Werner never saw a volcano in action. If he had, perhaps he would not have come up with such an unlikely theory. It must be noted, however, that the idea of the heat of volcanoes being due to the combustion of some flammable substance was very ancient, and was no more implausible than Hutton's idea of a great mass of molten matter at the centre of his 'earth machine', the heat for which was supposedly provided by combustion and pressure.

A feature of the Wernerian theory which attracted particular attention was the question of the origin of basalt. In some places in eastern Europe, 'pancakes' of this rock may be found as vertical columns on the tops of hills. Werner regarded these as being a late product of the universal ocean, which could supposedly rise as well as fall. For Huttonians, this was nonsense. The basalt was the product of lava that had been extruded from the earth and had cooled and cracked in columnar forms. This idea did not appeal to Werner, for the basalts in his local area did not seem to have any connection with volcanoes and looked as though they might have crystallized from a fluid. At the time, the main focus of the Vulcanist–Neptunist dispute had to do with the origin of basalt, and perhaps one should restrict this naming of the dispute to this particular issue. However, the debate between 'aqueous' and 'fiery' theorists transcended the problem of the origin of basalt and is usually taken to have had wider connotations.

It was mentioned above that Hutton was particularly interested in the granitic veins that he observed in Glen Tilt. It was exceedingly difficult to account for such veins or for dykes of basalt by means of the Wernerian theory. Werner did his best by supposing that what we would call igneous dykes were produced by infilling of basaltic matter from an overlying solution into cracks in the earth's surface. If this occurred twice at the same place, one might even hope to account for the inclusion of fragments of 'country rock' in a dyke. But this *ad hoc* hypothesis could hardly work for observations such as Hutton made in Glen Tilt, where there are numerous, variously shaped, veins. On the other hand, as we shall see in Chapter 9, there are many places where veins suggest that some kind of dissolved matter has penetrated rocks in an 'aqueous' fashion. The Huttonian theory of great magma chambers intruding into the crust from below (see Figure 4.3) is certainly not successful in all cases.

The question of time has always been important in geology. To my knowledge, Werner did not publish specifically on this question. But it should not be thought that he was a '6,000 years' man. Guntau (1984: 71) has referred to a manuscript of Werner in which it was stated that the universal ocean might have covered the earth for a million years. So while it is true that some 'Wernerians' such as the Irish chemist Richard Kirwan (1799) wanted to construct a biblical geology out of Wernerian theory, Werner himself did not take this approach. Rather, he was a deist, like Hutton.

Nevertheless, the idea of a universal ocean capable of holding vast quantities of granitic matter, or whatever, in solution will probably seem absurd; and it certainly appeared so to Werner's opponents in Scotland. But two points in particular should be mentioned. First, many kinds of mineral matter are soluble in alkalis, and there was no reason in particular why Werner's hypothetical ocean should not have been strongly alkaline in character. Second, it was a known fact that the sea levels seemed to be falling in some parts of the world. De Maillet had mentioned this. Other observers in the eighteenth century, particularly in the Baltic region, had noticed definite evidence for falling shorelines (Wegmann, 1969).[32] So the idea of a receding ocean was by no means implausible.

In fact, so far as Werner was concerned, the great virtue of his theory was that it was firmly based on empirical evidence. Indeed, he felt so strongly about this point that he coined a new term – 'geognosy' – to refer to his science, seeking to differentiate it from ordinary natural history, mineralogy or geology. As we have seen, the term geology was just coming into use at the time that Werner was in his prime. But, in his view, the writings going under the name of geology tended to be too speculative. His science was to be different – firmly grounded in facts.

This was indeed possible, if one left aside the speculative ideas about the universal ocean and such matters. Each stratum, recognizable by its lithological characteristics, was called a *Gebirgsformation* (or 'formation'). From their outcrops in Saxony (and Thuringia), it appeared that the different formations were arranged in some definite order, as had previously been shown by Lehmann, Füchsel and others; and this was taken to indicate a 'discoverable time sequence' (Werner, 1971: 20). This was the important feature of Werner's work so far as the history of geology was concerned. It meant that Saxony could serve as a 'paradigm' for researches about the earth in a truly Kuhnian fashion (Kuhn, 1962). Werner's students had a clearly defined and, to some degree, accomplishable task. They could take the Saxon sequence of strata and use it as a 'lens' through which to examine and comprehend strata in other parts of the globe. Thus (to continue to use Kuhnian terminology) geognosists were faced with a series of 'puzzles' – the fitting of the strata in different parts of the globe into the Wernerian framework. Sometimes a puzzle could be solved by noticing that a particular formation, expected according to the Saxon paradigm, was missing; but the overall order was maintained. If, however, the data didn't fit at all, there was an 'anomaly'. And, if faced with enough anomalies, the Wernerian paradigm might eventually have to be abandoned (as Kuhn's account of scientific change might lead one to expect). This happened to some extent as the students of Werner began to travel the world and came into closer contact with regions of igneous activity (see Chapter 8). But the Wernerian theory was modified as much as abandoned.

It is an important question whether Werner's geognosy should or should not be regarded as a historical science. We have noticed that Werner talked of 'formations'. Strata were not simply to be classified according to their appearances, their compositions or even their fossil constituents. They were entities formed during the course of time; and they had a geometrical arrangement which – by reason of Steno's principle of superposition, deployed along with Werner's Neptunist theory – indicated their order of accumulation (the youngest uppermost). Such considerations have led some historians of geology (e.g. Laudan, 1987: 94) to take the view that Werner's work did indeed mark the beginning of historical geology and was, therefore, an important new phase in the history of ways of thinking about the earth.

Even so, it appears that chronology was of rather little concern to the Wernerians and that they were chiefly concerned with working out the spatial relations of strata. This, of course, is just what might be expected within the context of a mining school. Thus, while there was indeed a temporal dimension to the Werner's thinking, it was not dominant in the application of his theory. For this reason, I prefer to use the term 'genetic' rather than 'historical' to describe Werner's geognosy. Werner had reached the conclusion (on an empirical basis, to be sure) that there was an ascertainable general sequence in the layering of the strata. This was so because there had been a specific sequence of changes in the universal ocean and then in the different ocean basins. The history of the globe was rather like a gramophone record: once it started playing it naturally followed a definite sequence. This I call a 'genetic' theory.

But Werner's geognosy was not an unqualified success as a genetic theory, in that he soon found rocks of similar appearance or lithology that occurred at different stratigraphic horizons.[33] This should not have been the case if a particular formation was characteristic of a particular period of the earth's history. So Werner was driven to the deployment of *ad hoc* hypotheses to account for the recurrence from time to time of similar conditions. (For example, as mentioned above, he had to assume occasional risings of the level of the oceans instead of a constant gradual diminution.)

In any case, according to Carozzi (1988), it is an error to conflate Werner's notion of a formation with the modern concept of the same name. The modern notion of formation does not, Carozzi says, have a temporal connotation. 'Formations' today have to do with lithostratigraphy, and they only acquire a temporal and historical connotation when linked with fossils and biostratigraphy or with datings based upon radiometric measurements.[34] In Carozzi's view, it took a 'long and traumatic time' for 'geologists to extricate themselves from the Wernerian concept of the "formation" '. Thus he does not see Werner's work in such a positive light as does Laudan.

I do not think it took geologists quite as long to sort things out as Carozzi suggests. Indeed, it seems to me, that the study of historical geology was well established by the early years of the nineteenth century. Cuvier and Brongniart, for example (see Chapter 5), who are rightly taken to be 'earth historians', used the term 'formation' in their stratigraphic work in the Paris basin. But it is indeed true that historical geology, as practised in the nineteenth century and through to the present, was and is something different from Wernerian geognosy. Its practice is analogous to that of the ordinary historian, who collects materials from archives and tries to piece together a persuasive story as to the sequence of events that led to the materials in the archives being the way they are. And, just as the historian needs to knit together the separate histories that are discovered from the information in the different archives, so too does the historical geologist seek to knit together the stratigraphic information collected in different geographical areas. Both processes are shaped by theory. In the case of the historian, the relevant theories are derived from political science, social psychology or whatever. The geologist uses theories about processes of erosion, sedimentation, volcanic activity, etc., and particularly ideas about the distributions of fossils (and ultimately their biological evolution). Both for the student of human history and for the geologist, the present is used as a rough, but not infallible, guide to the past.

This is not to diminish the importance of Werner in the history of ways of thinking about the earth. Quite the contrary. Laudan (1987) has described the spread of Werner's students across the globe – expounding, applying and testing his ideas – as the 'Wernerian radiation'. The term is most apt. The Wernerian radiation gave the study of the earth its first paradigm, marking a separation from mineralogy. It was the means of escape from a pre-paradigm condition.[35] But this is not to say that there was an immediate transition from mineralogy to historical geology as a result of the development of Wernerian geognosy. Indeed, historical geology could not manage without mineralogy.

During the early years of the nineteenth century, there was something of a battle royal between the followers of Werner and those of Hutton,[36] with the former at first much in the majority. From our perspective, it seems that Hutton was in the right: the notion of a universal ocean was absurd. But on the ground and at the time it seemed quite otherwise. If the reader will look back at Figure 4.3, it will be seen that Hutton's remarkable section of Arran can be construed from a Wernerian perspective if one is so minded. So too can a geological map of Arran. Indeed, as has been shown by Rudwick (1962), when the leading English geologist George Greenough, founder of the prestigious Geological Society of London, made a geological tour of Britain in 1805, he was unable to determine satisfactorily on the basis of the field evidence whether the Huttonian or the Wernerian theory was to be preferred. In consequence, there was a kind of theoretical agnosticism – a kind of 'plague on both your houses' – in the early papers of the Geological Society, and a general attempt to eschew theory, so dissatisfied were people with the heat generated by the Wernerian–Huttonian dispute.

It is clear, then, that in its day the Wernerian theory did not seem to be wrong in some transcendentally obvious way. The two theories can, in a way, be compared with the Ptolemaic and the Copernican theories in astronomy, one being a kind of mirror image of the other. There are two quite different ways of 'thinking about' a place like the Isle of Arran. A limited range of evidence did not, in itself, compel belief one way or the other.[37]

Some historiographical points may be made in relation to the view of Werner that has tended to prevail in the English-speaking world. This, I suggest, largely has to do with the influential history of geology, *Founders of Geology*, by Archibald Geikie, who wrote:

> [N]ever in the history of science did a stranger hallucination arise than that of Werner and his school, when they supposed themselves to discard theory and build on a foundation of accurately-ascertained fact. Never was a system devised in which theory was more rampant; theory, too, unsupported by observation, and, as we now know, utterly erroneous. From beginning to end of Werner's method and its applications, assumptions were made for which there was no ground, and these assumptions were treated as demonstrable facts. The very point to be proved was taken for

> granted, and the geognosts, who boasted of their avoidance of speculation, were in reality among the most hopelessly speculative of all the generations that had tried to solve the problems of the theory of the earth (Geikie, 1962: 212).

Geikie went on to accuse Werner of basing his system on knowledge of a limited area of the earth (namely Saxony) and of having a rigid 'onion-skin' model, which, however, became very flexible when faced with contrary evidence. Also, according to Geikie, Werner had a 'repugnance to writing', although he was (as Geikie willingly acknowledged) an outstanding lecturer and teacher. And, according to Geikie, the Wernerian or Neptunist theory rapidly fell into disrepute, even before the death of its author.

Such judgements provide excellent material for the connoisseur of anachronistic/Whiggish history of science.[38] But Alexander Ospovat, who has done much to restore Werner's reputation in the English-speaking world, has shown that the theory was not a simple 'onion-skin' model. Werner's ocean went up and down (though downwards overall) and was at different periods stormy or calm (Werner, 1971: 22). This may be thought of in two ways: the 'articulation of a paradigm' in Kuhn's sense; or the invocation of *ad hoc* hypotheses in a manner regarded as reprehensible by one such as Sir Karl Popper. Either way, it would not impugn the scientificity of Werner's work, although in a Popperian sense Werner's geognosy might not appear to be 'good' science.

Ospovat (1980) has further shown that Werner's geographical knowledge was much more considerable than Geikie implied, and that Werner published substantially more than Geikie led readers to believe. More importantly, it should be observed that various forms of Neptunism, though largely displaced from British geology by the 1830s, persisted longer (in modified form) in Germany, and in France and Canada too. This chiefly had to do with efforts to understand the nature of metamorphic rocks such as schists and gneisses, mineral veins and ore bodies. Metamorphic rocks show laminations or foliations, and sometimes seem to grade into sedimentary rocks or into crystalline rocks such as granite. Thus the idea of granite being simply the product of a large mass of magma cooling and crystallizing under the surface of the earth, as Hutton believed, was an oversimplification. This issue will be examined more fully in Chapter 9.

Leaving aside such historiographical issues for the present, we may enquire about the fate of the Huttonian–Wernerian debate in the earlier part of the nineteenth century. It has generally been thought that the Vulcanist theory won. In a sense this is true, although, as will be shown in Chapter 9, many ideas that had their origin in 'aqueous' theory continued to be used and developed in the later part of the century. The Vulcanist 'win', if that is what it was, had two particular supports, other than field observations such as those that might be made in the Auvergne.

First there were the experimental investigations of Hutton's friend, Sir James Hall (1761–1815) (1805, 1812, 1826). One objection to Hutton's theory was that marble could not be produced by the action of heat on limestone, as carbon dioxide (as we now call it, rather than 'carbonic acid air' or some such term) would be driven off, leaving quicklime. Hutton contended that under pressure the gas would not be able to escape, so the limestone could melt and be converted to marble. Hall more or less succeeded in doing this, by heating powdered chalk in a sealed gun barrel, obtaining a stony mass which might by a stretch of the imagination pass as marble. Or, as William Whewell (1832: 125) put it rather nicely: 'Sir James Hall drew from his oven a marble loaf made of chalk flower [*sic*]'. The Neptunists were not convinced. Perhaps chalk could be consolidated in this way, but that did not mean that that was the only way of making marble. An aqueous origin had not been excluded.

In another experiment, inspired by an accident at a glass factory in Leith, near Edinburgh, in which a large mass of glass that had escaped from a furnace cooled slowly and produced a crystalline mass, Hall took lava and cooled molten samples of it both rapidly and slowly. Rapid cooling yielded a glass, whereas slow cooling produced a crystalline product. Again, the Neptunists had an answer. After repeating Hall's experiments, Jameson argued that some alkaline component was driven off by the heat, so that the substances were not chemically the same before and after heating (Sweet and Waterston, 1967).

In his third series of experiments, Hall heated together a mixture of sand and salt water and succeeded in obtaining something approaching a consolidated sandstone. For Huttonians, this might have been expected to occur to the sediments laid down on the seabed when they were acted on by subterranean heat. The Neptunists, however, objected that, with the salt being there, a chemical process was involved and that a hot primeval brine might accomplish the same task. The debate was not, therefore, resolved simply on the basis of Hall's experiments, and perhaps the moral of the story is that experimental or laboratory geology does not necessarily succeed in overriding geological theory or geological fieldwork. Hall's experiments were, however, important as being (almost) the first examples of 'experimental geology'.[39]

The other line of evidence which certainly tended to support Hutton's theory was that directed towards showing – using thermometers in mine shafts – that the earth does indeed get hotter as one approaches its interior. This had been the opinion of early theorists such as Kircher, Descartes and Leibniz, but became overshadowed by the success of the 'aqueous' paradigm in the eighteenth century. Werner and his followers held that the main sources of terrestrial heat were the sun's daily input, the store from the primeval ocean and the occasional superficial volcanic actions due to local combustions of coal, etc. But the matter was examined more closely in an empirical way by several French investigators in the early nineteenth century (d'Aubuisson de Voisins, 1803, 1807; Cordier, 1827). French theoreticians such as Lagrange, Laplace and Fourier also held that astronomical and physical theory should lead one to expect that the earth's interior would be hot.

The empirical results of the French mining engineer, Louis Cordier (1777–1861), which he published in many journals (e.g. Cordier, 1827), were thought to be particularly reliable, although much work of a similar nature was also carried out in the mines of Cornwall and reported in the transactions of the Geological Society of that county. Taking his thermometer down various French mines and lodging it in the actual rock rather than in the open passages of mine shafts, Cordier found an extraordinarily high rate of increase of temperature of 1°F per 25 metres. From this, he extrapolated to the conclusion that the interior of the earth had to be liquid, and that it was merely covered with a thin solid crust of the order of 100 kilometres in thickness. This was consonant with the earlier ideas of Descartes and Leibniz. That the thin crust was flexible was, Cordier thought, evidenced by the occurrence of earthquakes. The earth was, however, thought to be cooling and contracting. This led to the shattered condition of its external crust. But, losing heat more quickly than the material of the interior, the crust exerted an inward pressure, which was released from time to time by the action of volcanoes. Such a theory, however, hardly gave a satisfactory account of the stratification of the earth's crust, and it was hardly a cyclic theory, as required for a Huttonian system of geology.

Cordier's system was thoroughly directionalist, as was most French geology at that time, following the tradition of Buffon. Cooling was a one-way process. Thus, while on the one hand Cordier's theory did support a Vulcanist theory of fire, rather than a Neptunist theory of water, from another point of view the theory was anything but Huttonian. To be Huttonian, it had to be a cyclic theory, and Cordier's was not. It meshed, rather, with the theoretical account given for the cooling of the earth by the French physicist Joseph Fourier (1768–1830) (Fourier, 1820, 1824), which was developed in association with the cosmological theories of Laplace[40] (see Chapter 12). Indeed, Fourier's work (certainly not that of the provincial Huttonians in Edinburgh) was the stimulus to Cordier's enquiries. On the 'output' side, Cordier's work was linked to the important and influential theory of mountain building developed by Elie de Beaumont, which relied on a cooling and shrinking earth. This will be discussed further in Chapter 8.

At the time of the debates between the Neptunists and Vulcanists – which were of considerable intensity, as has been shown with *élan* by Gillispie ([1951] 1959) – the great issue seemed to be whether fire or water should have precedence in a theory of the earth. Neither side can be said to have 'won' the debate. Victory was claimed by the Vulcanists, but it was decisive only on quite a narrow front, namely the origin of basalt. On larger issues, including such grand questions as the origin of granite, debate has continued through to the present. One might say, however, that the great divide among geologists – between those who specialize in hard (igneous and metamorphic) rocks and those who spend their time with soft (sedimentary) rocks – goes right back to the early days of geology and the realms of Vulcan and Neptune. And twentieth-century debates between 'migmatists' and 'magmatists' (see Chapter 9 and Figure 9.3) may also, with some stretch of the historical imagination, be seen as related to the distant dispute about fire and water. Which was the most appropriate agent in one's theories, when thinking about the earth's past?

CHAPTER 5

The Earth Surveyed and Geologically Mapped: the Territorial Imperative

We have seen from Chapter 3 that geological mapping began in the eighteenth century in the work of men like Guettard and Lavoisier and in maps such as those of Füchsel.[1] For the most part, however, such early maps were 'mineral' rather than stratigraphical (Greenly and Williams, 1930; Eyles, 1972; Taylor, 1985). They gave the outcrops or locations of rocks, minerals and ores, chiefly of economic significance, and had little to do with telling a history of the earth.[2] For this purpose, one needed some criterion or criteria for recognizing rocks that might be of the same age but different appearance. And one needed a suitable technique for the presentation of the information, so that it might readily be taken in almost at a glance.

As mentioned, the beginnings of the collation of information in the form of maps was to be found, for example, in the work of Füchsel. But the use of colour for this purpose was a great advance, and was, so far as is known, first utilized by Friedrich Gottlob Gläser (1749–1804), a mining engineer trained at the Freiberg Academy, in a map representing the area round Suhl and by the Werra River in Saxony (Gläser, 1775). The map[3] showed the locations of villages, industries, mines and ore outcrops. Three rock types were delineated: granite (red), sandstone (yellow) and limestone (grey). The boundaries were shown very approximately, essentially with three blocks of colour. But a large 'vein' of granite was mapped as passing right through the sandstone.

Three years later, Wilhelm von Toussaint Charpentier (1780–1847), a colleague of Werner at Freiberg, produced a simple map of Saxony, coloured according to suggestions made by Werner and displaying eight of the lithological subdivisions recommended in Wernerian theory (see Figure 5.1).

Four years later, the Frenchman Pierre Bernard Palassou (1745–1830) illustrated an extensive essay on the geology of the Pyrenees with the help of eight mineralogical charts, which showed dips and strikes of strata (Palassou, 1781). From an examination of Palassou's maps, it is evident that he had travelled

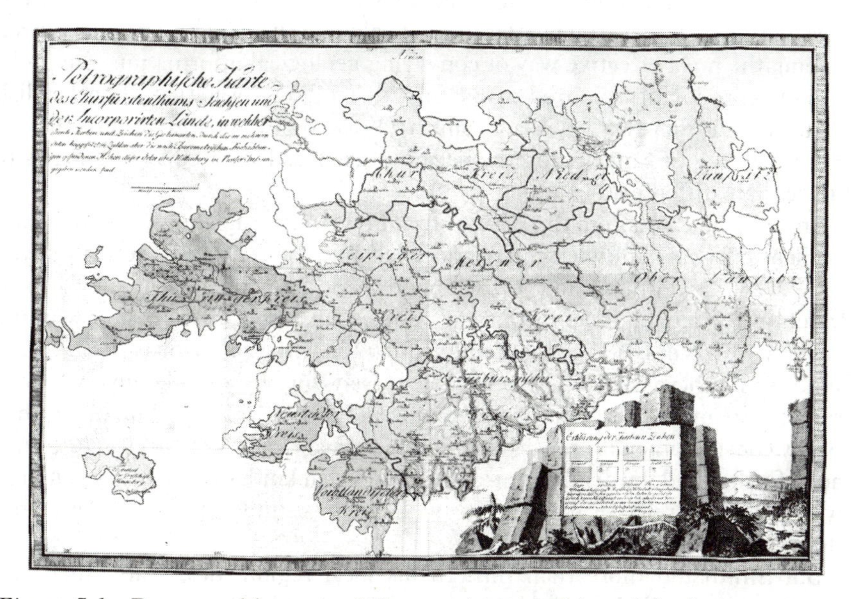

Figure 5.1 Petrographic map of Saxony (coloured in original), according to Johann Friedrich Wilhelm von Toussaint Charpentier (1778, Plate 1) (by permission of the Natural History Museum, London)

up the various valleys cutting into the Pyrenees and had recorded the different kinds of rocks and metallic ores to be found therein. He then linked the various exposures on his charts with (straight) lines, across hillsides where he had not, I venture to suggest, actually walked. Thus he was interpolating his data, as do modern map makers. From today's perspective, Palassou did his job in a crude and simplistic fashion. No modern map-maker would dare draw in his or her boundaries in quite such a cavalier fashion. However, it should be observed that Palassou was not mapping clearly delineated sedimentary strata, but the occurrence of different ore bodies in a mass of granite. His book did seek to represent numerous folded strata in section.

In 1801, the Italian Scipione Breislak (1748–1826) illustrated an essay on the volcanoes of Italy with the help of a large map of the Naples area, which depicted volcanoes and lava flows in red (Breislak, 1802). In America, a Scotsman, William Maclure (1763–1840), produced a coloured geological map of the eastern part of the country (Maclure, 1809), and another version in 1817 (Maclure, 1817), which shows six 'Wernerian' lithological subdivisions, including the 'Old Red Sandstone' or '1st Sandstone Formation'. Coloured geological maps of two parts of India were published by Benjamin Heyne (1814).[4]

However, the most important of the early efforts at mapping the terrain and representing different strata with different colours – which was soon recognized as being the most effective way of conveying geological information, the problems of reproducing coloured maps notwithstanding – was that of an English engineer and surveyor, William Smith (1769–1839). His achievement was particularly significant because he began to use fossils as well as lithologies to differentiate different rock units.

Smith, who came from the Bath district of western England, where there are excellent exposures of what today are called Jurassic rocks, was self employed as a surveyor and drainer in the great boom period for the building of canals in Britain in the late eighteenth century. With such an occupation, and working over many parts of Britain, he became acquainted with the various types of rocks found in southern Britain, which – fortunately for the foundations of stratigraphy – are mostly distributed in a simple and orderly fashion. Moreover, they display considerable differences in lithology – chalk, clay, sandstone, coal, shale and so forth – and the distribution of these different kinds of rocks can be entered on a map with relative ease, either with various kinds of shading or, more effectively, in different colours.

But Smith did more than this. In his local region, near Bath, there are exposures of so-called oolitic[5] limestone, rich in fossils. But the fossils were not all the same. Indeed, one subset of the oolitic limestone seemed to be characterized by one assemblage of fossils and the other subset had a different assemblage. Other rocks, of different lithologies, such as the chalk, had their own characteristic fossils; but the chalk too could be subdivided according to its fossil contents.

The hills of the Bath area are deeply dissected by valleys, which allow the surveyor to walk in among the rocks, so to speak, and gain a good idea of their three-dimensional structure. So, at least by 1799 and almost certainly earlier, Smith had enough information, collected as he engaged in his surveying for canals and other enterprises, to be able to draw up a list of the various strata of the district, from the coal measures to chalk, along with their thicknesses. The several kinds of fossils that might be found in each kind of rock were also tabulated. This table, though not immediately published,[6] circulated in handmade copies. Soon thereafter, Smith began to prepare materials for a coloured geological map of England and Wales – the precursor of all subsequent geological maps of Britain. Early drafts dating from 1801 are known, and even earlier ones are thought to have existed at one time but have been lost or destroyed (Judd, 1897; Cox, 1942). However, the first full production, with an accompanying memoir, did not appear until 1815 (Smith, 1815a, b).

Smith's map (which appeared in gradually emended form in the many copies that he made over the years, as his empirical knowledge increased) was remarkable in a number of respects. First, it was largely a single-handed effort, although he did receive substantial assistance from a polymathic surveyor friend, John

Farey (1766–1826) and from Smith's nephew, John Phillips, who subsequently became one of England's most distinguished geologists in the mid-nineteenth century, eventually taking the chair at Oxford. Second, Smith used a very clear, albeit tedious, method for the delineation of the strata. Each type of rock was allocated a characteristic colour (some of Smith's colours are still retained in modern maps), and he deepened the coloration towards the base of each

Figure 5.2 William Smith's geological map of Cumberland (1824) (hand-coloured in original) (by permission of the British Library)

stratigraphic unit so that they stood out from one another with considerable clarity.[7] The map's information was very detailed in the regions Smith knew well, such as Somerset and Gloucestershire, while it was hazy in less familiar and more complex areas such as Anglesey.

As said, Smith produced a map of England and Wales, and parts of Scotland, almost single-handed, and he issued various emended versions with more detail gradually added as the years went by. These maps cannot be reproduced conveniently here, as they are too large. But, as a sample of Smith's work, Figure 5.2 shows one of his County Maps series, that of Cumberland (Smith, 1824). This series was issued by Smith and Farey chiefly for commercial purposes, being small portions of Smith's main map and attractive to purchasers in the various counties of England and Wales.

While Smith is justly renowned for his grand principle, that fossils can serve as 'indices' of strata, it has been argued by Rachel Laudan (1976) that he did not always practise what he preached. For there are some places on his map where he coloured what we now know (by using Smith's own principle of characterizing rock units according to their fossil contents) to be different limestones with the same colours; and conversely some strata that we know by their fossils to belong to the same division of the chronostratigraphic column Smith depicted with different colours in different places.

Laudan's argument that Smith did not always rely on fossils finds support if we consult Smith's first table of strata, drawn up in Bath under his instructions in 1799.[8] It lists twenty-three rock units, from coal upwards to chalk. In each case, the estimated thicknesses are stated, the lithologies described and some fossils are named. The unit labelled 'Red Ground' (now regarded as Triassic) bears the comment 'No fossils known'. Yet this was one of the main units delineated on a surviving manuscript geological map of the Bath area, prepared in 1801.[9] Thus it is obvious that Smith first recognized his stratigraphic units by their lithologies. Indeed, logically speaking, he could hardly have done otherwise.

This is not to say, however, that Smith did not establish or use the principle that strata might be identified by their characteristic fossil contents.[10] As mentioned, in the Bath area, for example, he distinguished two lithologically similar units, an upper and a lower oolitic limestone, on palaeontological grounds. He had his principle in mind, even if he did not always use it or was not in a position to do so. Moreover, he performed a major task for British geology by sorting out the order of strata for the major units of southern Britain. This is well shown by Smith's impressive published hand-coloured figure, a *Geological Section from London to Snowdon* (Smith, 1819), which showed the relative thicknesses and spatial relationships of the upper parts of the stratigraphic column much as they are understood today. It was only when the section reached the complex terrain of Wales that Smith might be said to have been 'hand-waving'; for there the intensely complicated 'transition rocks', so called in Wernerian theory, were not sorted out satisfactorily, even in a preliminary way, and certainly not using the Smithian technique of arranging strata by their fossil contents.

For modern geologists, the geological map and geological section provide essential tools whereby the geological history of an area may be gauged. It is a nice, question, however, whether Smith was particularly interested in the history of the earth, as revealed by its strata. He was, after all, a surveyor, and I think it was essentially the geometrical arrangement of the strata that concerned him, for practical and economic reasons, as had earlier been the case for the Wernerians in Saxony. Smith believed that the strata 'were formed much as they are and that since their formation the crust of the Earth has not been much distorted or disturbed except at the Deluge and then only superficially'.[11] The general history of the globe was a question for others to examine.

This task was undertaken with great enthusiasm by the growing band of geologists in the nineteenth century, with geological mapping and palaeontological correlation being the main tools. In Britain, the mapping of the country as a whole was carried out separately from Smith, though using his work as an essential starting-point, by the gentlemanly geologists of the Geological Society of London, founded in 1807 under the leadership of George Bellas Greenough (1778–1855). Their map appeared in 1819 (Greenough, 1819) with an accompanying memoir (Greenough, 1820). A second edition was issued twenty years later. Greenough's map was specifically intended to be an 'objective', theory-free compilation of empirical facts. It contained a good deal more information than Smith could generate single-handed.

Greenough's map was a useful counter to the speculative theorizing that had accompanied the debates between the Neptunists and the Vulcanists, particularly in Edinburgh.[12] But in fact, if a map is to convey much meaning to the person who is using it, it is in some senses better that it be prepared with theory in mind. Otherwise, the map will carry little 'meaning'. Smith's own particular theory was remarkable. He imagined that there had been successive convulsions somewhere to the east of Britain which had spread successive layers of sediments across southern Britain (rather like successive layers of foam on a beach), yielding the distribution of rocks that was revealed by his mapping procedures. He also held to the Wernerian notion of 'universal' formations. Such ideas may not attract us today, but they did not stand in the way of Smith producing maps that were remarkable for their beauty and clarity, and gave concrete expression to his geological thinking.[13] In map-making, while objectivity is of prime importance, some theory is better than none.

It should be noted here that, although (so far as Britain is concerned) most historical attention has been directed towards the large early maps of Smith and Greenough, there was also another 'tradition' of more detailed mapping, dating from the same period, which has passed almost entirely unnoticed. In a recent paper, Torrens (1994) has shown how some detailed geological survey work and mapping of small areas were undertaken by 'practical men' such as Smith, Farey and others, generally on behalf of landed gentry who wished to know what might be the mineral riches of their estates.

One such gentleman-patron was Sir Joseph Banks, powerful President of the Royal Society. Torrens has located a remarkably detailed coloured geological map of Banks's Ashover estate in Derbyshire, dated 1812, prepared by Farey on Smithian principles.[14] If one did not know the date, one might easily suppose by its appearance that the map was a late nineteenth-century production. This recent discovery suggests that there was probably a much stronger tradition of geological cartography in the early nineteenth century than has previously been supposed. But it cannot be said to have had much influence on ways of thinking about the earth by the educated public at large. Of lower-middle-class origins, Smith was not welcomed into the early Geological Society. Banks had the requisite social qualifications and made considerable efforts on Smith's behalf, supporting some of his publications. But Banks, who was scarcely pleased to see the Geological Society rising alongside 'his' Royal Society, did not join the 'Geologicals'. So the early union of interests and practice between the gentlemanly geologists and the practical men, which Banks might have facilitated, did not occur as a general phenomenon. It was the ideas of the gentlemen that chiefly prevailed in the public mind, although Smith was later made much of, has come to be regarded as the founder of stratigraphy and in his old age was dubbed the 'father of English geology' by Adam Sedgwick.

In France, important contributions to geological cartography were made by the mineralogist Alexandre Brongniart (1770–1847) and the zoologist Georges Cuvier (1769–1832). Cuvier, the most influential French scientist in his day, is remembered particularly for his 'catastrophism', which will be discussed in Chapter 6. For the moment, we may note that in his work with Brongniart on the Paris Basin Cuvier endeavoured to imagine the circumstances which obtained at the time when the various sedimentary rock units (which, significantly, the authors called 'formations') were laid down. For example, fresh-water and marine deposits were distinguished (by analogy with modern deposits), and so the authors could provide readers with a geological history of the Paris region: when the sea advanced and when it retreated, when the waters were swarming with shells and when they were barren. Fossils were used to discriminate strata that looked similar and yet were chronologically different. That is to say, Smith's stratigraphical principle was deployed. For example, gypsums that had been described by Werner were distinguished from those of the Paris area by considering their relationships with certain shell beds. All this could be done without the provision of direct testimony from humans who had witnessed the actual geological events (Cuvier and Brongniart, 1808, 1811, 1822). Earlier, as we have seen, Robert Hooke thought it necessary to consult ancient records, myths and legends, in order to know something about the earth's history. Rocks and fossils were not thought sufficient in the seventeenth century.[15] On the other hand, Lavoisier's work provided a precedent, close in both space and time, for the activities of Brongniart and Cuvier.

Also in the early nineteenth century, work was done towards linking up with one another the geological observations made in different regions. An example of this kind of endeavour is provided by Thomas Webster (1773–1844), a Scottish-born architect who worked in London, becoming museum keeper, librarian, draughtsman, one of the secretaries to the fledgling Geological Society of London, and eventually in 1841 the first Professor of Geology at University College, London. In a classic paper (Webster, 1814), which dealt with the geology of the Isle of Wight, and the Hampshire and London basins, the work of Brongniart and Cuvier was followed, with the recognition of alternating marine and fresh-water sediments. Subdivisions of the chalk were made on palaeontological grounds, and the memoir was illustrated by coloured maps and sections. More importantly for our present purpose, Webster sought to correlate the rocks that he was investigating in southern Britain with those across the Channel that Brongniart and Cuvier had described in France, comparing graphically (in his Plate 10) the sedimentary basins of Paris, London, and the Isle of Wight (Hampshire).[16] Webster also used evidence from the high-angle tilts of the strata in the Isle of Wight and the Isle of Purbeck, and from appearances of faulting, to argue for occurrence of great earth movements in the past, and subsequent widespread erosion.

In Webster's investigations, we have, I suggest, the beginning of work that continues to the present: the mapping of the earth's strata and the effort to interpret the results so as to be able to recount its geological history. I call the emergence of this practice the 'historicizing' of geology.[17] The earth's sedimentary strata may be thought of as a kind of palimpsest and the task of the geologist – a historian of a kind – is to identify the elements of this palimpsest and to correlate the fragments and arrange them in temporal sequence. Then the history of the whole may be read from the record in so far as the characters of the fragments indicate their modes of origin. Any given formation is never universal in extent, but its lateral extensions should be determined as far as possible. Lithologies can and do change in space at the same time (e.g. a sandy beach not far from a muddy estuary), and so too do the fossils; but lateral correlations may be possible at least to some degree with the help of the palaeontological evidence.[18]

Thus, between the time of Werner in the 1780s and about 1815, geology became 'historicized'. The 'onion-skin' model of Wernerian theory (as construed in its most simple and unsophisticated form), which, however, had a definite discoverable order for the strata formed over time, gave way to the historical

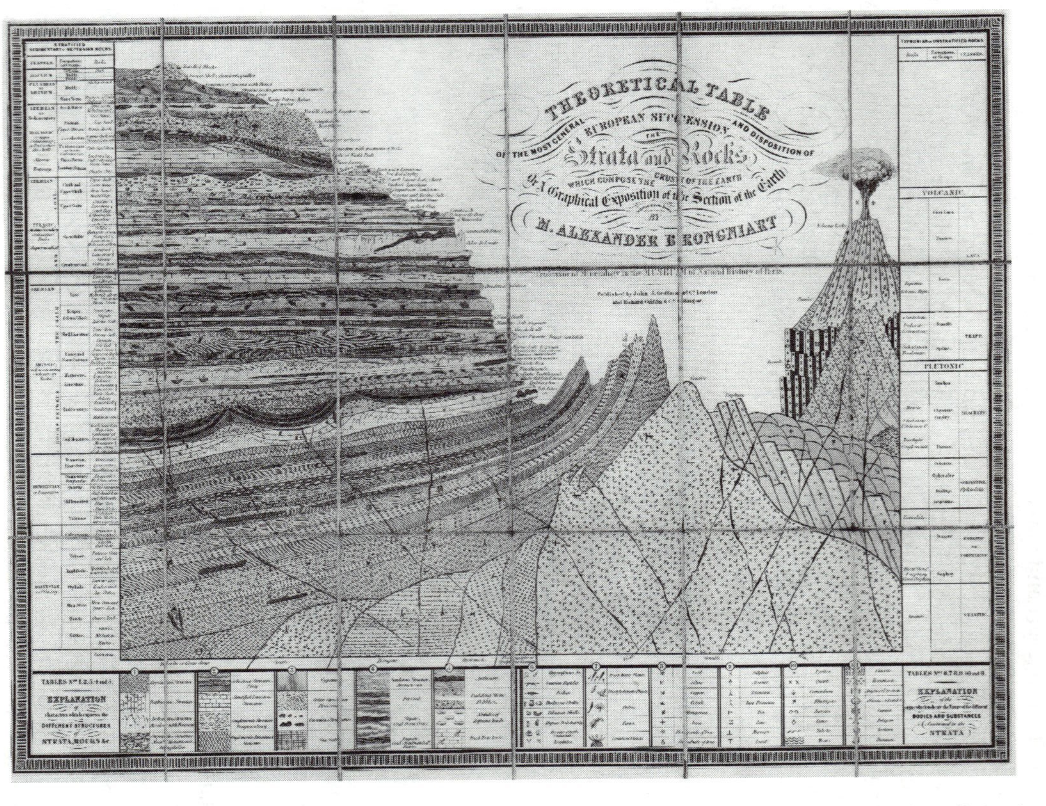

Figure 5.3 Section showing the European succession of strata in generalized form (coloured in original), according to Alexandre Brongniart (by permission of the Natural History Museum, London)

investigation of the sedimentary palimpsest. The results of such endeavours are well displayed in a remarkable section prepared by Alexandre Brongniart (n.d. [1829]), which well shows the understanding of the general history of European strata that had been achieved by the third decade of the nineteenth century (Figure 5.3).

It might be thought that this was an 'onion-skin' model, such as might have been produced by a sophisticated Wernerian, or even an advanced follower of Lehmann and Füchsel. The difference, I suggest, is that for the nineteenth-century Brongniart the chart was a composite diagram. It represented a kind of 'average' of what was known of the geological history of the various parts of Europe. It was a synthesis of several individual or local histories. Füchsel's diagram, by contrast, served for Thuringia alone. Werner's sequence, though based on the kind of information that Füchsel and others provided, was highly theoretical – more so than was necessary for the practice of historical (or historicized) geology. Even so, the Wernerians would have been happy to subscribe to the composite result that Brongniart was able to provide. But Werner himself, for all his theory, did not accomplish such a result. The stratigraphical knowledge had to be gained in the field.

That this was so is scarcely surprising. It should be noted, however, that the historicization of geology had to be a cooperative affair, with some institutional backing. It could not be accomplished single-handedly and overnight. Smith had a map, a (redundant) theory of the genesis of the strata and a clear understanding of the geometrical arrangements of British strata. He did not provide a history of the rocks in all their intricate detail. This did begin to emerge, however, in the early years of the nineteenth century. Brongniart and Cuvier were the leaders, but less well-known figures such as Webster played their part.

With the historicizing of geology, a new and attractive prospect opened up for geological research. Countries had to be mapped, and this needed to be done in some kind of coordinated manner. With the mapping of a country, its geological history would be determinable, and at the same time the very process of detailed mapping would reveal something of the country's mineral wealth and hence its economic prospects. With the burgeoning of the Industrial Revolution, national geological surveys became increasingly attractive and were pursued more and more assiduously in the nineteenth century and until the recent past.

In the first three decades of the nineteenth century, some individuals were bold enough to suppose that they might map a whole country. As we have seen, William Smith undertook the mapping of England, Wales and parts of Scotland. Richard Griffith (1784–1878) worked in Ireland (see Herries Davies and Mollan, 1980; Herries Davies, 1983, 1995) and had a preliminary version of his map on display as early as 1814, although a published version did not appear until 1838.

In 1808, a crude geological map of Scotland was submitted to the Geological Society by the Genevan Louis Albert Necker (1786–1861), who studied in Edinburgh in 1806 and spent his later years in Skye. Of greater importance was the work of John Macculloch (1773–1835), who single-handedly mapped the whole of Scotland, his map being published posthumously (Macculloch, 1836; Cumming, 1981). Although he started his work as a private venture, Macculloch eventually obtained financial backing from the government (in 1825), so that in effect he was a one-man official survey. His theory was largely Wernerian in character. Smaller local areas of Britain were also mapped successfully, quite early in the century – for example, J.S. Henslow's map of Anglesey (Henslow, 1822; Oldroyd, 1993).

The official survey of England and Wales also began as a one-man show, the showman being Henry (later Sir Henry) De la Beche (1796–1855) (see McCartney, 1977; Rudwick, 1985). Coming from a well-to-do military family and having studied at the Military College, Great Marlow,[19] De la Beche lived in Jamaica for a period, where he inherited a sugar plantation powered by slave labour. He studied the geology of the island and became the authority on the area. Returning to England in 1824, he settled in the West Country and pursued his geological interests.

Although at that time he was what has become known as a 'gentlemanly geologist' (Rudwick, 1985), De la Beche's financial position was not secure (nevertheless he found time for geological tours on the Continent in the late 1820s[20]). So, having begun the geological mapping of Devonshire as a matter of personal scientific interest, he soon turned to the government for financial support, as Macculloch had done before him. This was forthcoming from the Ordnance Survey, and so in 1832 De la Beche began what might be called a piece of 'contract work' for the government – namely the provision of geological maps of Devon, for a fee of £300. Impressed by the successful completion of this task, the Board of Ordnance appointed De la Beche in 1835 as a full-time employee. And thus the British Geological Survey was established. By 1839, a Museum of Economic Geology was established in London and by the mid-1840s a staff of about ten had been built up. The full-scale survey of the British Isles was under way.[21]

It should be noted that De la Beche was a man with some military background, and that his ability to undertake geological mapping was in part a result of the training that he received in mathematics and topographical surveying at his military college. Also, the department under which he came to serve was established in the first instance for strategic military purposes. Thus it should not be surprising that the Geological Survey was, as first conceived and as initially established, something like a military organization. The Survey 'officers' of the early period wore a military-style uniform. And the very ethos of the institution was, I suggest, military in character, in that, as the ground was gradually surveyed,[22] it was thought of as ground 'captured' or brought under scientific control.

But there was another aspect to this question of 'control'. How was the stratigraphical column to be subdivided? Concomitantly, how were the maps to be coloured? Werner had left his successors with the concepts of 'transition' and '*floetz*' (layered) rocks. The latter ran roughly from the coal-bearing formation of what we call the Carboniferous up to the chalk.[23] Below the coal measures there was the Old Red Sandstone (so named by Werner[24] and called the *Grès ancien* in France). And below that there was the great mass of transition rocks, whose fossil contents had received little systematic investigation before the 1830s. There was thus work to be done, and the question of 'control' came to modulate this work to a considerable degree.

The main subdivisions of the stratigraphic column, as they are known today, are given in Table 5.1, with the names of the persons who originally proposed the subdivisions and the dates when the names were bestowed (Berry, 1968). It

Table 5.1 *Subdivisions of the stratisgraphic column*

Quaternary (Desnoyers, 1839)	Pleistocene	Lyell	1839
Tertiary (Arduino, 1760)	Pliocene	Lyell	1833
	Miocene	Lyell	1833
	Oligocene	von Beyrich	1854
	Eocene	Lyell	1833
Secondary (Arduino, 1760)	Cretaceous	d'Omalius d'Halloy	1822
Mesozoic (Phillips, 1840)	Jurassic	von Buch[a]	1839
	Triassic	von Alberti	1834
Palaeozoic (Phillips, 1841)	Permian	Murchison	1841
	Carboniferous	Conybeare/Phillips	1822[b]
	Devonian	Sedgwick/Murchison[c]	1839
	Silurian	Murchison	1835
	Ordovician	Lapworth	1879
	Cambrian	Sedgwick/Murchison[d]	1835
Precambrian (Jukes, 1862)			

[a]As early as 1795, Alexander von Humboldt had identified a *Jura-Kalkstein*, but this only represented part of the modern Jurassic and the diagnostic fossils were not given.
[b]Earlier (1808), J.J. d'Omalius d'Halloy (1783–1875) had referred to a *terrain bituminifère*, with approximately the same meaning as that of the Carboniferous of Conybeare and Phillips. The Carboniferous unit is equivalent to the Pennsylvanian and Mississipian (taken together) in American geology.
[c] William Lonsdale also played a significant part in the establishment of the Devonian System. See Rudwick (1985).
[d]Although Adam Sedgwick established the concept of the Cambrian through his fieldwork in Wales, it was Murchison who coined the name for the system.

will be observed that the geologists contributing to this nineteenth-century subdivision of the stratigraphic column were chiefly British, and it is not unreasonable to construe the whole process as a kind of extension of the domains of the British Empire. It is, of course, unusual for a country as small as Britain to have within its boundaries such a high proportion of the geological systems named above – only the Miocene is missing in Britain. But a remarkable proportion of the systems we recognize today were established on the basis of work done in Britain or by British geologists overseas. In consequence, the modern stratigraphic column is not necessarily the one that would be produced by any set of nineteenth-century geologists. It is a highly contingent taxonomy. If, for example, the geology of the globe had first been worked out in New Zealand, no geologist would have placed a boundary between the Cretaceous and the Tertiary (Oldroyd, 1972c), for sedimentation continued in that part of the world, whereas there was a major break in Europe.[25]

Thus it was through the stratigraphical column that a far-flung European hegemony in geology was established, with what was to a large extent a British taxonomy imposed on the world in an almost imperialistic manner. Thinking of the earth as a kind of empire, geological mapping being analogous to colonial acquisition, is most clearly seen in the work of another gentlemanly geologist, Sir Roderick Murchison (1792–1871).[26] Like De la Beche, Murchison was originally associated with the army. Indeed, he fought in the Napoleonic wars. After this active period of military service, he married and settled down to the life of a country squire. But, soon tiring a life of fox hunting, at his wife's suggestion he took up geology, joined the Geological Society and learnt the craft of fieldwork among the more ancient rocks of Britain with the Reverend Adam Sedgwick (1785–1873), the Cambridge Professor of Geology.

The story of the relationship between Sedgwick and Murchison has been told many times and need not be detailed here.[27] There was initially an alliance between them against De la Beche, concerning the interpretation of the rocks of Devon. In this battle, the two amateurs[28] proved the victors for the most part, with the establishment of the Devonian system, the recognition that rocks of similar age might be very different in different areas (the sediments of Devon being equivalent in time to the lithologically different rocks designated the Old Red Sandstone in other parts of Britain) and the reliance on fossils as stratigraphic indicators.[29] But, with regard to the rocks of Wales, the two former colleagues fell out and eventually became not only geological opponents but personally estranged. It all had to do with territory.

Putting the matter in simple terms, Sedgwick and Murchison sought to subdivide the rocks of central and north Wales and the Welsh border region (particularly Shropshire) on the basis of palaeontological evidence. It turned out that these rocks – previously designated 'transition rocks' according to Wernerian theory and thought to be somewhat uninteresting – were often richly fossiliferous. In the 1830s, Murchison worked westwards from the Welsh border region, establishing what he called the Silurian System, coining this term after a tribe, the Silures, who once lived in that part of Britain. The Silurian rocks had their own distinctive fossils, with corals, crinoids, brachiopods, trilobites and the like, and several reliable subdivisions of the system were also established on palaeontological criteria.

Sedgwick, in contrast, worked outwards from the difficult mountainous regions of north Wales, where fossils were less common, and so he had to rely more on lithological and structural criteria. His system, the Cambrian, was not satisfactorily defined by its fossil content (or so Murchison later claimed). In their earlier collaborative work, the two geologists endeavoured to draw a boundary in Wales between their respective systems, but as has been well shown by Secord (1986) the job was not performed satisfactorily, so that in their later investigations the two geologists began to 'poach' on their respective territories. I suggest that the dispute was chiefly fuelled by Murchison's territorial ambitions for his system: the Silurian. His mentality is manifested in the extended subtitle of his great treatise, *The Silurian System* (Murchison, 1839), which read: *Founded on Geological Researches in the Counties of Salop, Hereford, Radnor, Montgomery, Caermarthen, Brecon, Pembroke, Monmouth, Gloucester, Worcester, and Stafford, with Descriptions of the Coal-fields and Overlying Formations.* It was as if Murchison were ticking off the terrain which had fallen under his sway. Subsequently, he became known as 'King of Siluria' (Woodward, 1908: 169), and as has been well shown by Stafford (1989) Murchison showed great zeal through the latter part of his career in getting his ideas accepted in as many parts of the world as possible, and he took delight in seeing the gradual spread of his Silurian domain.[30] This was his way of thinking about the earth.

Within Britain, Murchison's predilection for territory was manifested in two particularly obvious ways. First, he tried to extend his 'kingdom' downwards into Sedgwick's terrain, annexing all the fossiliferous rocks to the Silurian and leaving Sedgwick only with an unfossiliferous residuum. Thus Murchison's second major treatise on the Silurian rocks (Murchison, 1854) was entitled *Siluria: the History of the Oldest Known Rocks Containing Organic Remains.* Sedgwick strenuously resisted this encroachment and tried to push his Cambrian as high up the stratigraphical column as he could, having to fight every inch of the way. Eventually, the Cambrian was accepted as a definite geological system, with its own characteristic fossils. But the breach between Sedgwick and Murchison was never healed in this world, and the way out of the impasse was proposed by an 'amateur' geologist, Charles Lapworth (1842–1920) only in 1879, when he suggested the Ordovician system to accommodate the Welsh rocks for which Murchison and Sedgwick had earlier been contending.[31] Even then, well after Murchison's death,[32] the British Survey did not accept the Ordovician proposal, and only adopted Lapworth's system at the beginning of the twentieth century, after Murchison's protégé Archibald Geikie had retired.

The other egregious exemplification of Murchison's territorial ambitions occurred in Scotland, as I have described in detail elsewhere (Oldroyd, 1990a). In the north-west Highlands, there is a band of sedimentary rocks running from the north coast down to Skye. Murchison was aware of the occurrence of fossils in one of these bands (the Durness limestone) and took them to be Silurian. Above the limestone there were large areas of schistose rocks, and above these again were fossiliferous Devonian rocks. Murchison regarded the schists as having been metamorphosed *in situ*, and thus felt empowered to call them Silurian too, although they were devoid of fossils. As a result of this theoretical move, he was able to construe vast areas of the northern and central Highlands of Scotland as Silurian, so that the geological maps of Scotland of the 1860s and 1870s were painted predominantly in Silurian colours (*e.g.*, Murchison and Geikie, 1861).

I mention these matters in a little detail as they exhibit, I think, an attitude towards the earth that was characteristic of the nineteenth century. The planet was a place to be conquered and subdued, for territorial expansion was accompanied by wealth and power. Geologists like Murchison (he was no doubt an extreme case) took pride in seeing the expansion of the territories of their own countries. The expansion of geological knowledge held out the promise of increased power, through the discovery and acquisition of mineral wealth. The knowledge gained by geologists, displayed in maps, was not, of course, private knowledge. On the contrary, there was a keen desire to make the knowledge known through the publication of maps and accompanying memoirs. And, when international geological congresses were begun in 1878, one of their main functions was to try to ensure stratigraphical coherence and transnational stratigraphic correlations. Nevertheless, countries gained power through geological and geographical knowledge. And Murchison was exceedingly active in trying to promote his country's interests through increase of knowledge of this kind. Then, at another level, he sought his own aggrandisement by the expansion of the regions with which 'his' systems were associated. Finally, through his powerful social position, he sought to exercise his (and Britain's) hegemony by the encouragement of geological and geographical exploration and by his patronage of those favourable to his views in the geological surveys that were established in other parts of the world in the Victorian period (Stafford, 1989).

I find it helpful to see all this in Foucaultian terms. That is, one may interpret the process that I have been describing above in the light of the arguments proposed by Michel Foucault in his *Order of Things* (Foucault, 1966, 1970). The world (and its inhabitants) were dominated in the nineteenth century by considerations of power. Foucault has written of 'discipline' and 'punishment'. Humans became institutionalized in new ways in the nineteenth century by secular bureaucracies, whether it was in hospitals, prisons or systems of public education. Governments sought power through knowledge of the inhabitants under their sway. They were counted and documented. So too it was with the earth. It was brought under human control, and the various geological surveys that were instituted in the nineteenth century bear clear witness to this process.

The process of the gradual establishment of bureaucratic structures for the mapping of the world may be gauged by the data in Table 5.2.[33] It should be noted, however, that in some cases important geological maps were produced privately – as Smith had done for Britain – before official surveys were established. For example, there was Maclure's map of the eastern part of the United States (Maclure, 1809), Griffith's map of Ireland (1838), Dufrénoy and Elie de Beaumont's map of France (1841), or d'Orbigny's map of Bolivia (1842). In the German-speaking part of Europe, a considerable amount of mapping was carried out prior to the establishment of the official state surveys (see Koehne, 1915).

The list is Table 5.2 though regrettably incomplete, provides interesting information about the character of geological surveying in the nineteenth century. It was to a large degree conducted on a worldwide basis by people of British stock. De la Beche's survey provided a model for the planning and organization of survey work, and supplied the initial staff for many of the first colonial surveys, although the Americans didn't need much help in this regard. In the Americas, geological survey work was closely associated with initial topographical surveys and with determining the mineral wealth of the separate states. As a broad generalization, there seems to have been some correlation between the degree of social and economic development of a state and its interest in establishing a geological survey. Thus, as with Murchison in Britain, the American geological surveys were linked with the establishment of bureaucratic control of a domain (state). In Africa, it was the British and French colonies that were mapped and surveyed in the nineteenth century. In Spanish- and Portuguese-speaking South America – parts of two declining empires – little of note was done. The Muslim world did not take its own affairs in hand so far as cartography and geological surveying were concerned. It was not interested in the exploitation of the globe with the aid of geological knowledge. On the other hand, Japan – a would-be emergent military and industrial power – began its formal survey programme only ten years after the Meiji revolution. And China began its survey in the early years of the twentieth century as it endeavoured to build up military and economic strength to combat Western and Japanese imperialist threats. It was the powerful countries that interested themselves in matters geological; and so it has continued to the present day (with the power now projected under the oceans and towards other regions of the solar system).

I have sometimes thought that historians of geology have been too inclined to give attention to 'English-language geology'. Yet this seems to be warranted to some degree by the information compiled in Table 5.2. There were certainly important Continental geologists who did work overseas in the nineteenth

Table 5.2 *Bureaucratic structures for the geological mapping of the world*

Country	*First survey director or first government surveyor*	*Date*
Europe		
Britain		
England and Wales	De la Beche	1835[a]
Ireland	James	1845
Scotland	Geikie	1867[b]
Germany		
Bavaria	von Gümbel	1849–50
Baden	Sandberger	1860
Saxony	Credner	1872[c]
Elsaß-Lothringen		1872?
Prussia[d]	Hauchecorne	1873
Hessen	Ludwig	1882
Baden	Sauer and Schalch	1888
Mecklenburg	Geinitz	1889
Württemberg	Sauer	1903
Oldenburg	Schucht	1908
Thuringia		1923
Netherlands	Staring	*c.*1858
Norway	Kjerulf	1858
Sweden	Erdmann	1858
Switzerland	Studer	1859
Austria–Hungary	von Haidinger	1849?/1865?
Hungary	von Hantken	1869
Finland	Moberg	1865
France	Elie de Beaumont	1868
Portugal	Ribeiro	1869[e]
Spain	de Castro	1873[f]
Italy	Giordano	1877[g]
Belgium	Jochaus	1878[h]
Denmark	Johnstrup	1888
Bosnia		1898[i]
Croatia/Slavonia		1909?
Austria	Geyer	1919
Czechoslovakia	Purkyne	1919
Poland	Morozewicz	1919
Greece	Ktenas	1919
Middle East		
Turkey	Gencer	1935
Africa		
Northern Africa		
Algeria		1893 (?)
Egypt	Lyons	1896
Sudan	Barron	1905
Morocco	Barthoux	*c.*1921

Table 5.2 *(continued)*

Country	*First survey director or first government surveyor*	*Date*
Central Africa		
Southern Rhodesia (Zimbabwe)	Maufe	1910
Gold Coast	Kitson	1913
Sierra Leone	Dixey	1918
Nyasaland	Craig	1918
Uganda	Wayland	1918
Nigeria	Falconer	1919
Southern Africa		
South Africa	Kynaston	1910
Orange Free State		1878?
Cape of Good Hope	Seeley	1889
Cape Colony	Corstorphine	1895
Transvaal	Molengraff	1897
Natal	Anderson	1899
The Americas		
North America		
Canada	Logan	1842
United States (Federal Survey)	King	1879[j]
North Carolina	Olmsted	1824
South Carolina	Vanuxem	1825
Massachusetts	Hitchcock	1830
Tennessee	Troost	1831
Maryland	Ducatel	1834
Connecticut	Percival and Shepard	1835
New Jersey	H.D. Rogers	1835
Virginia	W.B. Rogers	1835
Georgia	Cotting	1836
Maine	Jackson	1836
New York	Mather/Emmons/Conrad/Vanuxem	1836
Pennsylvania	H.D. Rogers	1836
Delaware	Booth	1837
Indiana	Owen	1837
Michigan	Houghton	1837
Ohio	Mather	1837
Kentucky	Mather	1838
New Hampshire	Jackson	1839
Rhode Island	Jackson	1839
Vermont	Adams	1845
Alabama	Tuomey	1848
Mississippi	Millington	1850
Illinois	Norwood	1851
California	Trask	1853
Missouri	Swallow	1853
Wisconsin	Daniels	1853

Table 5.2 *(continued)*

Country	*First survey director or first government surveyor*	*Date*
Iowa	Hall	1855
Arkansas	Owen	1857
Texas	Shumard	1858
Kansas	Mudge	1864
Minnesota	Hanchett	1864
Nevada	(Appointment not made)	1865
Louisiana	Hopkins	1869
Colorado[k]	Smith	1874
Washington	Bethune	1890
Nebraska	Barbour	1891
North Dakota	Babcock	1895
West Virginia	White	1897
Wyoming	Beeler	1901
Florida	Sellards	1907
Oklahoma	Gould	1908
Idaho	Thomson	1919
Montana	Clapp	1919
New Mexico	Wells	1927
Utah	(No director)	1931
Oregon	Nixon	1937
Hawaii	Herscheler	1953
Alaska	Williams	1959
South America		
Argentina		1885
Mexico	Aguilera	1891
Brazil	Derby	1907[l]
Uruguay	Lammo	1912
Venezuela		1936?
Colombia	Biester	1938
Chile	Füller	1957
Bolivia	Danaso	1965
India, Far East and Pacific region		
India	Oldham	1851
Japan[m]	Wada	1882
Thailand (Siam)		1892?
Federated Malay States	Scrivenor	1927[n]
China	Ting	1912 or 1913
Korea	Park	1946
Fiji	Skiba	1951
Australasia		
Australia		
Victoria	Selwyn	1855
Tasmania	Gould	1859
Queensland (North)	Daintree	1868

Table 5.2 *(continued)*

Country	*First survey director or first government surveyor*	*Date*
Queensland (South)	Aplin	1868
New South Wales	Wilkinson	1875[o]
South Australia	Brown	1882
Western Australia	Maitland	1896[p]
Northern Territory	Nye	1935
New Zealand[q]	Hector	1865
Otago	Hector	1861
Canterbury	von Haast	1861

[a]De la Beche did survey work at an earlier date before the official foundation of the Survey.
[b]It should not be overlooked, however, that Macculloch's survey began in 1825.
[c]A good deal of mapping had been done earlier in Saxony by Werner's former students, and in 1835–45 by Naumann, von Cotta and Geinitz.
[d]This Survey was responsible for mapping most of the smaller German states as well.
[e]An earlier geological commission began work in 1857.
[f]An earlier commission was formed in 1849.
[g]Earlier work began in the Florence area in 1868.
[h]An earlier map was made by Dumont 'by order of the Government' (1836–54).
[i]Parts were apparently mapped earlier (1884) by the Austro-Hungarian Survey.
[j]From 1867, a number of nationally organized surveying teams operated prior to the establishment of the US Survey in 1879. See Merrill ([1924] 1969). The list of dates given for the various state surveys is not exact. Some semi-private surveys were conducted under state auspices before the dates given. On the other hand, some of the surveys whose dates are listed were discontinued, and some of these were later re-established.
[k]This was a territory, rather than a state until 1876. The State Survey began under George in 1907.
[l]An earlier commission was in operation from 1875 to 1877.
[m]Including Korea in the Japanese empire.
[n]Scrivenor was first appointed Government Geologist in Malaya in 1903.
[o]Earlier survey work was conducted by Samuel Stutchbury in 1850–5.
[p]Von Sommer was briefly 'government geologist' in 1847.
[q]Limited geological explorations were undertaken earlier (1859) in the provinces of Auckland and Nelson by Hochstetter and von Haast. These were partly private in nature but reports were made to the provincial governments.

century – such as the Austrian Ferdinand von Hochstetter (1864) in New Zealand, or the German Ferdinand von Richthofen (1877–1912) in China. Nevertheless, there is a case to be made that there was a significant correlation between British colonialism and English-language geology. As said, the earth was thought of in the nineteenth century as a source of wealth and power, which might be generated by investment of time, money and effort in geological research. It is not surprising, then, that there was a kind of geological empire building that went on within geology, as we have seen in the work of Murchison.

The construction of a geological map is a process involving both analysis and synthesis. As the geologist walks over the ground s/he has to abstract from the 'bloomin', buzzin' confusion' of experience certain observations that are to be entered on a topographical map (available in advance if possible). That is, certain rocks have to be noted and classified, either there and then or later in the laboratory with the help of microscopic inspection. The choice of the classification is not transcendentally obvious. Indeed, it can be both theory-laden and controversy-laden and may be quite different for different observers, as is shown in Figure 5.4. Here, to illustrate the point, we see the same area of north Wales mapped at about the same time by different nineteenth-century British geologists, John F. Blake, and Catherine Raisin and her mentor Thomas G. Bonney, who were then engaged in quite an acrimonious dispute about the rocks of that region (Oldroyd, 1993). The details of the controversy need not be recounted here, except to mention that the dispute had to do with the question of whether there was or was not an unconformity below the conglomerate. Blake's map suggested that there was such an unconformity; Bonney and Raisin's map suggested otherwise. The point, so far as the contestants in the dispute were concerned, was that the presence or absence of a break in the stratigraphic sequence, marked by an unconformity, was thought to have relevance to the age determinations of many of the rocks of north Wales, and more particularly whether Precambrian rocks were to be found in the region, and if so where. The point, so far as we are concerned, is that the rocks themselves did not 'speak', saying how they were to be mapped. Indeed, it will be seen, from a comparison of the two maps, that the lithological subdivisions used by the contesting geologists were significantly different, and so their boundary lines were drawn in quite differently. Yet the rocks themselves did not change at all. It was their representations that were different.

Of course, quite apart from the question of differences in theory, no geologist can cover every inch of the ground, and interpolations and extrapolations always have to be made in the making of a map of a given area on the basis of a finite number of observations. And the data that are collected are, I hold, influenced to some degree by the geologists' theoretical expectations and according to the needs of the controversies in which they are engaged. Attention will naturally be focused on rocks that seem to support the viewpoint of the particular controversialist. Then the data have to be synthesized. Such points are well illustrated by the maps of Blake and of Bonney and Raisin.

Thus the classic methodological procedures of analysis and synthesis, discussed for centuries by writers on the methodology (or methodologies) of science, are nicely exemplified in the processes involved in the construction of geological maps. The analytic part involves the observation and classification of the rocks in view and the selection from the totality of experience of certain 'data', which are then recorded on a map. Then the various items on the map have to be connected by drawing boundary lines, faults, etc. This is an act of synthesis. It is likely to be performed in part with ideas in mind about the

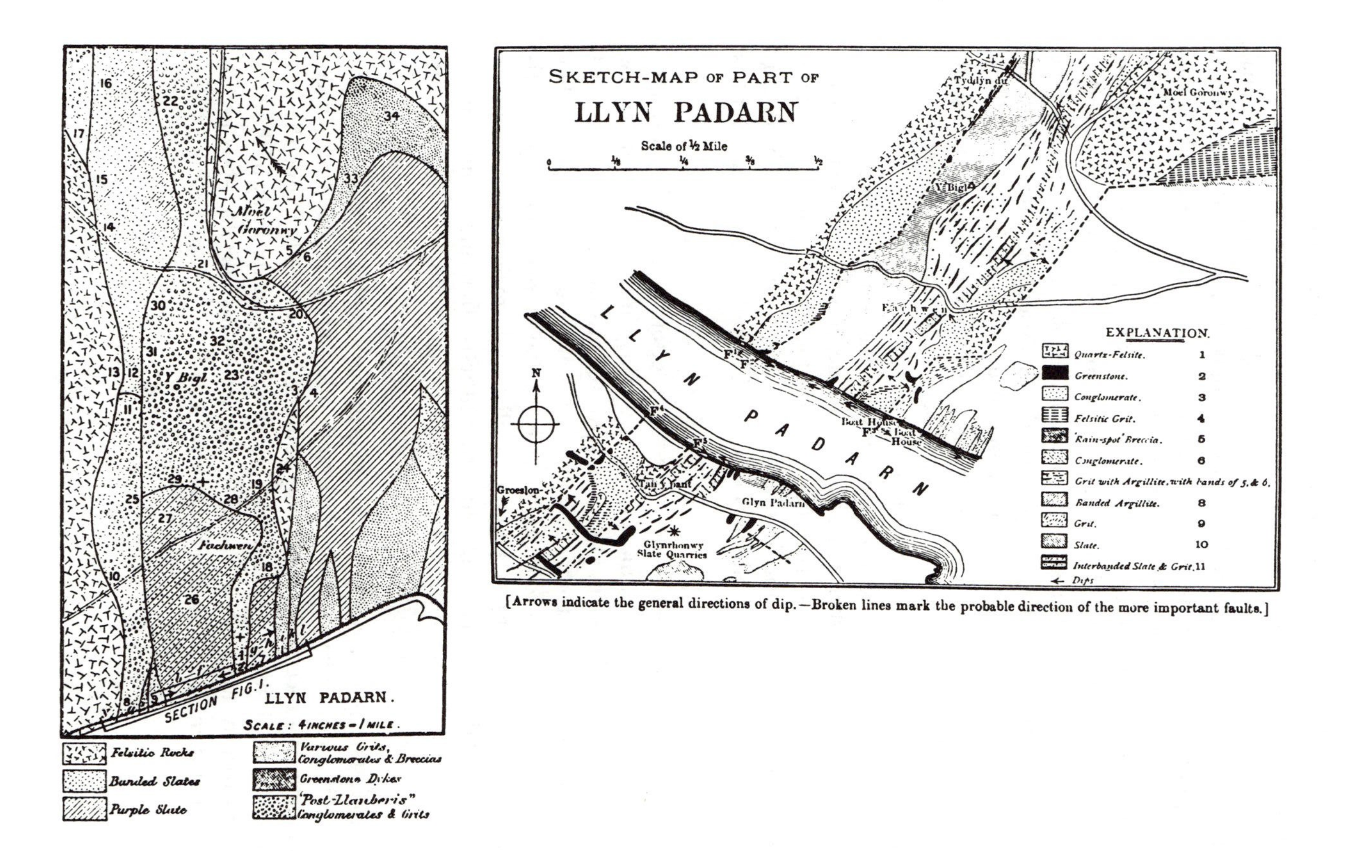

Figure 5.4 The same area of North Wales, mapped according to different theories by geologists engaged in a theoretical controversy (a) Blake (1893: 447); (b) Bonney and Raisin (1894: 594)

geometrical relationships of the strata and the means by which they were formed. But to some degree the geometry and the history can only be unravelled when the outcrops of the strata have all been entered on the map and rational cross-sections can be constructed. Thus the geologist is involved in a kind of 'boot-strapping' process, and the thought processes involved in the making and interpreting of maps are necessarily complicated.

One thing is certain, however: the construction of maps is an essential part of geologists' work. Indeed, it is one of the first and most important tasks that have to be accomplished when studying a region. Geological maps are of prime importance, for they have the complex role of representing four dimensions (three of length and one of time) by means of two-dimensional figures.[34] So map work has often been the source of controversy in the history of geology. And, as Figure 5.4 illustrates, it is a task that is perpetually subject to revision as theories change and new data are gathered.

But, important (and theoretically complex) as map work has been and is in geology, there is a danger that it can become a kind of fetish – an end in itself, rather than a means to other more important goals in theorizing about the earth. Referring to the situation in the first half of the twentieth century, the distinguished French geochemist and cosmochemist Claude Allègre (1988: 22) has written :

> Mapping absorbed about nine-tenths of the efforts of geologists throughout the world. It was a fundamental and indispensable step, but no new concepts emerged from it. However, some scientists made what amounted to a religion out of it and still practice it without knowing why they continue to do so. As each region was more and more completely investigated, the areas mapped became smaller and smaller. The 'mapping mentality' that began as a road toward synthesis often turned into a mania for more and more minute analysis. A means became an end in itself; generations of geologists wore themselves out in pursuit of it.

Perhaps nine-tenths is an exaggeration, and in modern geology detailed mapping can and often does play a vital theoretical role – in understanding palaeoenvironments or processes of mountain building, for example. None the less, Allègre's point is, I think, worth making, even if it represents a way of thinking about the earth by a representative of the modern school of black-box practitioners rather than the traditional 'man in the field'.

CHAPTER 6
Geological Time and the Tempo of Geological Change

In his biography of James Hutton, Playfair described the emotions felt when, together with Sir James Hall, they looked at the rocks at Siccar Point (see p. 95) and realized what the unconformity there implied for the age of the earth. As Playfair (1805: 73) put it: 'The mind seemed to grow giddy by looking so far into the abyss of time.'[1]

Not many in the eighteenth century looked at rocks and thought thus. Yet, by the mid-nineteenth century, most geologists had looked into the abyss and overcome their vertigo. The men who wrought this change included great theorists such as Georges Cuvier and Charles Lyell (1797–1875) and lesser figures such as George Poulett Scrope (1797–1876). In this chapter, we shall examine something of the lines of reasoning involved in bringing about such a major shift in the way geologists and educated people generally thought about the earth with regard to time.

Two striking facts may be noticed about the stratigraphic record. On the whole, the underlying rocks are more crystalline, consisting of materials such as granite, syenite, schist and gneiss, and these rocks are devoid of fossils. The crystalline rocks often look very old too. This may seem to be a very theory-laden statement. But I think most people would probably agree that the gnarled schists of Anglesey or the hard gneisses of the hills to the north of Montreal do look older than the tenacious clay of the London region, the friable soils of the North Island of New Zealand or the dusty loess deposits of China. And in stratigraphic order, the soft rocks usually lie over the hard ones.

The second point to note is that, as a rule of thumb, the fossils in the 'soft rocks' resemble the creatures living today more and more closely the higher the fossils are in the stratigraphic sequence. For example, the first fishes appear in the Silurian; but these creatures lacked jaws. In the Devonian, they have jaws, but the scales are hard and provide a kind of exoskeleton to the animals, whereas (as in modern sharks) hard internal bones are not found. In the Triassic, however, the fish remains are not remarkably different from the bony types found today.

Again, in the sediments near the bottom of the fossiliferous[2] part of the stratigraphic column there are creatures like trilobites or graptolites that are quite unlike any extant creatures, and there are no fish, no birds, no mammals. In the Jurassic, there are the famous coiled ammonites, now extinct; but these bear some resemblance to the modern nautilus, found today in the Pacific Ocean. In the Tertiary rocks, there are mammals aplenty and, as one ascends the stratigraphic column, they get more and more like those we see today.

However, the fossils do not form a continuously graded succession from the bottom of the sedimentary series to the top. On the contrary. Consider, for example, the well-known Gault clay of southern England. It is rich in ammonites and other fossils, but these do not grade into one another as one ascends a cliff-face made of the Gault (as for example at Folkestone). Different ammonite species are clearly recognizable, and the different types are used to define different 'zones' – which is essentially the deployment of the stratigraphic principle proposed by William Smith.[3] Moreover, one could hardly claim that the uppermost ammonites are in some way more modern in appearance than those at the bottom.

Nevertheless, although such phenomena were well recognized by the early years of the nineteenth century – that is, there seemed to be a certain jerkiness in the stratigraphic record – on the whole the fossil evidence seemed to display a general 'direction'. Now, within the stratigraphic record, there were appearances of minor jerks, such as are manifested by the ammonites in the Gault clay. And there were also major palaeontological changes, particularly at the top of the Permian and the top of the Cretaceous. How were such observations to be interpreted?

Empiricism was the predominant scientific philosophy at the beginning of the nineteenth century, and the initial response was to take the fossil evidence at face value. That is to say, the breaks in the palaeontological record were taken to represent actual breaks in the sequence of deposition. And the bigger the fossil break, the bigger the break in the record of the earth's history. Indeed, the major breaks were regarded as indications of major catastrophes in the earth's history. The matter was put with clarity and power by Georges Cuvier (1813: 24): '[T]he thread of operation is . . . broken; the march of nature is changed, and none of the agents which she now employs were sufficient for the production of her ancient works.'

It should not be thought that Cuvier adopted this 'catastrophist' view[4] because he was trying to squash the whole of geological history into a biblical time framework of a few thousand years. On the contrary, just as astronomers had burst the limits of space, so, he thought, geologists should magnify their perception of time: 'Would it not also be glorious for man to burst the limits of time, and, by a few observations, to ascertain the history of this world, and the series of events which preceded the birth of the human race? (ibid.: 4).' No, it was the stratigraphic record which suggested the occurrence of great catastrophes. Such a view was also supported by the surface appearance of the earth in northern Europe and the Alps. As will be discussed more fully in Chapter 7, it is littered in many places with large boulders and a layer of clay with all sorts of pebbles and fossils mixed therein (today called 'boulder-clay' and thought to be the residue of the ice ages). This evidence suggested that there had been some kind of great flood fairly recently in the earth's history. Other, earlier catastrophes were suggested to Cuvier by the major palaeontological breaks and major unconformities that occurred from time to time in the stratigraphic column.

Cuvier's ideas were perhaps never worked out perfectly. He suggested that there had been great catastrophes from time to time, but they were not worldwide. So, if the animals in some regions were wiped out by devastating floods or whatever, those regions might subsequently be colonized by other animals from regions that had escaped the devastation. Thus, Cuvier supposed, the general appearance of the stratigraphic column might be accounted for. His system was 'directionalist', though non-evolutionary.

Cuvier was ambivalent as to whether new creations occurred from time to time, after each catastrophe. But he did accept (ibid.: 148) that there had been a major catastrophe – the one recorded by Moses in the Bible – less than five thousand years ago.[5] And this was the message that many people took away from Cuvier's geological work.[6] In other words, it provided a rejuvenated physico-theology. Although the earth might be very old, the findings of the new historicized geology could be meshed with Judaeo-Christian history. The only requirement was that the six days of creation had to be regarded as six indefinitely long periods of time. The overall history of the earth and its inhabitants was progressive. It led towards the conditions of a world conducive to man's well-being. Mankind was 'specially created', and recently. Cuvier did not think that there was satisfactory palaeontological evidence for the existence of humans who had lived before the time of the Deluge or for persons who had been killed by the Deluge.

But, besides the stratigraphic interest and the interesting related theological issues,[7] Cuvier's work was of considerable importance for philosophical reasons. When he spoke of the thread of nature's operations being broken, and when he asserted that the processes of erosion, deposition, etc. that we see operating today were incapable of accounting for what we actually see around us (such as the boulder-clay), he was making a substantive claim – that things in the past were not the same as they are today, and that past geological processes were also different – certainly in degree and perhaps even in kind. There was a lack of uniformity in nature's operations.

This was, as I say, a substantive claim. But it had methodological implications.

If nature were radically different in the past from the present, how could the geologist use his knowledge of present-day processes to understand the earth's past? Worse, if the Deluge (which was regarded by many in the early nineteenth century as a supernatural event) were responsible for the earth's most recent catastrophe, then it could well be that Cuvier's other catastrophic episodes in earth history were also of supernatural origin. And if this were so – although Cuvier did not draw this conclusion – then it placed certain aspects of the earth's past outside the domain of regular scientific inquiry. Cuvier would have shrunk from this conclusion himself, but it was, I believe, implicit in his catastrophist theory. So if geology adopted Cuvierian principles it was different from the other natural sciences: in effect it had to deal with miracles.

Such a conclusion was not, perhaps, entirely uncongenial to some geologists in the early decades of the nineteenth century. At that period, there was almost a mania for geological research. It was considered a quintessentially manly or sportsmanlike kind of science. As said in Chapter 5, there was an acquisitive territorial imperative to nineteenth-century geology. Healthful, energetic and an occupation that could be pursued in almost any part of the world, it could also lock in with the religious preferences of many gentlemen of the time. Thus, in Britain at least, geology was the science of choice for a considerable number of clergymen.[8] It allowed a pleasing blend of science and 'muscular Christianity'. None the less, it proved possible to do perfectly respectable geology with metaphysical assumptions that might be regarded as quite inappropriate today. Much valuable work was done by the early nineteenth-century geological clerics.

But not all early nineteenth-century geologists saw matters thus; and it was those who followed the intellectual tradition of Hutton[9] who provided the main opposition to catastrophism and a revitalized physico-theology.[10] Among the 'dissenters', the leading figure was Charles Lyell. Lyell's original intended career was that of a lawyer, and he attended Oxford University with the object of pursuing that profession. But while at Oxford he came under the influence of the Reverend William Buckland (1784–1856), who lectured in mineralogy and geology and was Canon of Christ Church. Buckland was a most unusual man – a 'character'. Said to be somewhat coarse in his habits and certainly with a liking for the bizarre, he was nevertheless making a great impression in Oxford, both for himself and for geology, at the time that Lyell was a student at Exeter College. More specifically, Buckland was endeavouring to show how geology (essentially Cuvierian) and Anglican theology could be nicely reconciled. His inaugural lecture (Buckland, 1820), *Vindiciae geologicae*, sought to accomplish precisely this task. It went down well, and his lectures were exceedingly popular.[11] Under Buckland's influence, Lyell became strongly attracted to geology, and being a man of independent means he eventually devoted his whole career to geological research and writing, exerting a major influence on the development of the subject during the nineteenth century and, in some ways, up to the present.

Although it was from Buckland that he learned something of the craft of geological fieldwork, and conceived a love for geology, Lyell was not happy with the theoretical assumptions of Cuvierian geology that Buckland espoused. Rather, Lyell (who was of Scottish origin) was minded to revive the geological principles of Hutton and Playfair, and also to use some alternative ideas that were being developed in France by the wealthy (English) amateur geologist Scrope. Following in the steps of Desmarest, Scrope had done important work in the Auvergne district (the Massif Central), showing how there was evidence in that region for a long history of lava flows, some of which had largely been eroded away and separated from their vents by erosion, while others were relatively fresh in appearance and could be connected to their vents. Although the total picture was complex, it appeared to Scrope that the volcanic history of the district could be worked out, provided that one accepted a very large time-scale for the processes and assumed that lava flows in the past behaved essentially the same as those generated by volcanic action today.

Scrope published his ideas in two important memoirs in 1825 and 1827. His ideas have been skilfully analysed by Rudwick (1974), who has shown that for Scrope the Auvergne was rather like a Huttonian perpetual-motion machine, with lavas being successively poured forth and then eroded away, giving a landscape that was continually changing in detail through time, but remaining much the same overall.[12]

Not long after, Lyell visited the Auvergne and was convinced by the interpretations that Scrope offered (which were essentially Huttonian in character). Lyell then travelled on to Italy to examine the volcanic regions of that part of Europe. He was particularly influenced by what he saw near Naples and in Sicily. At Pozzuoli, round the bay to the north-west of Naples, Lyell visited an interesting Roman building (formerly called the Temple of Serapis). Close to the sea, this edifice still had (and still has) three columns standing, and some distance up the columns there were marks (of mollusc borings into the stone) which indicated that the columns had, since Roman times, been partially submerged. This strongly suggested that within about two thousand years the land at Pozzuoli had sunk and then risen again 20 or 30 feet. And this had been such a gentle process that the columns had not toppled over.[13] Lyell concluded, rightly, I think, so far as the empirical evidence was concerned, that significant geological changes could occur in a 'gradualist' manner and need not involve great catastrophes such as Cuvier envisaged.

In Sicily, Lyell made a close study of the great volcano of Etna and of the historical records of its eruptions. The mountain was growing, with the successive additions of layers of lava. Knowing the approximate rate of increase of the mountain and its total height, one could make a rough estimate of its age. It appeared to Lyell that it was at least several hundred thousand years old.[14]

Now below the lava flows at the edge of the mountain Lyell discovered fossil shells which were obviously geologically recent, for they were virtually identical to shells of molluscs swimming in the Mediterranean today. This meant that 'young' fossils – ones recent in appearance – were many thousands of years old. Thus the earth's youngest strata were very ancient in relation to a human time-scale. This was an extremely persuasive argument for the great antiquity of the earth as a whole.[15]

It might reasonably be objected that Lyell's argument about Etna and the age of the earth depended upon the assumption that the rate of formation of the mountain was roughly constant – that geological processes in the past occurred at about the same rate in the past as they do today. This was, in fact, precisely the argument that Lyell did adopt. Indeed, it became the hallmark of his geology. It was the methodological rule that he laid down for himself, namely that one should use observations of currently occurring processes to account for what had happened to the earth in the past. This rule was soon dubbed 'uniformitarianism' by William Whewell (1832).

On returning from his French and Italian journeyings, Lyell began work on a major geological treatise, which was eventually published in three volumes in the years 1830 to 1833: the famous *Principles of Geology*, subtitled *An Attempt to Explain the Former Changes of the Earth's Surface, by Reference to Causes Now in Operation*. This subtitle expressed exactly the method of thinking that Lyell sought to introduce into geology.[16] Lyell wanted geology to be a bona fide science, with a respectable methodology. The very title of his book reminds us of the *Principia* of Newton (1687). It was an important consideration for Newton that scientific investigations should utilize 'true causes' (*verae causae*), not speculative hypotheses.[17] Causes should not transcend the realms of the knowable and the known.

Now, in relation to geology, it is evident that if Lyell were to follow this principle he had to explain the unknown (the earth's past history) in terms of the known (what we can see happening today). We must use the present to understand the past. It was this 'uniformitarian' premise that Lyell argued for with considerable success in his *Principles*. It can be seen how his procedure was exemplified by his reasoning in favour of the great age of the earth, on the basis of his study of Etna.

We may think of Lyell's methodological principle from two points of view – the methodological and the substantive[18] – although the two intertwine with one another in a manner that is philosophically disconcerting. If one is to use the principle that 'the present is the key to the past',[19] as Lyell did in his reasoning about Etna, and as seems to be necessary for nearly all science (perhaps not for theoretical cosmology?) in the sense that one must argue from what is known to that which is unknown and in need of explanation, then, in the geological case, one is almost committed to the conclusion that the earth was in its distant past essentially the same as it is today. Certain historical, contingent details would differ from time to time, but broadly speaking everything would stay the same. Thus the geology that Lyell propounded – which had much in common with that of Hutton and Scrope – had the methodological assumption of uniformity, the further assumption of gradualism (as opposed to catastrophism), the idea that the laws of nature were the same in the past as they are today and, most boldly of all, that the world was broadly speaking the same in the past as at the present.[20]

How, then, did Lyell account for the main features of the fossil record, in accordance with his philosophy of geology? His suggestion was that the principal groups of animals (fish, molluscs, mammals, etc.) had always been on the earth, but that new species appeared from time to time and also became extinct from time to time. They were produced (by a mechanism about which Lyell could say nothing) in such a way as to be suited to the physical environment that might prevail on the earth at different times. For Lyell believed that the different parts of the earth's surface were constantly rising and falling, though slowly and gradually (as at Pozzuoli). This could cause climatic changes, for on some occasions the high land might be nearer the poles, while at other times it might be nearer the equator. Also, according to the disposition of the continents at different times, the ocean currents might shift and change. Thus the climates could change (within limits), and the changes would be tracked by the appearance and disappearance of different species. Overall, the changes were without direction. Thus Lyell was not a 'progressionist'. Rather, as emphasized by Gould (1987), his system was cyclic – non-directional. Whereas Buffon, for example, thought that the coal measures indicated that the earth had formerly been warmer and was steadily cooling, for Lyell the massive plant deposits merely represented a passing phase of warm climate in the northern hemisphere when the general altitude of that part of the globe happened to be low.

So Lyell could contemplate quite significant changes in the climatic circumstances in accordance with his theory. Indeed, in his later years, he somewhat reluctantly accepted the new glacial theory (see Chapter 7), assuming that his theory of climate – which supposedly depended on the fortuitous disposition of the high and low land – was sufficiently flexible to allow for the occasional occurrence of ice ages.

Lyell's theory may be described as a form of environmental determinism, although it was not fully naturalistic. Species were directly linked to the environmental circumstances. Thus Lyell (1830: 123) seriously suggested that if the right conditions recurred the great dinosaur, the iguanodon, at that time only recently discovered and the object of considerable public interest, might again roam the earth.

Lyell's theory could give a plausible account of the great breaks in the fossil record which had led Cuvier to propose the occurrence of great catastrophes. The idea was simply that, with a slow but steady turnover of species and places on the earth where there was no deposition for long periods (although deposition would be occurring elsewhere), it would be possible for sediments to show great changes in fossil form from one horizon to the next. This need not betoken a great catastrophe, but simply the non-deposition of sediment where the land in that region had been above sea level for an extended period of time.

Because of the supposed slow turnover in species, Lyell was able to develop a palaeontological method for the subdivision of the Tertiary rocks. These he classified according to the proportion of extant species that they contained: Newer Pliocene, 96%; Older Pliocene, 35–50%; Miocene, 17%; Eocene, 3.5%. This procedure accorded with his general philosophy of geology. But it could not be carried down below the Tertiary strata because there were so few extant forms below the Eocene.

Lyell's geology was persuasively argued and had a major influence, on English-speaking geologists at least, turning several important figures such as Sedgwick away from diluvialism or 'Flood geology'. Lyell was long regarded as one of the most important nineteenth-century theorists, setting geology on the appropriate methodological road. His influence on Charles Darwin was particularly important, as will shortly be discussed. Lyell's adroit historical introduction, designed to show that his predecessors who had ideas akin to his own were generally on the right track whereas others were not (Rudwick, 1970; Porter, 1976; Gould, 1987), carried many readers along with him. He suggested that geology should be clearly differentiated from cosmogony and this proposal was generally well received, and was surely a sensible suggestion at that stage in the development of geology. And, as said, by introducing his methodological precepts, Lyell seemed to be giving geology an approved scientific status.

There were, however, tensions in Lyell's argument. In his view, mammals, for example, ought to be discoverable in the very earliest fossiliferous rocks, the Cambrian. But there was not the slightest empirical evidence that this was so.[21] Second, he had no present evidence for the production of new species, such as was supposed to have occurred in the past.[22] Thus his own methodological preaching was not practised in this regard. Third, there was the unique case of man, who was supposed to be divinely created. Fourth, while Lyell's ideas on geological processes, the subdivision of the Tertiary, the great age of the earth, the emancipation of geology from cosmogony and a good many other matters were well received, his ideas on non-progression in the earth's history – that (apart from the advent of mankind) things had always been much the same in the past as they are in the present – was accepted by almost no one (Bartholomew, 1976). And of course Lyell's rejection of palaeontological progression made it excessively difficult for him to accept Darwin's evolutionary theory in the later part of his career.

Nevertheless, Lyell offered ways of thinking about the earth that were of the highest importance. Its historical past was to be understood by recourse to observation of its present processes and circumstances. Geology was a natural science, not a kind of physico-theology. It was distinct from cosmogony. Earth movements were chiefly up and down, and the continents were therefore constantly changing their configurations; and climatic conditions altered accordingly. So too did the forms of living organisms. The earth was of immense age, so that vast physiographic changes might be wrought by slow but inexorable processes. Evidence could be found for the slow but considerable erosive action of rivers.[23] There was no need to invoke catastrophes to account for geologists' observations. The catastrophist doctrine had arisen because of the lack of imagination as to the extent of historical time.[24] Nature moved with dignity. A 'gradualist' perspective was encouraged. The proper task of the geologist was to write the earth's history using uniformitarian guidelines.

As mentioned, Lyell thought that species appeared and disappeared from time to time, but he had no clear idea of the mechanism for the production of new species. In Volume 2 of the *Principles*, he seriously considered the theory of evolution that had been propounded by Lamarck in the early years of the nineteenth century, but decided against it. Lyell collected a large amount of information on the question of species (Lyell, 1970), but for a long time he was unwilling to accept the evolutionary hypothesis. Some might suppose that this had to do with Lyell's political conservatism and dislike of revolutionary change.[25] At a deep psychological level, this may indeed have been the case. But he had good empirical and methodological grounds for the rejection of the transmutationist hypothesis. One could not see species changing into other species at the present time, although there was clearly significant variability within species, which might be magnified by the artificial activities of plant or animal breeders. This was as far as Lyell was willing to go – at least in his early years.

When he did eventually accept evolution, Lyell did so with minimum change to his theory. He was eventually pushed to acceptance, Gould (1987) suggests, as a result of the long failure to find Palaeozoic mammals, and the recognition of human development, evidenced by stone-age implements and the discovery of Neanderthal man. Though accepting evolution, Lyell could still hold on to uniformity – uniformity in the rate of change (no catastrophic extinctions). But, says Gould, Lyell's intellectual shift obliged him to abandon his earlier conflation of methodological and substantive uniformitarianism, which he had successfully sustained for long years through successive editions of *Principles of Geology* by means of his powerful rhetoric.

It is a nice question as to the precise influence of Lyell on geology. The scientist/historian, Archibald Geikie, for example, gave little space to Lyell in his influential *Founders of Geology* ([1905] 1962). In contrast, modern historians, such as Rudwick and Hodge, have given much attention to Lyell, and this has no doubt seemed appropriate since the influence of Lyell on Darwin was unquestionably very great indeed. And so, with the tremendous attention devoted to Darwin in recent years, it is possible that Lyell has been given a bit more consideration than might have seemed warranted to the geologists of his own day, thinking of the world as a whole. Thus, the stature of Lyell has perhaps become unduly magnified simply by reason of his close association with Darwin, with whom, in my opinion, historians have become so obsessed that it has become difficult to see the true shape of science in the Victorian era. Nevertheless, the position of Darwin is, of course, of great importance, and we are perfectly justified in examining his part in the story of how thinking about the earth has developed. In any case, Lyell's works were undoubtedly read widely, being translated into several languages.

Although Darwin did not meet Lyell before embarking on his celebrated *Beagle* voyage, he familiarized himself with the *Principles of Geology* (Lyell, 1830–3) during the course of his journey, and endeavoured to use the new uniformitarian principles and the idea of endless changes of the relative levels of land and sea, with the land slowly moving up and down in different places, rather like a yo-yo.

Darwin's most successful application of this idea was in relation to his theory of coral reefs. For some islands in the Pacific it is clear that the land has been rising in recent times, for one may find large blocks of dead coral well above sea level.[26] In other places the coral forms fringing reefs, with the coral below sea level for most of the time. Darwin suggested that this state of affairs could be understood by supposing that in such places the land was sinking with the coral growing upwards at such a rate as just to keep pace.[27] In his book on coral reefs (Darwin, 1842), he provided a map with vast areas of the Pacific region coloured red and blue, according to whether the land was rising or falling. This gave just the kind of support that Lyell's theory required.

On the other hand, when it came to the application of Lyell's theory of species and his account of the fossils in the stratigraphic record, things were much less satisfactory. Indeed, Darwin found significant anomalies in the Lyellian theory. According to Lyell's view, newly created species were supposed to be fitted to their conditions of existence, and species became extinct if and when they could not accommodate themselves to changing environmental circumstances. But, in South America, Darwin found the remains of giant mammals that were evidently extinct, but whose disappearance did not seem to be attributable to any obvious changes in the environment. Or, as is well known, he found differences in species (e.g. of tortoises, finches and mockingbirds) in the Galapagos Islands which did not seem to be linked to significant differences in the environmental conditions of the different islands (although some correlations of form and habitat could be recognized).

Such observations set Darwin thinking, and eventually, after he had got back from his voyage in 1836, he reached the conclusion (the following year) that his observations suggested that living organisms had 'transmuted', or, as we would say, they had evolved. Then, in 1838, he devised the first version of his theory of transmutation, with the idea that organisms gradually changed, adapting themselves to changing circumstances as a result of the best adapted forms surviving, there being insufficient resources for all organisms that are 'born' to survive. Subdivision of species into new species was ascribed at first chiefly to geographical isolation, and later to some other mechanisms.[28] Darwin's theory was eventually published in book form, as late as 1859, in the celebrated *Origin of Species*.

As is well known, the publication of *The Origin* caused a great furore, and to this day there are many people who reject the notion of evolution, or if they accept it they will not allow that it is a naturalistic process as Darwin envisaged. Or they may deny that the mechanism proposed by Darwin (natural selection) is adequate to its task, even when supplemented with all that we now know about genetics, molecular biology and so forth. Ignorance of or a negative attitude towards Darwinian evolutionary theory is particularly strong in the United States, although much of the world's most important technical and philosophical work on evolutionary biology is conducted there.

But such controversies need not concern us here. The thing to note is the way in which the stratigraphic column became intelligible when viewed in the light of the Darwinian theory. Broadly speaking, the simpler and more primitive organisms appeared at the lower levels, and became gradually more complex, such that 'advanced' creatures could be found in the upper strata. Thus molluscs were to be found in the Cambrian, fish in the Silurian, mammals in the Jurassic and so on.[29] This gave sense to the leading stratigraphic principle that William Smith had enunciated at the end of the eighteenth century.

Darwin's theory was, however, a way of 'reading' the fossil record. For the trends and divergences that his theory might have led one to expect to find in the fossil record were hardly to be discerned in the rocks themselves. Darwin's solution was simply to plead the incompleteness of the record. To take a simple example, within the Cretaceous (e.g. in the Gault clay, mentioned above) there are numerous clearly defined species of fossil ammonites without known intermediaries. This is of benefit from the point of view of the stratigrapher, as the sharp distinctions between the different species facilitate the palaeontological correlation of strata. But it is something of a mystery as to why the fossil types are so clearly delineated one from another.[30] All Darwin could do was to assert that the stratigraphic record was very incomplete. Others, such as Murchison, contended that the evolution/transmutation theory was simply wrong. In the later years of the nineteenth century, and particularly in the United States, Darwin's explanation of transmutation came in for considerable criticism, and various alternative mechanisms for change were canvassed. This has led Peter Bowler, for example, to write extensively on the question of the 'eclipse' of

Darwinism (Bowler, 1983). But such issues lie chiefly within the domain of the history of biology and need not be examined further here. For the present purposes, we will content ourselves with the thought that Darwin's theory made the stratigraphic column intelligible – more so than did Lyell's theory. And Darwin's theory was wholly naturalistic.

There is, however, the all-important question of the age of the earth to be considered. As we have seen, Cuvier had spoken of how 'glorious [it would be] to burst the limits of time'. And Lyell, with his arguments on Etna, the slow cutting back of the Niagara Falls and other evidence, and in accordance with his metaphysical preferences, had envisaged an almost limitless age for the earth. On the question of geological time, Darwin followed Lyell without hesitation and made some rough estimates of his own as to the order of magnitude of the age of the earth. For example, Darwin's property in Kent overlooked the valley of the Weald, and in *The Origin of Species* he estimated that it had taken about three hundred million years for the valley to have been excavated – by the action of the sea, cutting the valley at the rate of about one inch a century. Today, geologists think that this is a considerable overestimate, for the oldest rocks in the centre of the valley (the Hastings beds) are themselves only thought to be about one hundred and forty million years old, and their folding and subsequent erosion occurred some considerable time after this. Nevertheless, Darwin was surely justified in regarding the earth as extremely old. Needless to say, to be feasible his evolutionary theory required enormous 'drafts of time'.[31]

Yet the physicists of Darwin's day did not allow that the earth was as old as Darwin desired. The challenge came chiefly from the Glasgow professor, William Thomson, Lord Kelvin (1824–1907). This distinguished physicist had been considering such questions as the source of the sun's heat from the earlier part of his career, but in 1862 he published a semipopular article on the question (Kelvin, 1862) which explicitly attacked the Darwinian theory, with its limitless 'drafts of time'. Kelvin attributed part of the sun's heat to the energy generated by meteoric impacts, but thought that such impacts could not be sufficient to maintain solar heat indefinitely. Hence he deduced (by an argument that involved several insecure assumptions) that the sun was cooling and was perhaps between one and five hundred million years old, with a lower limit of ten million. In further papers, published in 1863 and 1864, Kelvin considered the time it might have required for the earth to cool from a molten state to one that was solid throughout, but with a temperature gradient and heat loss the same as those currently observed. This gave him a figure of between twenty and four hundred million years. Kelvin's theorizing was also at odds with Lyellian uniformitarianism, since if the earth were cooling it must have been hotter and more geologically active in the past than at present.

A further argument appeared five years later, in 1868 (Kelvin, 1871), and was more plausible than the previous two. It had been accepted since the geodetic survey work of the eighteenth century that the earth was an oblate spheroid. Presumably it had acquired this shape when it was wholly molten and prior to its solidification. Then, considering the shape of the earth, it was possible to calculate its rate of rotation at the time of solidification. But, Kelvin argued, this rate had subsequently been decreasing because of the frictional effects of the tides. If this rate of decrease could be calculated, then one had an estimate of the time taken for the earth to solidify and then slow down to its present state, cooling the while. While acknowledging that the calculations could only be approximate and that there were a number of factors to be taken into account, such as the presumed change in the earth's moment of inertia due to its contraction on cooling, it appeared that a reasonable estimate for the time since solidification was one hundred million years. Again, this was incompatible with the uniformitarian geology of Lyell and Darwin.

The whole episode has been closely studied by Burchfield (1975), and the details need not be retraced here. Kelvin returned to his arguments and his calculations many times during the later part of his career, with his numbers gradually coming down – to about twenty million years – as he refined his calculations. Most geologists, with limited expertise in physics, had no means of countering the arguments, and the non-mathematical Darwin was quite unable to offer a challenge. All he could do was hang on and hope that something would turn up. In any case, no one really knew how much time was required to make the Darwinian theory admissible. Even today, it is hard to form a mental picture of one, ten, a hundred or a thousand million years. So, with some unease, geologists in the late nineteenth century came to accept that the earth might only be about one hundred million years old. Whether that was time enough was hard to say. But the days of unlimited 'drafts of time' seemed to be over.

Then something did turn up. It came in quite an unexpected way with the discovery of the phenomenon of radioactivity, and the recognition that there were radioactive materials in the earth that were generating heat, thus nullifying Kelvin's calculations about the earth's rate of cooling. In 1903, Ernest Rutherford (later Lord Rutherford) and H.T. Barnes announced the heating effect produced by the disintegration of radium, and the following year Rutherford spoke at the Royal Institution, arguing that the earth need no longer be thought of as a mere cooling body as it had an '*internal* supply of heat' (Eve, 1939, p. 107). The idea was quickly picked up by Charles Darwin's son George Darwin (1903), a distinguished mathematician/astronomer/geophysicist, who had for long been trying to 'problematize' Kelvin's arguments, but with only limited success. Now radioactivity came to the geologists' rescue in that George Darwin could plausibly suggest that a heat-generating substance (radium) might be present in the sun and in the earth. We no longer think that the sun is kept hot by radium. However, radioactivity helps to keep the earth warm. Anyway, since the beginning of the present century geologists and biologists have felt able to draw as many 'drafts of time' as they may reasonably require without fearing that they are in contravention of physical principles as they do so.[32]

Moreover, as is well known, radioactivity provides techniques whereby the age of the earth may be estimated (see Dalrymple, 1991). In the early years of the century, workers such as Bertram Boltwood of Yale, Arthur Holmes in Britain and the Irishman John Joly, together with many others, did work on radioactive minerals and their decay products, so that by 1907 Boltwood, following the method of calculating the ages of minerals suggested by Rutherford, had a figure of 2,200 million years for a Precambrian rock from Ceylon; and in 1911 Holmes had a figure of 1,640 million years for the same material by the uranium/lead method. Such investigations have been carried out ever since, and the present estimate for the age of the earth as a whole is about 4,540 million years (Dalrymple, 1991: 402). Alhough estimates have increased significantly over the last thirty years or so they seem to have settled down of recent years, and at the moment I do not anticipate any major change on this front. This seems to be the amount of time that our minds and our theories must deal with when thinking about the earth. It should be emphasized, however, that the time since hard-shelled animals first appeared on the earth is but a fraction of this. Current thinking is that the Cambrian began about 570 million years ago, although this figure may be reduced somewhat in the next few years, according to recent reports.

CHAPTER 7

Thoughts About Climate, Glaciation and Carving the Earth's Surface

The catastrophism discussed in Chapter 6 was, in part, the result of a geological 'accident' – a quirk of the particular epoch in which mankind evolved, that is during and not long after great periods of glaciation. As noted, the science of geology emerged in northern Europe in the eighteenth century. And much of the terrain of this part of the world is swathed in so-called boulder-clay. There are also the remains of glacial moraines and (in Ireland particularly) structures called drumlins[1] and eskers,[2] which are thought to have been produced by glaciers. There are intriguing scratch marks on rocks, now thought to have been produced by the gouging action of rocks pushed over a rock surface by glaciers. There are loose blocks of rock lying all over the place, far distant from the nearest place where they are found *in situ* (glacial erratics); or there are large pieces of rock standing loose on the tops of hills as if some great bird had dropped them there (*blocs perchés*). There are raised beaches around several coastlines. There are loose gravels, sometimes mixed with fossils, near the tops of certain mountains or hills. There are marks of ancient lake margins in Scottish glens. And there are valleys where the rivers in them seem much too small to have carved the valleys, no matter how much time is allowed for the process – indeed, some valleys, such as those cutting into the chalk hills of southern England, are without rivers altogether.

Today, such phenomena are accepted as evidence of the occurrence of great ice ages in the fairly recent geological past. But the 'reading' of the evidence mentioned above is very much a theoretical process. It could easily be, and initially was, read in quite a different manner. That is, the evidence seemed to indicate the occurrence of a great flood (or floods) that had once swept over the world, depositing great quantities of debris in the process.

Such 'diluvialist' interpretations can be traced back well into the seventeenth century. As we have seen (p. 54), John Woodward imagined that all the materials of the earth's crust were dissolved or held in suspension at the time of the biblical

Deluge, and were then deposited in layers, according to their densities, once the waters receded. John Ray (1692a, b, 1693, 1713) likewise thought that there had been a great flood responsible for much of what might be seen on the earth's surface today. The Irish chemist, Richard Kirwan (1799), was content with the idea that fossils had been emplaced by the action of the Noachian Flood. Indeed, he embraced the idea with enthusiasm. And well into the nineteenth century, standard texts such as that of Conybeare and Phillips (1822) took it as axiomatic that there was clear geological evidence for the Flood.

There were two broad reasons for this. First, as we have seen, early geologists thought that the earth was a young object, so that much could be learnt about its past on the basis of written testimony, and most particularly that of the Bible – although there were also pagan authors, such as Plato and Ovid, and ancient records from Babylonia, that recorded catastrophic floods. Myths and legends from other parts of the world likewise seemed to sanction some kind of universal flood. Second, the interweaving of geology and theology (in physico-theology) was a peculiarly attractive occupation, favoured until well into the nineteenth century, as we saw, for example, in the previous chapter when considering the work of William Buckland. Even so, it was not necessary to have one's geology built on a theological foundation to arrive at a catastrophist geology. Catastrophism seemed to be warranted for the likes of Cuvier and Buckland by clear empirical evidence (Hooykaas, 1970).

There was, in fact, a range of positions that might be taken between the catastrophist views of Cuvier and the slow-but-steady approach of the uniformitarians (Page, 1969). For example, the Great Flood might have been a relatively gentle affair. Perhaps there had indeed been a biblical Flood, but the waters had risen quite slowly and receded likewise. Indeed, this might seem the best approach, if one did not wish to invoke miracles to explain Noah's preservation.

But there were those – most notably Hutton and, later, Lyell – who rejected the idea of a universal Flood altogether. For the deist farmer Hutton, it seemed implausible that the great designer, God, would act so as to destroy the life he had created (Hutton, 1795, vol. 1: 273). Thus one of the major features of Hutton's theory was that, with limitless quantities of time supposedly available for geological processes, there was no reason why rivers might not excavate their own valleys. This position has been called 'fluvialism' (Herries Davies, [1969]).

But what could Hutton – a fluvialist – say about the various phenomena mentioned at the beginning of the present chapter? In fact, Hutton did not deal very satisfactorily with such matters. There are, none the less, a couple of interesting pages in his *Theory of the Earth* (Hutton, 1795, vol. 2: 218–19) where he mentioned the possibility of the former greater extent of the Alpine glaciers, thereby accounting for the occurrence of huge boulders well beyond the valleys in which the modern glaciers are still to be found. (He reached this conclusion after a visit to the Jura in 1794.) But Hutton did not infer a change in climate to account for such phenomena. Rather, he supposed that the Alps had formerly been higher than at present, producing larger glaciers. Hence the observation was yoked to Hutton's idea of the slow formation and erosion of mountains, which was an essential part of his general theory.

The idea of the former extension of the glaciers of Alpine Europe and ultimately the idea of the ice age(s) came, appropriately, from Switzerland.[3] Early suggestions that the glaciers might have extended significantly beyond the positions of their present terminal moraines were made by a clergyman, Bernard Kuhn, in 1787, and by a mountaineer and hunter, Jean-Pierre Perraudin, as a result of his observations in the Val de Bagnes in 1815 (Herries Davies, [1969]: 264). Their preliminary suggestions were taken up by the engineer Ignace Venetz (1837[4]) and Jean de Charpentier (1834, 1836, 1841), a graduate of the Freiberg Mining Academy, director of the salt-mines at Bex in the Canton of Vaud and honorary Professor of Geology at the Lausanne Academy. Charpentier's book (1841) presented a map showing the former extension of the Rhône glacier into the central plain of Switzerland, based on the distribution of erratic blocks from the region of the headwaters of the Rhône.

But these men were relatively minor figures, and it was not until Louis Agassiz (1807–73), Professor of Natural History at Neuchâtel and one of the leading naturalists of the nineteenth century, took up the cause that the idea attracted much attention. Agassiz is remembered particularly for his researches on fossil fishes, for his 'polygenism'[5] and for his writings on taxonomy. However, it is arguable that his grand idea of an *Eiszeit* (ice age) was his major contribution to science (although, according to Imbrie and Imbrie (1979: 28), the term was suggested by his botanist friend, Karl Schimper). It is believed that Agassiz initially wrote out a statement of his theory during the course of one night in 1837 (having spent the summer with Charpentier at Bex the previous year), and his paper was presented to the *Société Helvétique* at Neuchâtel the following day (Agassiz, 1837, 1838). A field-trip to the Jura was organized immediately after the Neuchâtel meeting. Leading geologists such as von Buch and Elie de Beaumont were unconvinced. Nevertheless, by 1840, Agassiz had written a full-scale book – *Etudes sur les glaciers* – as well as numerous articles on the topic. The *Etudes*, which were dedicated to Venetz and Charpentier, contained detailed evidence for the former extension of the Swiss glaciers. So too did Charpentier's book.

Although there were exceptions, such as Lyell, it was commonly and plausibly supposed in the nineteenth century that the earth was cooling after its formation. But if this were so, how could the earth have been colder in the past than at present, as required if one were to invoke the notion of an ice age? Agassiz did not really answer this question, but arbitrarily proposed that cooling occurred in a curious fashion. He supposed that the temperature dropped, rose somewhat and then remained constant for a time. It then dropped again to a new low, rose a bit and again remained steady. The process being repeated – and with an overall diminution of temperature – the time would come when the temperature was significantly colder than at present. This was Agassiz's 'ice age'. We lived at a time when the earth had temporarily warmed again, but one could look forward to an even colder epoch in the future.

The theory was, in a sense, a kind of catastrophist doctrine of a Cuvierian flavour (although it went beyond the empirical evidence, reliance on which catastrophists aspired to). Agassiz could give no reason how or why the earth's temperature should have fallen in such a peculiar manner. But, if the idea were accepted, then the observations of glacial erratics, etc., could be explained. Each dip in temperature, followed by a modest upturn and a period of stable conditions, might correspond to a Cuvierian catastrophe.

The reception of Agassiz's theory, which by 1840 had been developed to the notion of a full-blown ice age with a great ice cap extending all over northern Europe, has been examined in detail, so far as the English-speaking world was concerned, by Herries Davies [1969]. Agassiz took his ideas to the meeting of the British Association for the Advancement of Science at Glasgow in 1840. The great convert he wanted, of course, was Lyell. But Lyell was difficult to persuade.

In Volume 1 of his *Principles*, Lyell (1830: 299) had suggested that the material, interpreted today as glacial detritus, might have been emplaced by the melting of floating icebergs, carrying within them rocks, stones and mud. In Volume 3 (Lyell, 1833: 149–50), he offered a speculative suggestion to explain the occurrence of erratics in mountain valleys. There might have been small icebergs floating in lakes dammed by landslides. Then, when the blockage was cleared, there might be a great rush of water down a valley, and this might deposit erratics at the mouths of the valley or beyond.

In 1840, when he met and conversed with Agassiz and did some fieldwork with him in Scotland, Lyell was temporarily converted to the idea of an ice age. But the following year he had second thoughts and recanted. Agassiz was asking too much of Lyell. An ice age would mean that conditions were not essentially the same in the past as at present, and the uniformitarian principle would be flouted.

Lyell did, however, come round to the Agassiz's way of thinking, albeit reluctantly, and it was accepted in his *Geological Evidences of the Antiquity of Man* (1863). Lyell's solution to the causal problem was fully in line with his general theory. He supposed that if, at any given epoch, there happened to be more elevated land near the poles and less near the equator then the overall temperature of the globe would be reduced. Conversely, if there were more high land near the equator and less near the poles the world would be warmer (as he had proposed to account for the extensive vegetation of the coal measures). Thus in principle he could account for ice ages within the framework of uniformitarian geology. It was, however, difficult to see how there could have been such major climatic changes in geologically recent times by this mechanism, and the evidence in Europe pointed to recent glacial conditions.

Another significant difficulty had to do with lakes. The distribution of these is rather peculiar. Apart from lakes in rift valleys and volcanoes, most large lakes are found in 'shield' areas such as Canada or Scandinavia, or in mountainous regions such as the Alps or New Zealand. But the Himalayas are curiously deficient in lakes, except for a few places at lower altitudes such as the famous Kashmir lakes.

Now, the large U-shaped valleys of many mountain regions often carry relatively small rivers. The reason for this, according to the glacial theory, was that the valleys were mainly excavated by glaciers, now departed, not by the rivers – which in any case carve valleys that are V-shaped in cross-section. But this raised a severe difficulty for the glacial theory. For, if the valleys were carved out by the glaciers, how could there be lakes in the valleys? The problem generated considerable debate: could glaciers excavate rock basins, or would they inevitably cut away any obstacles so that there would be no chance of lakes forming as a result of glacial action? Indeed, as one of Hutton's contemporary critics, J.-A. Deluc (1790–1), had objected: if rivers cut their own valleys, how could there be lakes? Likewise, if glaciers cut their own valleys, how could there be lakes? Herries Davies [1969] has called this the 'limnological objection'.

In Britain, the leading exponent of the notion that glaciers could excavate rock basins and thus originate lakes was Andrew Ramsay (1862), a senior member of the Geological Survey. He maintained that if glaciers gouged out valleys in rocks that were of unequal hardness there might sometimes be situations where a glacier could dig downwards without necessarily pushing forward; or in riding over a hard obstacle it could form a hollow higher up the valley; or lakes could form behind moraines as glaciers retreated. Lyell, however, objected, arguing in his *Geological Evidences of the Antiquity of Man* (Lyell, 1863: 311–19) that in many cases lakes did not occur where they might be expected according to the theory of glacial origin. He proposed instead that flexures of the earth's crust ('unequal movements of upheaval and subsidence') might be held chiefly responsible. Today, of course, this is indeed thought of as a possible cause of the origin of a lake; or lakes may be produced by faulting (there is a good example at Lake George, near Canberra). But Ramsay's theory of rock basins is also thought to be sound in appropriate cases. Thus Deluc's old 'limnological argument' argument against Hutton was defused.

The theory of glaciation could also account for changes in sea level, which had long puzzled geologists (Wegmann, 1969). It could do this in two ways. The American geologist Charles MacLaren, in reviewing Agassiz's *Etudes sur les glaciers* (MacLaren, 1842), suggested that sea levels would be lowered if large quantities of water were locked up in the form of land ice. Also, T.F. Jamieson (1865, 1882) suggested that the sheer weight of ice might depress the earth's crust, but after the ice melted the land surface might slowly rise up again, perhaps sometimes in jerks. Jamieson's theory accounted for the formation of raised beaches, and also for the well-established fact that the land in the Baltic region was rising relative to the sea. MacLaren's idea, which had later supporters, such as S.V. Wood in his note to *The Reader* of 9 September 1865, proved to be of immense importance for later studies in biogeography. For with the sea level being generally lower in the glacial epoch there would be opportunities for the ready migration of animals across regions that are now covered by water. In particular, it would have allowed the migration of humans from Asia into the Americas and from South-east Asia into Australia.

The glacial theory provides a fascinating example of the phenomenon of the theory-ladenness of observations. British geologists had been walking over the mountains of their homeland for many years, but until Agassiz visited Britain in 1840 and introduced them to the idea of glaciated landscapes and small-scale evidence for glaciation, like the striations that may be found on otherwise smooth rock surfaces, such features had not been recognized as important, and were not recorded. After Agassiz's visit, people began to see evidence of glaciation everywhere. A good example has been noted by Herries Davies ([1969]: 263):

> Notice to geologists. – At Pont-aber-glass-llyn, 100 yards below the bridge, on the right bank of the river, and 20 feet above the road, see a good example of the furrows, flutings, and striae on rounded and polished surface of the rock, which Agassiz refers to the action of glaciers. See many similar effects on the left, or south-west, side of the pass of Llanberis.[6]

Today, it is easy to see the world through a similar set of theoretical spectacles, although most visitors to such mountainous regions do not, I think, envisage them as formerly covered by a mass of land ice, which carved out the landscape so that it takes on the appearance we see today. Geologists, however, immediately recognize a glaciated landscape when they encounter one, and they – and some members of the general public also – now see the world differently from their predecessors as a result of the establishment of the glacial theory.

But we are still left with a difficulty. In the jargon of philosophy of science, Agassiz's theory (at the time of its being put forward at any rate) was 'overdetermined' (Lugg, 1978). In some cases in science one can think of a number of different theories that may account for the evidence. In such a situation, the empirical evidence alone does not determine which theory is to be preferred. So any one of the theories on offer can be said to be 'underdetermined' by the evidence. But, in Agassiz's case, he really had no theory as to why there had been an ice age. So we might say (perhaps quixotically) that his theory was 'overdetermined'.

Once Lyell acquiesced with regard to the idea of an ice age, there was not much further objection to the idea of a former frigid climate and a period (or, as it soon came to be realized, several periods) of glaciation in the Pleistocene. Other geological eras, such as the Permian, were also found to display evidence of glaciation in other parts of the world outside Europe, for example in Australia. Such observations played havoc with Agassiz's speculative idea about the way the earth cooled: for, according to his model, one would scarcely suppose that the earth would have been cold enough to have suffered an ice age back in the Permian.

So what might be the cause of glaciation? It seemed that the cause had to be extraterrestrial in origin – perhaps due to alterations in the amount of energy generated by or received from the sun. But why should that occur?

The problem was tackled with energy and imagination by several theorists in the nineteenth century, and work continues to the present. It is one of extreme complexity and difficulty. The first person to try to work out a theory of ice ages according to astronomical causes seems to have been Joseph-Alphonse Adhémar, in his *Révolutions de la mer*, published in Paris in 1842.[7] Considering the motion of the earth, its axis makes a slow rotation (which would appear clockwise if one were looking down on the North Pole from above) in a period of 26,000 years. But the earth's elliptical orbit also rotates (anticlockwise) in space, and the combined movement gives the so-called precession of the equinoxes. The phenomena were analysed mathematically by the French astronomer and mathematician Jean d'Alembert (1749, 1754/1759), following which the overall precession period was calculated to be 22,000 years.

Because of the ellipticity of the earth's orbit and the precessional movements, the seasons are unequal in the different hemispheres and are also slowly changing. For half of the precessional cycle, one hemisphere has a slightly longer summer, and then it has a slightly longer winter. (At present, the southern winter exceeds the southern summer by 168 hours.) Adhémar suggested that whichever hemisphere had the longer winter would suffer an ice age. Rather fantastically, he imagined that a great Antarctic ice-sheet might attract water from the northern hemisphere to create a bulge of water in the southern hemisphere. Then, as the southern hemisphere climate ameliorated, the huge ice-cap would be eaten away at its base by the warming water, and would stand up as a kind of giant mushroom of ice. The collapse of this would lead to a great tidal wave sweeping northwards, laden with icebergs (Adhémar, 1842: 316–28). This was yet another example of nineteenth-century catastrophism. But in Volume 4 of his *Cosmos* von Humboldt (1883: 460) pointed out (correctly) that, although the lengths of the seasons would change due to the precession of the equinoxes, the actual amount of heat received annually in each hemisphere would be the same: there could be a shorter hotter summer or a longer, less hot summer. It should all balance out.[8]

Nevertheless, the astronomical theory did receive further attention from an unlikely source: a Scottish autodidact, James Croll (1821–76).[9] The son of a stonemason, Croll worked as a carpenter, but an injury led to his attempting non-manual work, at which he was unsuccessful. Eventually he was reduced to the position of janitor at Anderson College, Glasgow. But in that post he had access to the library, and it was there that he came across Adhémar's strange book. This led to Croll trying himself to produce an astronomical theory of the ice ages, and he succeeded in getting several papers published in the *Philosophical Magazine* (Croll, 1864, 1867a, b). His ideas so impressed Archibald Geikie that Croll was taken on to the staff of the Scottish Branch of the Geological Survey in 1867, and published a major book, *Climate and Time*, in 1875. That year, he was elected a Fellow of the Royal Society and later he received an LL D from St Andrews.[10] Croll's theory was given popular exposition by Sir Robert Ball in *The Cause of an Ice Age* (1891).

I shall discuss Croll's astronomical theory in a little detail. It is both interesting and important, and very probably correct in its essentials, though still controversial among those who think about the earth. His basic idea was that, in addition to the precessional movements mentioned above, the shape (ellipticity) of the earth's orbit is slowly changing because of gravitational interactions with the other planets in the solar system. Sometimes the orbit is almost circular while at other times it is more elliptical. Consider, then, a situation when the earth's orbit is markedly elliptical, as shown in Figure 7.1. *X* and *Y* represent the positions of the equinoxes. Now, as Poisson, Herschel and von Humboldt

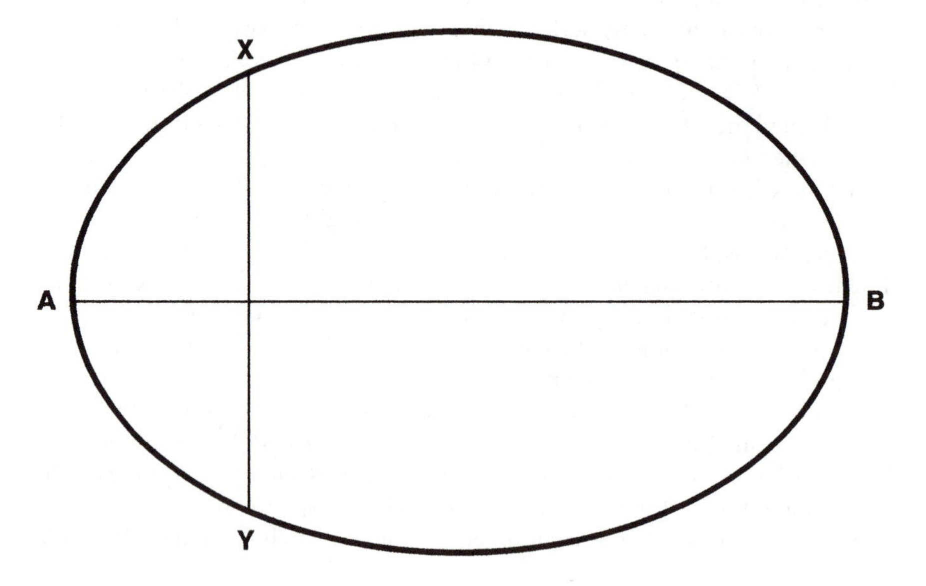

Figure 7.1 Earth's orbit round the sun (ellipticity exaggerated)

pointed out, whatever the elliptical shape of the orbit, the total heat received by the earth while it passes through *XAY* is equal to that which it receives as it passes through *YBX*. However, the heat it receives in the northern and southern hemispheres is not equal in any given half-year, owing to the inclination of the earth's axis to the plane of the ecliptic. And with the earth's inclination (the 'obliquity of the ecliptic') being about 23° 27′ it is the case that, of the total amount of heat received from the sun on a hemisphere of the earth in a given year, 63 per cent is received during the summer and 37 per cent during the winter (Ball, 1891: 90). Now according to the hypothesis of Croll, the line joining the two equinoctial points (*XY*) is not fixed. It could be as shown in Figure 7.1. Or the line could be along the major axis of the ellipse (*AB*), or at any intermediate position. That is to say, the direction of the earth's axis of rotation relative to its elliptical orbit could slowly change. And, as said, the ellipticity of the earth's orbit might also change. So there could be a complicated set of geometrical relationships, slowly changing over time. The heat distributions would vary according to the geometrical configuration at any given time. Such changes might occur as a result of the secular changes of all the spatial relationships between the orbits of all the planets in the solar system.

Consider again the case of Figure 7.1. Suppose the earth's axis is inclined so that it is summer in the northern hemisphere for part of the orbit *XAY*. Then the earth will have a short hot summer in the northern hemisphere and a long cold winter; but the southern hemisphere will have a long mild summer and a short cold winter. Even so, the total heat received in each hemisphere in a year is the same.

We may, then, consider the situation where we have two major variables: the ellipticity of the earth's orbit and the direction of the line joining the equinoxes (*XY* in Figure 7.1). If these are both slowly changing, there will be complicated interactions, giving complex changes in the earth's climate. Essentially, for periods when the ellipticity of the earth's orbit is small – that is, it is almost circular – then alteration of the equinoctial line is not going to make much difference, and conditions are going to be pretty much the same in the northern and southern hemispheres. But, if the ellipticity is great, then the long cold winter in any given hemisphere is not compensated by the short hot summer, and so glacial conditions may ensue in that hemisphere. This is because of the properties of water. It takes time for snow and ice to melt. In contrast, the other hemisphere will enjoy a warm climate overall.

Croll's theory envisaged occasional periods of high ellipticity when one or other hemisphere might be glaciated. Thus we might explain the occurrence of the Pleistocene and the Permian glaciations. But one could also envisage that the line between the equinoxes (*XY* in Figure 7.1) could swing round so that it might sometimes be at *AB*, or at any intermediate position. If aligned along *AB*, the heat distribution between the northern and southern hemispheres would be the same and glaciation would not ensue. If at *XY*, however, there could be glaciation in one or other of the hemispheres, according to which way the earth's axis was tilted, assuming that the orbit was particularly elliptical at that time.

It may be seen, then, that Croll's theory allowed for the occasional occurrence of glacial epochs, at periods of high ellipticity. But these would be interrupted, in any given hemisphere, by interglacials, as the equinoctial line made several of its swings during a period of high ellipticity. At periods of low ellipticity, such swings would hardly affect the climate. It should be noted that Croll supposed that a period of glaciation in the northern hemisphere (say) was accompanied by a warm climate in the southern hemisphere, and vice versa.

Croll's theory was highly ingenious and in the nineteenth century it seemed to offer exactly what was needed to account for the empirical evidence of glacials and interglacials. Moreover, it offered a pleasing integration of geological and astronomical theory. It was objected that the astronomical effects, though real, were too small to be significant. Croll's response was that the astronomically induced changes would be amplified by feedback mechanisms, and in fact much of his book (Croll, 1875) dealt with such matters. If a particular hemisphere were cooler, the trade winds would blow harder, and more moisture would be transported, adding to the precipitation and the glaciation. In this suggestion, he later received support from the influential American geologist T.C. Chamberlin, whom we shall meet again in Chapter 12. Chamberlin (1899) had the idea (from the physicist John Tyndall) that increase in carbon dioxide in the atmosphere would tend to produce a global warming (by what is today called the 'greenhouse effect'), and this would give rise to more water vapour in the atmosphere, which would then, however, block off the sun's radiant heat. There might, then, be a cycling of temperatures, which would interact with the fundamental cause of it all – the alterations in the earth's orbit and its axis of spin. Chamberlin (1898b) also had the idea that the amount of carbon dioxide in the atmosphere would be related to the carbon locked up in limestones. Thus he was one of the first to consider the theme of sedimentary cycling, which we shall also treat in Chapter 12.

Croll's theory was persuasive to his colleague James Geikie, in the Scottish Survey, who (later becoming Professor of Geology at Edinburgh in succession to his brother Archibald) specialized in glacial or Pleistocene geology and palaeoanthropology (Geikie, 1874, 1914). The theory obviously needed geological confirmation in the field as well as abstract theory, and it was suggested by Geikie that the occurrence of river terraces at different levels might provide indications of climatic cycles in the Pleistocene. Thus one might hope to find field evidence not just for one great *Eiszeit*, as envisaged by Agassiz, but a series of such 'ages'.

Evidence for such a series seemed to be provided by the work of the US Geological Survey under Chamberlin and Frank Leverett, who thought they had evidence for four major ice ages in North America, on the basis of a sequence of four glacial 'tills',[11] which were called the Nebraskan, the Kansan, the Illinoian and the Wisconsin, the last of these being the youngest (Imbrie and Imbrie, 1979: 114). In the 1880s, working for his *Habilitationsschrift* in geography at Munich, the geomorphologist Albrecht Penck (1858–1945) studied the river valleys running into the Danube from the north-eastern side of the Alps. He found (for example, in the valley of the Steyr) what appeared to be four sets of river gravels: one at the present level of the river and three at higher levels. Each gravel layer was thought to have been formed during a period of glaciation, when, with reduced vegetation, there was a high rate of erosion. Each could be traced up the valley into material that was evidently the former terminal moraine of a glacier. When the climate ameliorated for a time, the water flow in the river increased and carried away some of the previously deposited gravels. So the river cut a deeper bed, and also a wider bed because it would meander over the valley. Then, in the next period of glaciation, there would be a further deposition of gravel in the previously formed wide valley; and so on. The net result was a series of valleys within valleys, and a series of gravels at different levels, the highest supposedly being oldest, contrary to what one might expect according to the usual Stenonian principle. With his co-worker Eduard Brückner, Penck named his four supposed glaciations the Günz, Mindel, Riss and Würm, the first three names being tributaries of the Danube, while the Würm was a tributary of the Isar River near Munich (Penck and Brückner, 1901–9: vol. 1, 110).

Penck also estimated the ages of his glacial deposits. Looking at the rate at which sediments were being deposited in the Swiss lakes and at their thickness, it seemed that the process had been going on for about the last 20,000 years. By analogy, then, one could estimate the times for the other interglacials, 60,000 being suggested for the Riss-Würm and 240,000 for the Mindel-Riss (ibid.: vol. 3, 1169). Thus was born a theory of glacial stages that lasted until after the Second World War. It appeared to cohere with the American results and with the theory of Croll. But it proved difficult to find a correlated series of gravels and terraces on the southern side of the Alps.

However, while the theory of the glacials and interglacials proved robust for many years, Croll's astronomical explanation proved less resilient, for the reason that it appeared that the glacial stages of the Pleistocene were paralleled in the southern hemisphere, which was completely at odds with the astronomical theory of the ice ages, which had glacial conditions in one hemisphere opposed to interglacials in the other. But that was not the end of the story.

Another attempt to produce an astronomical theory of the ice ages was made by the Serbian Professor of Applied Mathematics at Belgrade University, Milutin Milankovich (1879–1958). He sought to determine accurately the geometry of each planet's orbit, and to show how it had evolved over the centuries This was necessary if one were going to make an accurate estimate of the history of the earth's orbital motion over time. He was then in a position to understand how the eccentricity of the earth's elliptical orbit, the obliquity of its axis of rotation to the plane of the ecliptic and the positions of the equinoctial points had all changed over time. Thus, after immense mathematical labours, Milankovich (1920) was in a position to publish the results of the calculation of the radiation received in the two hemispheres over the last several hundred thousand years.

Milankovich's book went largely unnoticed, but it did attract the attention of the famous German climatologist Wladimir Köppen,[12] and the two scientists entered into correspondence. Köppen thought that the decisive factor causing glaciation would probably be the amount of solar radiation received at high latitudes during the summer months. Accordingly, Milankovich calculated a graph for the radiation received at different latitudes during the last 600,000 years. And, to their pleasure, the minima for high latitudes appeared to correlate with Penck's Günz, Mindel, Riss and Würm glaciations.

Milankovich's work continued. He had a 41,000-year cycle for the obliquity of the ecliptic, which seemed to control the radiation received in the polar regions, and a 22,000-year precession cycle, which influenced the radiation received at the lower latitudes. He then sought to calculate the effects that would be produced by the changes of radiation, considering the various feedback processes that had been suggested by Croll. The problem was tackled by considering the known levels of the snowline at different latitudes. Finally, he published a graph which showed the estimated extents of the ice-sheets for any time during the last 650,000 years (Milankovich, 1938).

The fate of the Milankovich theory since the 1930s has been both interesting and complex, and full details cannot be given here.[13] With the emergence of the introduction of methods of radiocarbon dating there was the possibility of direct comparison of the ages of Pleistocene deposits with the radiation curves calculated by Milankovich. The results were not at first encouraging. In America, evidence for periods of warm climate were found in the middle of what should have been one of Milankovich's glacials. In general, when comparisons were attempted for the older deposits, the Milankovich results did not seem to fit. But how reliable were the radiocarbon datings for the older deposits?

Further techniques were brought into action: the analysis of deep-sea cores, with their planktonic contents; efforts to make correlations with the rises and falls of sea levels[14] (due to water being locked up in ice-caps during periods of high glaciation); determination of $^{18}O/^{16}O$ isotope ratios in Foraminifera[15] from the deep-sea cores as estimates of temperatures at the times the foraminifers were living. Marine terraces were examined at different latitudes, and it was suggested that the obliquity and the precession cycles might have different effects at different latitudes. The ages of coral reefs were examined closely. Work on geomagnetism, with the magnetic reversals[16] that were to be so important in the establishment of the idea of sea-floor spreading (see Chapter 11), came into the efforts to make age correlations, for the magnetic alignments of the rocks in the different parts of the cores could be determined and compared.

In 1948, an International Committee defined the beginning of the Pleistocene by the appearance of cold-water species in southern Italy. This occurred 1.8 million years ago, which was wildly different from the age of the Pleistocene crudely determined by Penck and Brückner. Also, the deep-sea cores showed nine or more glacial phases, not just the four of Penck and Brückner.

Then work carried out in a quarry in Bohemia by George Kukla showed alternation of loess[17] with soils from interglacials, the whole being datable by palaeomagnetic reversals. Kukla then went back to re-examine the terraces studied by Penck and Brückner. The terraces were there, but were found to contain interglacial materials in some places. Indeed, the Würm terrace was found to contain a rusty piece of a bicycle. The Penck theory had truly collapsed, and with it, incidentally, went a whole mass of theorizing in palaeoanthropology – but that is another story. The supposed correlation between Milankovich's cycles and the Penck glacials, envisaged by Köppen, also collapsed at the same time.

Further work (of which there has been a great deal since the 1960s) has been directed towards trying to get a match between the deep-sea cores – which can be dated radiogenetically, be compared palaeomagnetically and have their temperatures at time of formation estimated by the oxygen isotopic ratios – and the radiation curves calculated by the Milankovich method (which were reviewed by a Belgian astronomer, André Berger (1977)). By 1976, all the evidence seemed to be fitting together at long last (Hays et al., 1976). There were a 100,000-year radiation cycle, due to variations in the eccentricity of the earth's orbit, a 41,000-year cycle due to variations in the axial tilt, and 23,000- and 19,000-year cycles arising from the precession of the equinoxes. All these seemed to be discernible in the deep-sea cores, and matched with the requirements of astronomy.

Perhaps that takes us to the end of the story of the problem of the earth's climate; perhaps not. There is the further complication of all the changes due to the movement of the continents relative to one another (see Chapter 11), and the consequent changes in oceanic currents, etc., which can have dramatic effects on climate. And exceptionally large volcanic eruptions may also affect the climates, possibly even initiating the onset of glacial epochs. But, so far as the Pleistocene glaciations are concerned, the astronomical theory seems to be working satisfactorily. Geologists have found a way of thinking about the problem that makes sense as the many pieces of the jigsaw fit together consistently. According to a coherence theory of truth, to which I subscribe, this is strongly suggestive that the way of thinking is correct.

Other nineteenth-century developments in geomorphological theory besides those linked to the glacial hypothesis are of considerable interest and should be noticed here. Much of this work was accomplished in North America, where the great nineteenth-century surveys revealed so much new information about topography and land-forms. The terrain in the arid western states was and is particularly suitable for the study of land-forms, and it was survey work in such states as Colorado, Utah, Arizona and Nevada that led to the establishment of new ideas about the formation of landscapes and new terminology for their description and for their formation.

Among the many American scientific explorers and surveyors of the west, the names of John Wesley Powell (1834–1902), Grove Karl Gilbert (1843–1918), Clarence Edward Dutton (1841–1912) and William Morris Davis (1850–1934) stand out as exceptional in importance for their ideas about the development of land-forms and the changes of these over time. After a somewhat irregular education, with some science included, Powell enlisted in the Federal Army and participated in the American Civil War. He lost his right arm in the bloody battle of Shiloh in 1862 but continued to serve, rising to the rank of major. After the war, Powell secured a post as Professor of Geology at the Illinois Wesleyan University. Subsequently, he took up professional surveying and rose to the directorship of the United States Geological Survey in 1881.

In 1869, Powell made an epic boat trip down the Colorado River, through the Grand Canyon, single-handed so to speak, and a second expedition was undertaken in 1871–2. An account of these exploits was published in 1875. As a result of his observations, Powell concluded that the form of the land that he had been exploring had been produced chiefly as a result of the action of running water sculpting the landscape. (Thus, like Hutton, we may call him a fluvialist.) And the concept of 'base level' was introduced: '[N]o valley can be eroded below the principal stream, which carries away the products of its surface degradation' (Powell, 1875: 163). This 'base level' was not an actual physical feature observed in the field, but a kind of abstract concept – or 'imaginary surface', as Powell called it (ibid.: 203) – with the help of which one could interpret the observations and envisage the processes occurring to produce the observed landscape. The processes of erosion were working to reduce all land to a single base level, ultimately that of the sea, but obviously such a state of affairs was never reached because of the occurrence of earth movements from time to time. So subsidiary and transient base levels might be found in different places and at different times. Unconformities might be understood in this light, as can be seen from Powell's section of one of the walls of the Grand Canyon (Figure 7.2).[18]

Of particular importance was Powell's suggestion (1875: 153) that drainage patterns might sometimes be older than the mountains through which the rivers run. This state of affairs could come about by rivers cutting down at about the same rate that the land is rising, which can explain what are otherwise very odd features of landscape – for example, rivers that rise on one side of a mountain range may cut through the range and flow to the sea on the other side. Thus

Figure 7.2 Section of the wall of the Grand Canyon, according to John Wesley Powell (1875, facing p. 212) (by permission of the British Library)

one could talk about 'antecedent valleys' and 'antecedent drainage patterns' (ibid.: 163). Powell also referred to 'consequent valleys', which arose in the first instance as a direct result of flexures of the earth's surface. Various other terms for the classification of valleys were introduced.

Such ideas were, in a way, just developments of the old fluvialist doctrines of Hutton (more particularly as developed by Playfair in his *Illustrations of the Huttonian Theory* (Playfair, [1802] 1964)). But they were extremely important for adding support to the idea that subaerial erosion was a major factor in the carving of land-forms and that marine erosion was, in comparison, not such an important agency. Moreover, Powell taught geologists how to reconstruct the physical history of a region by studying its drainage system in relation to its rock structure. In fact, with the work of Powell one was seeing the emergence of a new interdisciplinary science: geomorphology, or physical geography.

G.K. Gilbert served on the Ohio State Survey, and later with G.M. Wheeler in survey work west of the hundredth meridian. Subsequently, he became one of Powell's assistants in the Rocky Mountains region. Gilbert joined the Federal Survey when it was formed in 1879, and continued there until the end of his working life. Studying the 'basin and range' country of the western states, he proposed that much of the topography had originated in 'block faulting' rather than by erosion acting on a folded surface, such as had been studied extensively by geologists in the Appalachians (Gilbert, 1875: 41). Nevertheless, like Powell, Gilbert placed great emphasis on the significance of subaerial erosion, and he introduced the important concept of 'grade', that is, the idea of a continuous descending curve of a stream channel, just sufficient for the current to carry its load of sediment. Thus a graded profile of a river will look as shown in Figure 7.3. But, if the resistance to erosion is unequal, then the gradient of the river will vary accordingly and energy will be concentrated at the resistant spots (as in rapids or waterfalls). Thus there is a constant dynamic adjustment between the forces of erosion and the topography of the earth. Putting the matter another way, when a river is running swiftly, with more energy than that required to carry its suspended load, it will 'corrade' its bed. A sluggish river deposits material. In the intermediate condition, when the river neither deposits nor

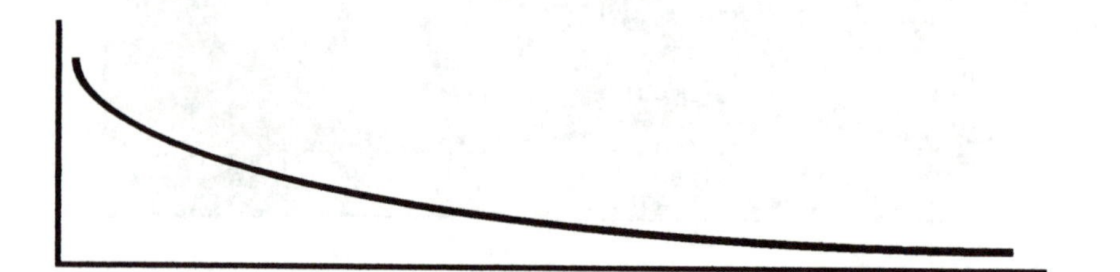

Figure 7.3 Profile of a graded river

corrades, it is at 'grade' or is graded. All rivers, suggested Gilbert (1877: 111–12), work toward this equilibrium condition. The principle could in fact apply to all transportational slopes.

The profile shown in Figure 7.3 also indicates the general shape of mountain divides, steeper towards the summits. This is particularly the form to be expected in well-vegetated regions where the rate of transportation is the limiting factor for erosion. On the other hand, in arid regions there is no accumulation of disintegrated material and the rate of erosion is limited by the rate of weathering. This will be dependent on the diversity of rock texture, and so the rock structure becomes embodied in the topographic form (Gilbert, 1877: 119). This is what is so commonly observed in the American West. As Gilbert put it (ibid.): 'With great moisture the law of divides is supreme; with aridity the law of structure.'

Gilbert thought about the earth in terms of equilibria and ratios, especially between forces and resistance to forces. Land-forms gradually emerged through time as a result of the interaction of forces. In a sense, such ideas had been implicit in the work of Hutton and Playfair, but now Gilbert was trying to discover the laws according to which the forces of erosion and deposition operated. So geomorphology could become a branch of science with its own special laws. The processes by which the face of the earth became sculpted were seen as law-like and hence susceptible to scientific description and analysis.[19] Thus a new aspect of the earth was brought under intellectual control. The symmetrical structure of a laccolith mountain,[20] Ellsworth Arch, provided Gilbert with a textbook example for the study of the development of a drainage system on a newly elevated terrain (ibid.: 145).

Dutton, like Powell, was also an officer in the Civil War. Afterwards, he became interested in geology and, with Powell's patronage, led a field expedition in Colorado, and later he became head of the Colorado division of the Federal Survey. It was Colorado, then, that provided the main inspiration for Dutton's theoretical ideas, which were, however, chiefly stimulated by Powell's work so far as thinking about land forms was concerned. Notably, Dutton (1880–1, 1882a, b) deployed Powell's ideas on antecedent drainages (which Dutton called 'persistence of rivers') and base levelling. Dutton was, like the other pioneer geologists of the desert areas of the American West, a thoroughgoing fluvialist. His thinking on the excavation of valleys by fluvial erosion is readily apparent from one of the illustrations from his report of 1882 (*Tertiary History of the Grand Cañon District* (Dutton, 1882a)) as shown in Figure 7.4.

In Dutton's illustration, the form of the canyon depended on the hardness of the rock. While the river was cutting through the upper unit (*A*), which was 'obdurate or very unyielding to the attack of weathering' (Dutton, 1882a: 251), a narrow canyon with almost vertical walls was formed, since the river could easily remove the small amount of detritus produced from the hard rock. But when the erosion proceeded as far as unit *B* ('notably softer and less

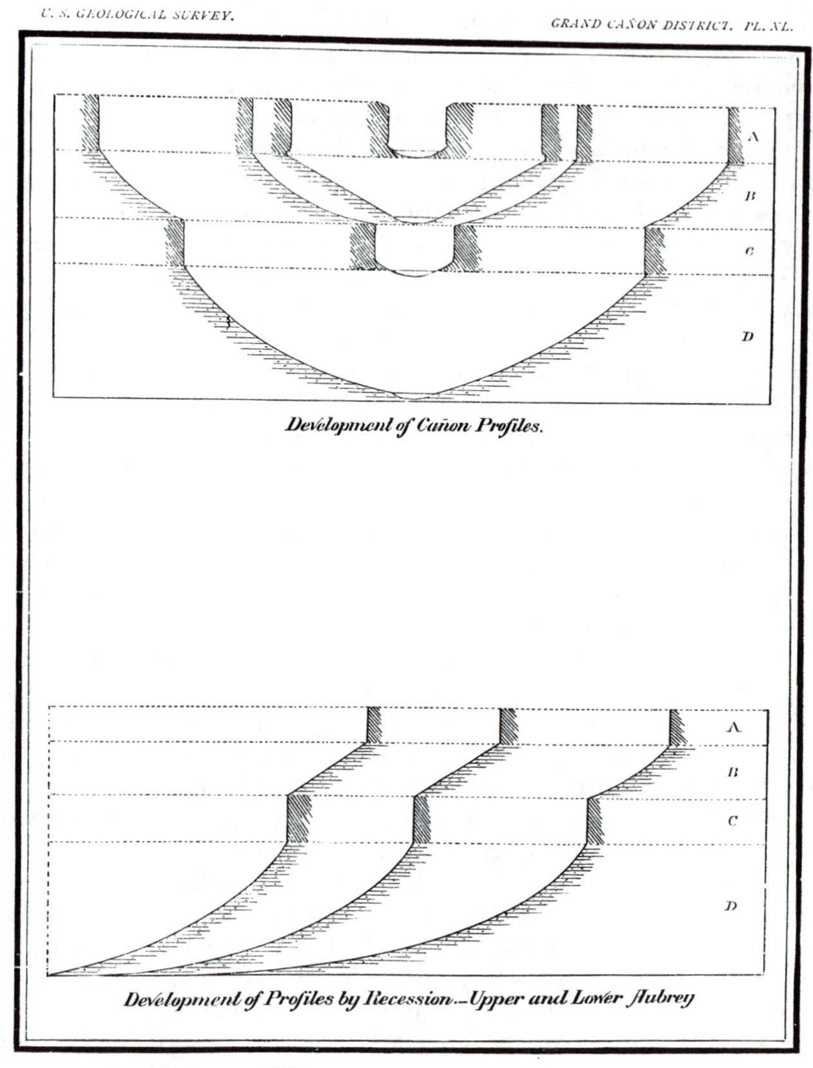

Figure 7.4 Development of canyon profiles, according to Clarence Dutton (1882a, Plate 40, facing p.250) (by permission of the British Museum of Natural History)

Figure 7.5 View of Kaibab Valley from Sublime Point, Grand Canyon region, by William H. Holmes (Dutton, 1882b, Sheet 15) (by permission of the British Library)

obdurate') the supply of debris increased and the rate of transportation slackened because of the reduced declivity of the river. So a smooth slope of talus was formed. On cutting down to units *C* and *D*, the land-form was, so to speak, duplicated. As Dutton pointed out, the talus or debris was the regulator of the cliff profile. These ideas were simple enough, but Dutton performed a useful service by putting them into the form of general principles. They can be seen in operation, for example, in the Blue Mountains quite close to my home in Sydney.

An idea of the marvellous terrain which so inspired the development of new ways of thinking about the development of land-forms may be gained by examining Figure 7.5, reproduced from one of the publications stemming from Dutton's explorations. This shows a magnificent drawing (coloured in the original) of Point Sublime, overlooking the Kaibab Valley in the region of the Grand Canyon. The artist was William H. Holmes.[21] It is scarcely surprising that such scenery stimulated new ways of thinking about the earth's surface (Pyne, 1982) and that the opening up of the American West led to the emergence of the new branch of geology – geomorphology.

Dutton regarded the notion that stratified rocks were derived from the degradation of the land as a proposition 'fundamental to geology'. Indeed, it was 'to geology what the law of gravitation is to astronomy' (Dutton, 1880–1: 95). In this regard, he was again giving expression to a view in geology that had been established in the literature since the time of Hutton and Playfair. But, again, the old ideas were given greater force by the wealth of empirical evidence afforded by the explorations and surveys of the western parts of the United States. Dutton also proposed (or gave the name to) the fundamental geological concept of isostasy, but discussion of this is deferred to the following chapter, where we shall consider nineteenth-century ideas on large-scale processes of mountain formation.

The several ideas relating to geomorphology and landscape formation that I have been discussing above were all brought together in a grand synthesis by William Morris Davis, who succeeded in bringing them into a coherent whole with the help of concepts from the evolutionary theory of the Darwinians, especially the 'German Darwin' Ernst Haeckel. Davis represented the development of landscapes as being analogous to the development of living organisms, and in this way, the earth was brought under the aegis of a developmental way of thinking that many nineteenth-century theorists found peculiarly attractive.

Davis graduated from Harvard with a Master of Engineering in 1870. This led in the first instance to a career in surveying, which included work for the preparation of the route for the Northern Pacific Railway through Montana. This period led him to develop ideas on geomorphology, and in 1885 he obtained a position as assistant professor at Harvard, eventually rising to the rank of full Professor of Geology in 1898.

The lens through which Davis thought about the surface of the earth was provided by the idea that it grew and evolved in a manner somewhat similar to living organisms: indeed, he liked to call his approach 'physiographic ontogeny'. (Such broad ideas were developed, however, with special reference to the rivers and valleys of Pennsylvania.) On a young mountainous land-surface, the valleys are V-shaped in cross-section, as the rivers carve relatively quickly into the high ground. But the valleys broaden as they grow older, so in time nothing of the original plain that is being dissected remains, except hilltops and crests of ridges. Finally, as a river runs into the sea, the land on either side may be almost entirely flat, with the river making great loops and perhaps running out through a delta region. Generally, there are estuarine mud flats. The first stage of development Davis called that of youth, the second maturity and the third old age or senescence. He also referred to an intermediate adolescent stage (Davis, 1889a: 203–13). So the whole river system 'evolved'. The picture was like the 'unfolding' of a bud into a flower, rather than something that took place by natural selection.

Davis's language was rich in organic metaphor. Rivers 'lived and survived'. They had 'life histories' and 'life cycles'. They did 'work' and 'completed their task' in eroding a terrain. The headwater of a river 'gnawed at its sources'. In 'infancy', a river might be 'embarrassed' by the original inequalities of the surface, so that, being at first insufficiently powerful to remove the inequalities, lakes might form in the depressions. Rivers could even be 'pirates', capturing other rivers on occasion (Davis, 1889b).

Moreover, a river might be 'rejuvenated'. This could happen if, at some stage in its development, the land surface were elevated quite rapidly compared with the cutting down of the river valley. Then a V-shaped valley could cut its way into a recently elevated plateau, and sometimes there would be a waterfall at the point where the rejuvenated valley, eroding deeply and quickly, met the old land surface of the plateau.[22]

But, if rivers could be 'revived' or 'rejuvenated' by land elevation, they could also be 'drowned' by land subsidence, giving valleys such as those of Sydney Harbour or San Francisco. Or, if an 'old' meandering river near the sea is drowned, then one might find a circular bay such as the famous Botany Bay to the south of Sydney. If a river falls directly into the ocean by a waterfall, the elevation of the land must be very recent. Such cases are uncommon, but I have seen a good example on the south coast of Java.

Davis made full use of Powell's concept of 'base level'. As a river gradually 'aged', it worked its way down to its base level. Ultimately, this must be the sea, but large streams could serve as temporary base levels for their tributaries. Likewise, Davis deployed the concept of 'grade': rivers erode until their currents are such that they just carry their load of sediment, neither depositing it nor cutting down further. The curve or profile of the river under these circumstances is its grade. When a region is eroded to a plain, Davis (1889c, p. 429) called the resultant land form a 'peneplain', a term used by geomorphologists up to the present.

In addition, Davis used the idea of 'antecedent drainage'. Indeed, in the map sketching the river system of Pennsylvania, the rivers (e.g. the Susquehanna) could be seen to cut right across the fold lines of the hills (Davis, 1889a: 188). Thus the rivers were older than the hills.

Eventually, Davis went the whole way on the Darwinian road, envisaging a 'struggle for existence' between rivers:

> [I]n consequence of elevation to a greater altitude, the streams have revived their lost activities, and set to work to sink their channels and open out their valleys in the process of reducing the land to its proper level again. . . . In the process of adjustment thus called forth, every stream struggles for its own existence . . . the steeper streams have gnawed more quickly into the landmass than the flatter ones, and the divide between the contesting streams has consequently been pushed in the direction of fainter descent (Davis, 1889b: 108).

In this fashion, the geomorphological ideas propounded by his predecessors such as Powell working in the western states were brought under the aegis of the general world-view of nineteenth-century evolutionism by the Yankee William Morris Davis. It was an appealing synthesis, which quickly attracted numerous adherents; and Davis has for ever left his mark on the language by means of which we think – perhaps unconsciously these days – about the earth.[23]

CHAPTER 8

The Making of Mountains and the Pulse of the Earth

We have already encountered a number of interesting theories about the way in which mountains are created. For some early theorists such as Buffon or Werner, mountains were primordial features, left over from the time when the earth first formed. For Hutton, on the other hand, mountains were formed on a regular basis as a result of upheaval of the earth's surface by the emplacement of subterranean magma, and then the surface of the high ground was carved by the forces of erosion, chiefly running water but also glaciers in some cases.

Such ideas were not based on a detailed knowledge of the disposition and extent of mountains worldwide, and what we might call a 'natural history' of mountains was needed before ideas about their origins could be developed more fully. By the nineteenth century, such information was generally available, as a result of the many explorations that had been made in the eighteenth century and which were prosecuted with even greater vigour in the nineteenth, colonial expansion from Europe being a major factor in the process.

Also, towards the end of the eighteenth century, and into the nineteenth century interest in mountain regions developed in various parts of Europe as one of the many manifestations of Romanticism. Thus we have the Lake Poets in Britain, who associated with geologists such as Adam Sedgwick (1846) and with the chemist Humphry Davy, who also lectured on geological topics (Davy, 1980). Alpine landscapes were described enthusiastically by Rousseau and Goethe; and in the work of Horace Bénédict de Saussure (1740–99) we find a man specifically interested in mountains from a scientific perspective, but who also initiated the sport of mountaineering. Before de Saussure's time, mountainous regions were thought to be chiefly the place for peasants rather than gentlemen or scholars. Indeed, mountains were often regarded with fear and were not necessarily regarded as beautiful. To an extent, their beauty was a Romantic 'invention' (Nicolson, 1959).

An important theory about mountain formation was developed by one of

Werner's most distinguished students, Christian Leopold von Buch (1744–1853). As a student of Werner (he actually resided with him) and with fieldwork experience in Landeck, Silesia (where he prepared a geognostic survey of the region for the Prussian government, by whom he was employed as Inspector of Mines), von Buch initially followed his teacher's Neptunist theories (von Buch, 1797, 1798, 1810a), but, as he widened his geological experience, things didn't seem to agree with the teaching of Werner. Granite was found in the Alps, but the rocks to the north and south were different, the dolomites of the south Tyrol having no equivalent to the north. Then, after examining the volcanic region of central France in 1802 (the Massif Central of the Auvergne region), von Buch (1802–3) moved somewhat toward the Vulcanist camp, though claiming that the Auvergne basalts were different from those he had previously seen in Germany and Sweden (ibid.: 256). It was suggested that the Auvergne basalts were first formed in the aqueous/Wernerian manner, and then heated and extruded so as to resemble igneous rocks. That is, there was supposedly a conversion of granite into trachyte (see p. 197) and then into basalt.

In the Auvergne, von Buch observed what we would call lava plugs (or tholoids), such as that of the famous Puy de Dôme, which appeared as hemispherical elevations, without craters. He also saw what appeared to be lava streams that had issued from the bases of apparent former volcanoes. He supposed that the outer parts of what we would call dormant volcanoes were formed from the country rock, which had been upheaved to form large caverns filled with lava.[1] Such structures were not, however, thought to be volcanoes proper, as they lacked either direct vents to the atmosphere from the earth's interior or craters. The theory was curious in that many of the Auvergne volcanoes (which, though quiescent at the moment, are very recent in geological terms) consist essentially of ash heaps. It is hard to think of them as upheaved country rock. Nevertheless, it was from this area that von Buch began to develop his 'crater of elevation' theory. His Auvergne theory was a kind of halfway house between Neptunism and Vulcanism. But, after visiting the Canary Islands in 1815, von Buch (1825) did move into the Vulcanist camp.

Detail of the theory of 'craters of elevation' is given in von Buch (1810b, 1836a, b). It might seem obvious that a mountain such as Etna was formed by successive emissions of ash and lava, forming a nicely shaped cone. This, of course, was the idea assumed without hesitation by Lyell and used by him to demonstrate the great age of the earth. But still von Buch did not accept what may seem to us the 'obvious' mode of formation of mountains such as Etna. He argued (von Buch, 1836a, b) that the angles of slope of such mountains as Etna, Vesuvius or Stromboli were too great for them to have been built up the way Lyell supposed. It was maintained, for example, that only when the angle of slope was as low as 3° could a mass of lava spread and accumulate to a considerable height (von Buch, 1836b: 109); otherwise it would just spread out. And Etna's angle of slope was far greater than 3°.

Von Buch's thought, then, was that mountains originated as a result of 'great display[s] of power from within' the body of the earth. One might think of them as blisters on the face of the earth. With the rise of magma, the overlying strata are heaved up to form dome-shaped structures, but only in some cases does lava break through to form a volcano.[2] For Etna, the height and bulk of the mountain had first been produced by a 'crater of elevation'; only subsequently had the volcano erupted to give lava flowing from an orifice and down the side of the mountain. The various Pacific islands, as well as the Peak of Teneriffe in the Canaries, also supposedly originated as craters of elevation.

The theory was Huttonian or Vulcanist in spirit, but as it involved presumed sudden and violent upheavals it belonged in the catastrophist geological camp, which set it apart from Huttonian theory. Von Buch (1824) also supposed that uplift could occur along linear axes, giving linear mountain ranges. Soon he began to recognize different mountain ranges according to their directions. And, according to the field evidence, it appeared that these could be of different ages. However, he did not pay much regard to the role of erosion or denudation in the carving or shaping of mountains.

Another writer of about the same period who considered the question of mountains from a general theoretical perspective, and in a way that was related to the ideas of von Buch, was the Frenchman Léonce Elie de Beaumont (1798–1874), Professor at the Ecole des Mines in Paris, Inspector General of Mines in France and, in 1832, successor to Cuvier's chair at the Collège de France. Elie de Beaumont also served as Permanent Secretary to the prestigious Académie des Sciences. Thus in his day he was a force to be reckoned with in French science.

Early in his career, Elie de Beaumont carried out fieldwork towards the preparation of the geological map of France, during the course of which investigations he began to develop some general principles about the geometrical distribution of mountain ranges and the several stages of sedimentary deposition revealed by the stratigraphical record, as that was coming to be known with the establishment of the geological systems such as the Carboniferous, etc. He was considerably influenced by the ideas of another of Werner's students, the polymathic explorer Alexander von Humboldt (1769–1859), who had travelled to South America (1799–1804) partly with the intention of examining the directions of mountain ranges. Von Humboldt thought it his task to look for geognostic patterns on a global scale, seeking analogies between geological formations and geographical features such as the directions of coastlines and mountain ranges in different countries. He thought he could see indications that mountain ranges were arranged in certain preferential directions, and that the dips and strikes of different mountain ranges were related to their ages – an idea called '*loxodromisme*' (von Humboldt, 1823a: 56). Moreover, they appeared to be produced by giant elevatory forces operating from within the earth (von Humboldt and Bonpland, 1814–29, vol. 6: 591–4), rather than the deposition of matter from above, as Werner's theory supposed.[3]

In his preliminary announcement of the results of his South American investigations, von Humboldt (Year 9/1801: 47–8) suggested that the apparent pattern of the primitive mountains might have originated as a result of the action of forces of attraction and affinity as the earth's crust first formed. It was almost as if the earth – or at least its visible exterior – had a structural pattern because it had 'crystallized' at its initial formation.

Following up such ideas, Elie de Beaumont (1829–30) suggested that the several 'revolutions' that had led to the apparent breaks in the stratigraphical record might be associated with particular episodes of mountain formation. A local uplift might have worldwide geological effects on marine fauna and hence the apparent universal character of the geological 'revolutions' (in Cuvierian theory) might be explained. The suggestion was attractive, for with the demise of the Wernerian theory that materials were deposited from a primeval ocean in a specific order[4] the correlations that might be made between strata at considerable distances from one another were no longer explained satisfactorily. With Elie de Beaumont's theory, however, there was some prospect of correlating faunal changes with major unconformities, which in turn could be associated with different episodes of mountain building. This idea, I suggest, could well have come to the mind of someone engaged in the technicalities of geological mapping. Also, with the help of the framework of Elie de Beaumont's theory, the respective ages of mountains could supposedly be deduced from examination of the unconformities between the different geological systems.

Thus the desire to have meaningful order in the stratigraphic succession, even in the period of 'historicist geology', did not die with the collapse of Werner's 'onion-skin' stratigraphy. For von Humboldt (1823b: 17), Werner's geognosy offered a continuing and extended research programme. He looked for 'identical formations' that were similar to one another in both composition and relative position within the stratigraphic column; and for 'parallel formations' that were 'geognostic equivalents', 'represent[ing] each other' (i.e. they occupied similar relative positions in the stratigraphic column but were mineralogically different). Likewise, Elie de Beaumont looked for large-scale parallels and correlations.

In 1829, Elie de Beaumont recognized only four major 'systems', or as we might say today 'orogenies': that is, periods of mountain building. In his later work, he increased the number considerably, to the extent that his catastrophism almost gave way to a form of gradualism. Be that as it may, Elie de Beaumont also had an explanation to offer as to why there had been the major periods of mountain building. The idea was that as the earth had slowly cooled it had contracted, and so its solid crust had buckled up from time to time. The popular analogue came to be that of the wrinkling skin of a drying apple. Simple as it was, this concept became the paradigm theory of the earth for most of the nineteenth century.

Although his later theorizing was more gradualist in character than his earlier views, Elie de Beaumont's conception of the process of the 'wrinkling' of the earth's crust was essentially 'catastrophist' in character. He thought that the 'crumplings' (orogenies) were quite sudden, so that there would be associated tidal waves producing widespread biological extinctions. Thus his geological theory, published at about the same time as that of Lyell, continued the Cuvierian catastrophist ideas in France, in opposition to the uniformitarian geology being developed by the Scotsman. However, in England De la Beche was sympathetic to Elie de Beaumont's theory and arranged for the translation of some of his publications.

Elie de Beaumont (1831), for example, is an interesting paper in that it shows that the term 'geological system' had not settled into its modern usage. We speak of a mountain system, but in so doing we think of the meaning of 'system' in quite a different sense from the way it is used in the phrase 'Silurian System' (or whatever), which connotes the time when certain rocks were formed or sediments deposited, rather than the time that they were uplifted. But, according to Elie de Beaumont's way of thinking, there were regularities or patterns, both temporal and spatial, in the earth's mountain ranges. Specifically, parallel mountain chains were thought to belong to the same orogeny (episode of mountain building) and were consequently of similar age.

Of course, a withering apple doesn't usually shrink so as to produce a regular geometrical pattern on its skin. But for Elie de Beaumont a zone of mountain formation would (ideally) form half of a great circle round the earth. There were, he thought, fifteen such half-circles, which crossed each other in such a fashion as to form a network of twelve pentagons on the surface of the earth, thereby giving the earth the approximate form of a pentagonal dodecahedron. Thus was developed the grand hypothesis of the *réseau pentagonal* or 'pentagonal network' (Elie de Beaumont, 1829–30, 1831, 1852) (Figure 8.1). As in von Humboldt's thinking, it was as if the earth were an object that was susceptible to crystallographic analysis. The theory was neat, but poorly related to the facts. None the less, helped by Elie de Beaumont's powerful position within the scientific community, the theory was utilized, particularly in France, for a number of years. It may be remarked that, although Elie de Beaumont's theory was radically different from Lyell's, they both had cyclic characteristics. But there was an important difference in that the Frenchman's doctrine, with its secular cooling of the earth, also had a directionalist 'arrow' (Gould, 1987).

It will be recalled that Lyell's theory, like that of Hutton, envisaged the wearing down of mountains and the deposition of sediment in the ocean basins,

Figure 8.1 Example of the 'pentagonal network' for the alignment of mountain ranges, according to Léonce Elie de Beaumont (1852, vol. 3, Plate 5) (by permission of the British Library)

followed by the elevation of rocks, the expansion of the underlying magma and the possible conversion of the pressured and heated sediments into igneous rock. But the cause of the vertical uplift remained obscure, especially for a globe that was presumably slowly cooling. Even Lyell, when he visited Switzerland in 1835, had difficulty in making sense of what he observed in terms of his own theory (Lyell, K.M., 1881, vol. 1: 453–4).[5]

Gradually, during the nineteenth century, the idea of lateral forces as being important in the formation of mountains came to the fore. Even in the eighteenth century, de Saussure had made observations (not published in his own lifetime) in the Alps, regarding what we would call 'thrust faulting' (Carozzi, 1989). Also in Switzerland, Arnold Escher von der Linth (1807–72) of the Zurich Polytechnic recognized the phenomena of foldings, overthrusting and inversions of stratigraphic order in the Glarus Canton to the south-west of Zurich (Escher von der Linth, 1839, 1841, 1846). After years of investigation and careful map work, he came to the conclusion that there had been manifest inversions of strata as a result of a colossal 'overpushing' or 'overshoving' (*Uebershiebung*) in the Glarus area (Escher von der Linth, 1841: 61).[6] In Britain, De la Beche (1846: 222) attributed foldings of the rocks of south Wales and parts of Ireland to the effect of lateral pressures arising from a contracting earth. In America, the brothers William and Henry Rogers (1805–81 and 1809–66) compared the structure of the Appalachian Mountains to what might be expected if strata had been thrown into wave-like structures by the action of undulatory movements in the earth's liquid interior, so that the layers of rock had turned over like surf on a beach (Rogers and Rogers, 1843). But the lateral movements supposedly arose from the (vertical) escape of molten and gaseous matter from under the mountains, through fissures in the core of the mountain system (Faill, 1985).

The idea of the Rogers brothers was an attempt to adapt the older geology to the new evidence furnished by the Appalachians. It was also applied in the European Alps (Rogers, H.D., 1858: 902). In a different style, the Swiss geologist, Albert Heim (1849–1937) of Zurich, who inherited the unpublished manuscripts of his teacher Escher von der Linth, published a major work which suggested how, under lateral pressures, folds could transform into thrust faults (Heim, 1878, Atlas, Plate 15, Figure 14). He also envisaged a huge double fold for the rocks of the Glarus Canton, which, however, required forces acting from opposite directions to produce the observed geometrical arrangement of the strata.

As has been shown by Greene (1982), there were exceedingly tangled debates in the mid-nineteenth century in America about the causes of mountain formation. Nowhere was the debate more complex and confusing – or confused – than with respect to the structure of the Appalachian Mountains. A major protagonist in these debates was the Yale professor, James Dwight Dana (1813–95). He envisaged contraction[7] by secular cooling of the earth, leading to the formation of the ocean basins (Dana, 1846: 353), but accompanying these depressions there might be fissuring of the crust at the edges of the area of depression, with intrusion of magma into the lines of fissuring and consequent elevation of mountain ranges at the margins of the ocean basins (Dana, 1847a, b, c, d). Dana supposed that there might be pools of fire (such as envisaged by Hopkins, 1842) underlying a sinking region where sediment might accumulate – later called a 'geosyncline' (see p. 174). As the geosyncline deepened, the fiery material might be displaced laterally under the adjacent region and, accumulating there, it might generate a 'geanticline'. Hence there might be evidence of lateral forces at work in such cases. The Andes provided an obvious example, generating the rock andesite (see p. xii). But it was difficult to see how the theory would work satisfactorily for complex fold belts such as that of the Appalachians, although that range provided ample evidence for lateral earth movements.

An alternative model was offered by James Hall (1811–98) of the New York Geological Survey. He sought a gradualist theory which, however, would also do something to explain the elevation of mountain ranges, including folded ones such as the Appalachians, which neither the theory of Hutton nor that of Lyell accomplished satisfactorily. Commonly, mountain ranges contain vast thicknesses of sediments, as can be shown by mapping in the field. But how can one explain a deposit that is, say, 40,000 feet thick,[8] when no oceans in the world are anything like as deep as this? Such a deposit could not have accumulated in accordance with uniformitarian principles simply by the collection of sediment in a stable pre-existing ocean basin.

Hall's solution to the problem of the accumulation of such huge thicknesses of sediments was to suppose that the crust collapsed under the weight of accumulating sediment, so that the trench or basin into which the sediment was deposited was continually deepening. Thus was born the idea of a 'geosyncline' (Hall, [1857[9]] 1883). The subsidiary foldings and faultings that one could see in the Appalachians (for example) were thought to have occurred more or less at the time when the sediments were deposited, rather than in some subsequent, perhaps catastrophic, deformation. Elie de Beaumont's lines of mountain elevation were, for Hall (1859: 86), 'simply lines of original accumulation'.

But, one might ask, how or why did the sediments collected in a geosyncline become elevated to form a mountain range? Hall's slogan was 'elevation is due to deposition'. Unfortunately, he never gave a satisfactory explanation of the elevatory part of the process. The point was not lost on his contemporaries. As Dana (1866: 210) complained, Hall offered a 'theory of the origin of mountains, with the origin of mountains left out'. In fact, so far as Dana (1873a) was concerned, Hall had got his causal arrow pointing in the wrong direction. For Hall accumulation of sediment caused subsidence, whereas for Dana subsidence made possible the accumulation of sediment.

The debate between Hall and Dana can perhaps be seen as a clash between two different philosophies of science. Dana was looking for a 'higher-level' cause for the formation of geosynclines. But for Hall, who might be said to evince a positivist philosophy in this matter, one should not be looking for ultimate causes. Field observations apparently demonstrated that there had been vast accumulations of sediment in trench-like structures. This was the primary fact. The reason why such accumulations had occurred was a subordinate question with which the geologist need not necessarily be concerned.

But, although positivism was a characteristic feature of the nineteenth-century philosophy of science and thus might seem to favour Hall's approach, so far as most theorists in the latter part of the nineteenth century were concerned, there was a high-level cause for the formation of geosynclines, and a simple one too. In the last analysis, the downwarpings, the elevations and the evidence of lateral pressures and lateral movements could all be explained in terms of the secular cooling and contraction of the earth. Again we have the apple theory of geotectonics: the earth was shrinking and growing wrinkled with age. But the theory, though seemingly rather obvious, has manifest disadvantages. If the earth cools, it must contract, and – by the principle of conservation of angular momentum – its rate of rotation should increase. But there is no evidence that this has occurred.[10] Further, one might expect the wrinkles of a cooling earth to form an 'all-over' pattern but, in fact, they are concentrated in particular bands such as the Andes.

Nevertheless, geosyncline theory was generally received with favour, and there was a proliferation of names as the geosyncline paradigm was 'articulated' in the Kuhnian sense.[11] In America, Charles Schuchert (1923) had 'monogeosynclines', 'polygeosynclines' and 'mesogeosynclines' (or mediterraneans). The German Hans Stille (1936a, b, 1941) invoked the idea of 'eugeosynclines', 'miogeosynclines' and 'orthogeosynclines'. The American Marshall Kay ([1951] 1963) went further, with 'autogeosynclines', 'zeugeosynclines', 'exogeosynclines', 'epieugeosynclines', 'taphrogeosynclines', 'parageosynclines', 'paraliageosynclines' and 'deltageosynclines'! This luxuriant terminology, which had to do with the various kinds of materials that might be found 'in' a geosyncline, their supposed relationship(s) to neighbouring stable continental areas ('cratons'[12]) and the depth of water in which the deposits were thought to accumulate, betokened a paradigm in some state of crisis – or one that was being 'articulated' to breaking-point.

If there were puzzles in America and for geosyncline theory, the situation was perhaps even more complicated in Europe, where the mountains of the Alps presented a great jumble. The Alpine conundrum was tackled vigorously by the great Viennese geologist, Eduard Suess (1831–1914), in his *Entstehung der Alpen* (1875). As we have seen, in the theory of von Buch, mountains were supposedly formed as craters of elevation, produced by massive intrusion of magma from below, which had forced up strata to their present heights. For Heim, in contrast, the Alps were the result of lateral thrusts or transverse crushing, such that two masses moving together had collided, producing huge overfolds and overthrusts. So far as field evidence in the Alps showed, the uplift theory seemed to Suess to be less plausible.

But Heim's pincer movement didn't seem right either. The Swiss field evidence, complicated though it was, suggested that there had been a 'one-sided' shove. It seemed as if the Alpine range had been formed by a general movement towards the north. Indeed, the folded and (in some cases) overturned strata seemed to have broken northwards like waves on a beach. But, unlike the Rogers brothers, writing about the Appalachians in the 1840s, Suess did not suppose that there had been a catastrophic motion. Rather, the 'waves' or folds had been produced slowly, with gradual deformation and accompanying metamorphism of the rocks. This remarkable movement was optimistically explained in terms of the contraction theory, in accordance with the 'apple paradigm'.

In the latter part of his career, from 1878 to 1904, Suess worked on a massive synthetic work which sought to provide a general synthesis of geological knowledge and endeavoured to propound an overarching theory of the earth: the great treatise *Das Antlitz der Erde* (Suess, 1883–1909, 1904–24). The author invited the reader to consider the general configuration of the earth, especially the shapes of the continents and the oceans (which for Lyellian theory, for example, were quite fortuitous) and the dispositions of the mountain ranges. And then, still utilizing contraction theory, which could supposedly yield lateral or horizontal movements as stresses due to contraction were relieved, Suess began to develop his own elaborate theory of the manner in which the 'face of the earth' had been formed. Particularly, he was concerned to show that the Lyellian 'yo-yo' (or oscillatory) theory of the rising and falling of different portions of the earth's crust was unsatisfactory.

The basic idea of Suess's theory (1904–24, vol. 1: 138, vol. 4: 622–3) was that contraction could give rise to subsidence, which would in some cases generate lateral or tangential forces. Mountain ranges would be the residual parts that had not subsided or would be formed as a result of lateral movements, rather than being the product of upward movement of material as in the old theories of Hutton or von Buch. The model might be thought of as follows. As the molten material of the earth's interior cooled it would contract, and then a solid portion of the crust might collapse into the underlying space. In such a situation, there would be lateral forces acting from the uncollapsed portions on to the collapsed parts, and these could give rise to the great thrusts of the mountain ranges. For example, parts of Germany such as the Black Forest might correspond to an area left standing which had otherwise been a region of collapse. And, as this region sank, material would spread northwards from the Alpine areas, thrusting over the collapsing 'foreland' strata, forming mountain ranges in the process. In the 'backland', such as the Mediterranean and the Adriatic seas, where collapse might also occur, there might be substantial igneous activity and extrusion of lavas.

Similarly, the Carpathian Mountains might be thought of as riding over the Russian foreland. In Asia, however, the general direction of lateral pressure seemed to be from north to south, as might be seen, for example, in the mountains of Yunnan, south China. So in Europe there was a northward advance, while the opposite was the case in Eurasia. Either way, the model was one of contraction, founderings and tangential thrusts.

For a Lyellian, one would expect to associate uplift with volcanic activity (Darwin had done this in South America). But, argued Suess, in the Mediterranean, where there was considerable volcanic activity, there was no historical evidence either of general elevation or subsidence. So the Lyellian theory was rejected yet again.[13]

An important feature of Suess's theory was the attempt to link his tectonic theory with the history of the globe revealed in the stratigraphic record. The proposal was that the collapse of a portion of the ocean floor would yield a regression of the seas (worldwide). But this would stimulate erosion on the more exposed land surface(s); and this would produce an increased supply of sediment, which would lead to the filling up of the oceans and the consequent transgression of the sea upon the land. Further contraction of the earth, followed by another collapse, would bring about another cycle in the long-term process. The theory offered a partial synthesis of uniformitarian (or gradualist) concepts with those of a catastrophist hue. Also, it allowed the linkage of tectonic theory and the stratigraphic record, which had likewise been Elie de Beaumont's grand aim. And again we have a cyclic theory.

Suess was thus proposing a powerful hypothesis as to why worldwide correlations of strata might be effected in a surprising number of cases.[14] It had to do with worldwide changes in sea level, yielding synchronized depositions or non-depositions of sediments round the globe (Suess, 1906: 535) and would explain, in a general way, why the different geological systems – Jurassic, Cretaceous and so forth – could be recognized worldwide. In thus seeking to link tectonics with stratigraphy, Suess (1888: 680, 1906: 538) introduced a new term, 'eustatic movement', and today we speak about the concept of 'eustacy'.[15]

The idea of general changes in sea level[16] had already been put forward in the nineteenth century in association with the development of glacial theory (see p. 149), as the oceans might have been locked up in the polar ice-caps. But Suess's concept was wider in scope than glacio-eustacy. It might, if true, help to explain the gross features of the stratigraphic column, at least since the Cambrian. We shall see below how the idea was developed by his successors.

In 1887, the French geologist Marcel Bertrand (1847–1907) discussed the evidence in Europe for the occurrence of several distinct phases of mountain formation: the 'Caledonian', 'Hercynian' and the 'Alpine' (Bertrand, 1886–7);[17] and he also envisaged links between the major phases of mountain building across the Atlantic, as shown in Figure 8.2. It will be observed that, in this figure, there are (very roughly speaking) two trend lines: mountain ranges that run east–west and those that run north–south. Five years later, Bertrand (1892) suggested a theory that was reminiscent of that of Elie de Beaumont. But, for Bertrand, mountain ranges formed a *réseau orthogonal* rather than a *réseau pentagonal*. Perhaps fortunately for Bertrand's reputation,[18] he soon gave up this

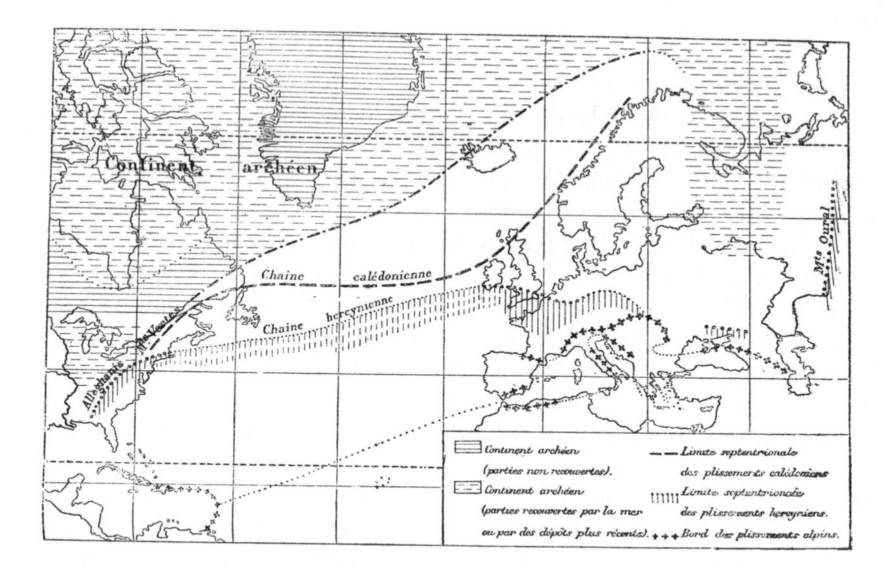

Figure 8.2 Different systems of mountain chains, according to Marcel Bertand (1886–7: 442)

fanciful notion, which was even belied by his own diagrams, as for example Figure 8.2.

However, the idea of looking for intercontinental correlations between lines of mountain building (or fold belts), for which task von Humboldt had been a pioneer, was followed up by other geologists. For example, the distinguished Alsatian geologist Emile Haug (1861–1927) sought to develop a picture of the earth that was a kind of cross between that of the American geosynclinists and that of Continental geologists such as Suess and Bertrand. Dana (1873b: 171) had envisaged the gradual 'expansion' of continents by the successive formation of geosynclines off the coasts of continents (marginal accretion). Haug (1900) wanted to link such ideas with those of Suess, effecting correlations between episodes of mountain building and marine cycles, and hence with the stratigraphic record. But, unlike Suess, Haug was willing to countenance once again the idea of upward and downward movements of the old Lyellian type.

In his grand *Traité de géologie,* Haug (1907–11, vol. 1: 157) distinguished

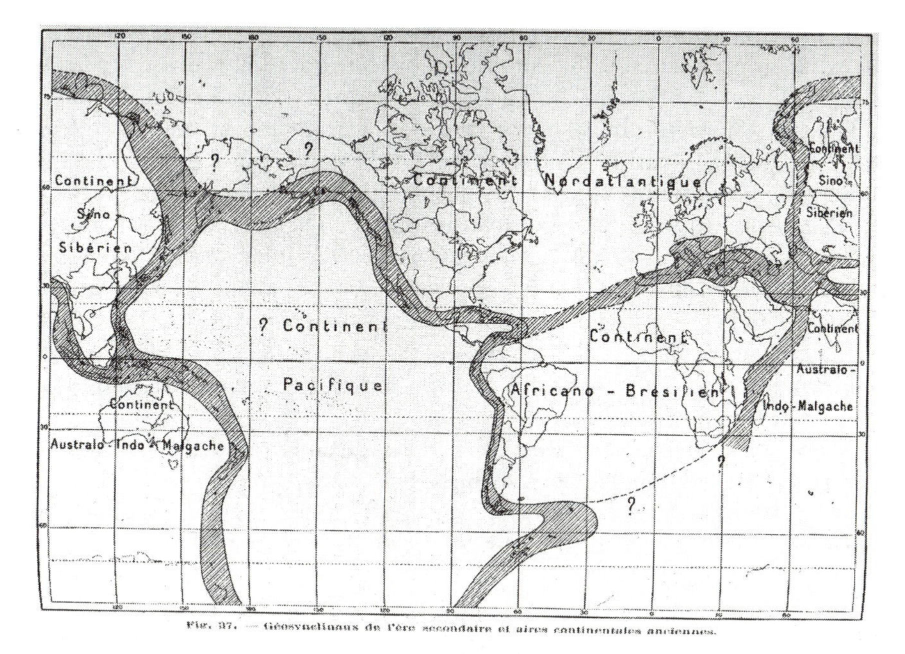

Figure 8.3 Arrangement of geosynclines and continental areas in the Secondary era (Mesozoic), according to Emile Haug (1907–11, vol. 1: 162) (by permission of the British Library)

between neritic (shallow water) and bathyal (deep water) deposits. The former, he pointed out, were usually relatively thin and variable in composition, while the latter were thick and uniform. Geosynclinal deposits (as he defined them) were of the latter type. Following Hall, Haug pointed out that mountainous fold regions were typically constituted of the thick and uniform materials. With this generalization in mind, Haug displayed the dispositions of Secondary (Mesozoic) geosynclinal deposits on a world map (see Figure 8.3[19]) and noted (ibid.: 164) that there was a coincidence between the tectonic zones and the zones of sedimentation.

In Figure 8.3, it should be noted that for the Mesozoic Haug was proposing a world that was largely continental in character, with a trench of geosynclinal deposition extending across the region of the present Atlantic Ocean. In fact, he attributed the present ocean areas to the breakup of the continents in a huge, relatively recent, geological event. Following Suess, he supposed that there had formerly been a fairly narrow sea (the 'Tethys') linking the present-day Mediterranean and Caribbean regions. The continents might rise, in which case the sediments would be deposited chiefly in the shaded regions of Figure 8.3. Or they might fall, in which case there would be great marine transgressions, and the continents would receive coatings of neritic deposits. The theory envisaged the existence of relatively stable continental platforms separated by fairly narrow, geologically active, mobile belts where geosynclinal deposits were received.

Haug was not active in promoting the idea of lateral movements. Nevertheless, they did form part of his theory. When the continents rose, space would supposedly be created in the intervening geosynclinal slots, where great thicknesses of sediment would accumulate. Conversely, when the continents fell, the sediments in the geosynclines would be squeezed and folded, to the extent that they would be pushed up to form fresh mountain ranges. There was, of course, a 'chicken and egg' problem. Did movement of the continents produce effects in the mobile belts? Or did activity in the belts cause movements of the continental masses? It was the old problem for geosyncline theorists, debated earlier by Hall and Dana.

In one sense, Haug was a 'fixist': the relative positions of the continents and the intercontinental mobile (geosynclinal) belts had supposedly been constant for long periods of time. However, it may be noted that his diagram was remarkably similar in general outline to those drawn later by plate tectonics theorists (see Chapter 11).

But this does not mean that we should think of Haug as an early continental-drift theorist. Rather, the phenomena to be explained were the same for both fixists and mobilists. Different theories could be devised to provide alternative ways of thinking about the same empirical phenomena. Haug, it should be emphasized, was turning away from the wrinkled-apple theory.

As has been shown by Greene (1982, ch. 10), there were good reasons, quite apart from the general considerations previously mentioned, why, by the time of Haug's theorizing, the contracting-earth theory was coming under question. Let us examine some of the ideas developed in the second half of the nineteenth century to show why this was so. One of the major scientific efforts of nineteenth-century colonialism was the work for the Great Indian Survey.[20] But unexpectedly this threw up an anomaly in that different distances were found between the cities of Kalianpur and Kaliana, 375 miles to the north, according to whether astronomical methods or standard surveying techniques on the ground were used.[21] An explanation was offered by J.H. Pratt (1809–71) (Pratt, 1855), Archdeacon of Calcutta and a formidable Cambridge-trained mathematician (third wrangler). His suggestion was that the discrepancy was due to an error in the astronomical method, caused by the attraction by the adjacent Himalayas of the plumb-line used in the determination.

Pratt's thoughts on this question attracted the attention of the British Astronomer Royal, G.B. Airy (1801–92), who began to think of the static forces involved so far as mountain ranges were concerned, if it were the case that the earth consisted of a fairly thin crust overlying a fluid interior, according to the theory of Cordier (p. 106). This model seemed incompatible with the existence of great mountain ranges such as the Himalayas, which might be expected to collapse into the underlying fluid unless there were some compensating force acting upwards from below. The difficulty could be overcome, Airy (1855) suggested, by supposing that mountain ranges were rather like icebergs. They floated on the magma below, but there was much more solid rock under the mountain range than might be seen above sea level: the ranges supposedly had 'roots'.

Then, in the 1870s, a British geophysicist, the Reverend Osmond Fisher (1817–1914), Rector of Harlton in England, a Fellow of King's College, London, and a former Fellow of Jesus College, Cambridge, developed significant arguments against the contraction theory. The amount of crustal shortening that might be expected according to the data of Lord Kelvin's theory of secular cooling was insufficient to account for the sum of all the folding that might be seen in so many parts of the globe. Fisher noted that the dark basaltic rocks, such as are found on sea floors, are generally more dense that the light, silica-rich granites, typically found in mountain ranges. So the comparative levels of the dark and the light crystalline rocks could be understood in terms of Airy's theory. That is, Fisher was accepting the general principle of there being 'roots' under mountain ranges dipping into a subcrustal fluid. In his *Physics of the Earth's Crust* (Fisher, 1881), he proposed the idea of convection currents within the earth's interior, with molten rock rising beneath the oceans and descending beneath the continents. However, Fisher did not, like Cordier, suppose that the greater part of the earth's interior was a fluid. His model was one in which most of the interior was solid – which accommodated Kelvin's physics – but there was a relatively thin fluid layer beneath the crust.

Fisher's ideas were taken up with approval by the more influential American geologist Clarence Dutton (1841–1912), who introduced the term isostasy (or 'equal standing')[22] and refined the concept considerably (Dutton, 1871, 1892). A spinning sphere such as the earth might be expected to be spheroidal if its composition were homogeneous; but if heterogeneous there would be depressions where the crust was more dense and bulges where it was less dense. Thus isostatic balance might be expected to obtain, unless disturbed for some reason – in which case isostatic adjustment might be expected. In general, the figure of the earth would have protuberances where matter was light and indentations where it was heavy.

Like Fisher, Dutton contended that a contracting earth theory was insufficient to account for the vast foldings evident in the earth's crust. Again, the Appalachians caused trouble, seeming to evidence lateral forces. But one might try to make sense of the observations by supposing that isostatic equilibrium was constantly being disturbed by erosion and deposition. The continental margins received sediment and thus, becoming heavier, tended to sink, whereas the continental uplands, being denuded of material, would become lighter and hence would tend to rise. Thus Dutton's thinking offered some explanation of geosynclinal phenomena, being more in line with the ideas of Hall than those of Dana. But Dutton had no satisfactory explanation of the elevation of continental masses as a whole or of the folded structure of the Appalachians.

Of course, the doctrines of isostasy and contraction/cooling were not necessarily at odds with one another, and it was open to geologists to try to combine the two hypotheses. This was the tack taken by Bailey Willis (1857–1949) in his paper 'The mechanics of Appalachian structure' (Willis, 1893). Material rising upwards by isostatic adjustment might be made good by the subcrustal inflow of material from the adjacent areas beneath the oceans. And as this material moved towards the continental regions it could readily be folded and plicated in the process. This eclectic theory had useful explanatory features, but it did not attract general attention or acceptance. It did, however, indicate the difficulties under which contraction theory was labouring. If one thinks of the contraction theory as a kind of Kuhnian paradigm, then Willis's theory might serve as an example of a hypothesis put forward when the paradigm was in a state of 'crisis'. It was one of the several hypotheses suggested during the last twenty years of the nineteenth century and into the twentieth century to account for the formation and structure of mountain ranges. None proved entirely convincing.

In the early years of the twentieth century, the American geologist with probably the highest standing was Thomas C. Chamberlin (1843–1928), who worked with the US Geological Survey, was later President of the University of Wisconsin and completed his career as head of the Geology Department at the University of Chicago.[23] We shall be concerned with his grand ideas about the origin of the earth in Chapter 12, but here we need to look at his ideas on eustacy and the diastrophic control of sedimentary processes, and hence of stratigraphic correlation (Chamberlin, 1898a, 1909).[24] These were an early example of 'pulsation theories' of the earth, which were important in the first half of the twentieth century.

It was partly as a result of his interest in glacial phenomena (which he encountered early in his career in Wisconsin) that Chamberlin became interested in recurrent or periodic geological phenomena. He believed that there were 'pulsations' in the earth's history: great earth movements, or orogenies, punctuating periods of relative quiescence. These would give rise to cycles of rapid marine regression, followed by slow periods of transgression, as the land masses were slowly worn down by erosion (base-levelling, in the language of the geomorphologists). There was, however, a general permanence of the areas of the oceans and the continents (in accordance with prevailing American thinking about the earth).

The processes of erosion and base-levelling were easily understood. But the old question of the cause or causes of uplift remained. According to Chamberlin's suggestion, the 'motive cause' (to use Aristotelian language) was still contraction, but this was chiefly restricted to the regions of the ocean basins. So, when oceanic subsidence occurred ('encouraged' by the deposition of sediment in the oceans), the less dense continental blocks were correspondingly squeezed upwards, according to Dutton's principle of isostasy. Mountain formation was, then, a massive corollary of movements occurring between oceanic and continental masses. Chamberlin's earth was solid to its core, so there was no fluid interior into which blocks of crust could founder.

However, the theory, while having something in common with Continental ideas such as those of Suess or Haug and drawing significantly on American theory such as the isostasy concept of Dutton, was ultimately rooted in Chamberlin's ideas on the origin of the earth, to be discussed in Chapter 12. The Continentals Suess and Haug were, incidentally, rather unsympathetic to the 'American' theory of isostasy.

Chamberlin's theory, then, was in brief as follows. There were periodic founderings of the ocean floors and associated 'up-wedgings' of the continents, typically to a level above that of isostatic equilibrium. Following such an episode, which would necessarily involve a fall in sea level, there would be erosion of the continents and sedimentation in the ocean basins. This would lead to a rise in sea level, which would be assisted by the small fall of the continents by isostatic adjustment after being wedged upwards.

There would thus be episodic regressions and incursions of the sea, and these would give rise to the main geological systems of the stratigraphic column, for the changes in sea level, fairly rapid when the collapse of an ocean basin occurred, would lead to stresses on living organisms with rapidly altering habitats. And so we might expect to see major biological changes evident in the rocks from time to time. There would, therefore, be a theoretical basis for what is found in the stratigraphic column, a causal explanation for the origin of apparently worldwide unconformities and an ultimate basis for stratigraphic correlation.

Various other interesting 'pulsation' theories were published subsequently in the twentieth century, such as Barrell (1917) and Haarmann (1930). Here we may look first at an interesting model proposed by the Harvard geophysicist David Griggs (1939), which sought to deal with the old enigmas of geosyncline theory. Then we shall examine a theory more closely related to classical stratigraphy (Grabau, 1940).

Of the ideas on the causes of geosynclines and episodes of mountain building available to geologists before the 'triumph' of plate-tectonic theory in the 1970s (see Chapter 11), the most attractive was that of convection currents in the earth's interior, investigated theoretically and with the help of mechanical models by Griggs. (But the existence of great convection currents within the earth had been proposed by William Hopkins as early as 1839.)

Griggs pointed out that if one had a closed pipe, in the shape of a rectangle and held vertically, if this pipe contained a liquid that was cooled at the pipe's upper horizontal arm and heated at its lower horizontal arm, and if there were a one-way valve on one of the vertical arms such that liquid could only rise upwards in that arm but not descend, then the system could exhibit periodic movements. Without the valve in place, the convection cycle starts to gather speed to a maximum, and then slows down, even though the supply of heat to the lower horizontal arm is constant. For, if the upward flow of the heated liquid is rapid, it may get driven beyond the cooling zone, and then warm liquid will be driven down to the lower horizontal pipe. Also, cooled liquid may pass the heater at the lower part of the pipe before it has time to get heated. Even so, after a time – after a few oscillations – things settle down into a steady convective movement. But, with the valve in place, things work differently. Flow does not begin until there is a convective driving force sufficient to open the valve. When the valve opens, there is a rapid movement for a short period. But this slows as the convective force diminishes, and the movement ceases when the force is insufficient to keep the valve open. Then, after a while the temperature difference between top and bottom is re-established and the flow begins again.

So far so good. The ingenious suggestion made by Griggs was that the valve simulated the behaviour of a plastic material within the earth which has a threshold strength below which flow does not occur. So his simple apparatus could serve as a model for convection currents that might act discontinuously within the body of the earth. Thus one might have a model that would allow one to understand the episodic nature of mountain building movements.

Griggs then sought to simulate the earth's internal convection currents to see what effects they might produce. He used a tank of glycerine (more dense) to represent the earth's interior mantle (see p. 341) and a mixture of oil and sawdust (less dense) to represent the external crust. The desired currents in the glycerine were generated mechanically, not thermally, by rotating horizontal drums in the lower liquid. On rotating the drums in opposite directions, one could see the 'crust' being dragged down between them, as it were in a 'geosyncline'. Then, when the motion of the drums was stopped, one could see the buoyant 'crust' material (oil and sawdust) rise up again – even above the general level of the crust – and spill over sideways.

Thus the model did just what was required of it. With heat in the earth's interior as the motive cause, episodic convection currents (in a material that only yielded when a certain force was exceeded) could produce periods of downwarp in the descending arm of a convection cell, allowing the accumulation of geosynclinal sediments. But this downwarp would eventually slow down and stop, being followed by elevation due to buoyancy effects, after the convection currents ceased temporarily; and lateral, 'thrusting' movements in mountainous regions might also be explained.

Next we may consider the fascinating work of the American geologist

Amadeus Grabau (1870–1946). In the earlier part of his career, he taught at Columbia University, but from 1920 he was a professor at Peking University. In his years in China his health prevented him from undertaking much fieldwork, but, despite his intellectual isolation, he developed a bold and elaborate theory of the earth, which culminated in the publication of his great treatise *The Rhythm of the Ages* (Grabau, [1940] 1978).

In his early years in America, Grabau became involved in a dispute with E.O. Ulrich of the US Geological Survey about problems in sedimentology. Grabau (1906) wanted to use the concept of 'facies[25] migration' from the German sedimentologist Johannes Walther (1893–4, vol. 3: 979);[26] but Ulrich (1911) believed that the concept made stratigraphic correlation virtually unworkable and that geological time units were sufficiently coarse for Walther's principle not to be relevant. As Johnson (1992) has pointed out, in the last analysis the debate was about whether local-to-regional earth movements dominated stratigraphic correlations, or whether (as Chamberlin thought) there were global events that involved worldwide changes in the levels of the sea relative to the land.

Moving to China, and thereby increasing the size of his stratigraphic database, Grabau believed that he could discern (or divine) evidence for worldwide changes in sea level from the stratigraphic records in different countries. He envisaged periods of rise and fall of the ocean floor, with consequent transgressions or regressions of the oceans. These changes could, he supposed, be correlated with the great geological systems. At the Sixteenth International Geological Congress held in Washington in 1933, Grabau presented his ideas in what came to be a famous paper, entitled 'Oscillation or pulsation' (Grabau, 1936). A diagram from this paper (Figure 8.4) shows how he envisaged a succession of marine incursions and regressions for the Palaeozoic strata. The figure was not prepared by anything so simple as the drilling of boreholes. It was produced by a vast amount of reading in the stratigraphic literature, supported by data collected in China by students under Grabau's instructions. If true, the diagram certainly supported the idea of continuous cycles of deposition and erosion, which might be controlled by worldwide fluctuations in sea level.

To account for movements of the ocean floors which might be responsible for the eustatic changes suggested by Figure 8.4, Grabau invoked ideas developed by the Irish geologist John Joly (1857–1933), of Trinity College, Dublin, who had interested himself in the thermal effects of radioactive changes occurring in the body of the earth (Joly, 1909, 1925). In fact, Grabau's theory of earth movements (which supposedly controlled changes in sea level) was in large measure similar to that of Joly.

Joly envisaged the gradual accumulation of heat in and under the continental parts of the crust as a result of the radioactive disintegrations slowly occurring

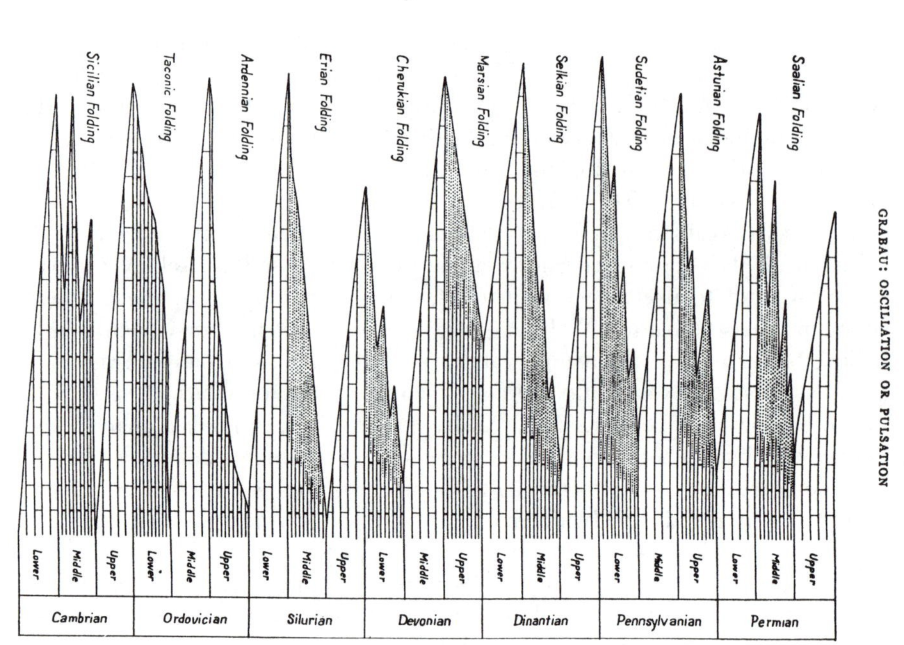

Figure 8.4 Figure illustrating Amadeus Grabau's pulsation theory (Grabau, 1936: 551)

there. He adopted the idea (which went back to Dana) of the essential constancy of the relative dispositions of the continents and oceans, but had the continents rising and falling, according to the following model. As heat accumulated under the continents, the dense basaltic matter there would eventually melt, expand and become less dense. In consequence, the continents would sink somewhat, according with the principles of isostasy. And thus there would follow a marine transgression over the continental areas, probably already worn down by erosion at this stage.

But the heating of the basaltic substratum and the basalt of the oceanic basins would be accompanied by earth expansion, and thus there would be great igneous eruptions at that stage. The loss of heat would then lead to contraction and solidification of the basaltic material and its concomitant increase in density. As a result, the continents would rise up again, causing a marine regression. Thus we have, in Joly's model, a constant 'earth machine', powered by the energy liberated by radioactive decay.

It should be mentioned that the moon also played a part in Grabau's theory. While the crust and subcrust were solid, no significant tidal effects were generated by the moon. But, at the period of melting of the basaltic material, tidal effects were supposedly produced such that the hot subcontinental basaltic matter could be shifted to the oceanic regions, thereby supplying the material for the massive volcanic activity. Such was Joly's grand theory. He estimated that a complete cycle might take about ninety million years.

Grabau's theory differed from Joly's in that it offered ideas about the very early stages of earth history, before the 'rhythm of the ages' was established. As we shall see in Chapter 11, an important idea developed by Alfred Wegener, the chief founder of the theory of continental drift, was that there was an original single primeval continent, termed 'Pangaea'. This slowly broke up into fragments during the course of geological history, eventually giving rise to the distribution of continents as we see them today.

According to Grabau, the primeval globe consisted of an iron/nickel core, above which there was a large 'pallasite sphere'.[27] Above this was a basaltic 'sima'[28] layer, and above this the light 'sial',[29] of which granite was a typical constituent.[30] The 'sial' was compared by Grabau ([1940] 1978: 6) to 'a scum of lighter substances, which crystallized out on the surface of the heavier foundation rock'. Above the sial sphere was the hydrosphere, and above that the atmosphere. Grabau believed that the whole surface of the globe was originally covered by sial, and there was also a complete covering of water. So the old idea of Werner's universal ocean was revisited.

But Grabau still faced a difficulty: how did Pangaea form such that the sialic 'scum' became concentrated in a single area of compressed and crushed gneisses and schists, leaving the underlying basaltic sima revealed in a large oceanic region? At this point, Grabau virtually invoked a miracle – or at least an exceptionally grand *ad hoc* hypothesis – suggesting that when the early earth was neatly covered by its layer of sial and its universal ocean, it had come into fairly close contact with some kind of stellar object, which had pulled the sial towards itself, forming a 'Pangaea' of intensely folded, compressed and metamorphosed rock, leaving the sima stripped bare of sial, but covered by ocean. Then this obliging star, or whatever, had taken itself off, leaving an earth that could evolve more or less according to the mechanism proposed by Joly.

Grabau's 'stellar body' was a gratuitous *ad hoc* hypothesis, for which he had no independent line of evidence. Nor did the hypothesis give rise to any testable consequences other than those that were originally required to be explained. Grabau did make a model, with a globe covered by a thin cloth. He then claimed (with a photograph to support the claim) that if the covering were drawn back (like a half-closed paper lantern) it would acquire folds similar to those anticipated according to an attempted reconstruction of the fold lines of Pangaea. (This was obtained by treating the continents rather like pieces of a jigsaw and fitting them together in a single supercontinent. Then one might see how the fold lines formed a pattern.) But this was scarcely convincing; much less convincing than Griggs' effort.

By moving as far forward as 1940, I have begun to tangle the nineteenth-century theories of mountain building, for which geosyncline theory and isostasy were major considerations, with later ideas of continental drift, to be discussed in Chapter 11. But they were just so intertwined in Grabau's theorizing. He regarded geosynclines as 'the incubators of mountain chains' (Grabau, [1940] 1978: 50). As he put it: 'Mountains are of interpulsation growth; the folding, when it occurs, takes place while the geosyncline is drained, that is during an interpulsation period. This is a fundamental law of the genesis of folded mountain chains' (ibid.: 51).

Grabau's theory had the merit of generality and it was also usefully eclectic, drawing together ideas from a variety of different theorists. However, it can hardly be said to have offered a satisfactory general synthesis, for he even invoked further 'extra-telluric' agents to help account for the major orogenies (Grabau, [1940] 1978: 56), rather than embedding them in basic physical theory, as did Griggs. Without Grabau's *ad hoc* cosmic intruders, the problem of the uplift of mountains was still not solved satisfactorily. On the other hand, the idea of a 'pulsation theory' to account for the existence of the major geological systems was certainly attractive.[31] Given that Grabau worked in considerable intellectual isolation and in a country where the political conditions were extremely unstable, I think his achievements were remarkable.

The cyclic character of Grabau's thinking about the earth is worth comment. (It was certainly more important than his *ad hoc* invocation of stellar agents interacting with the earth.) The idea of cycles was certainly not new. It can be found in ancient ideas about the cosmos, including those of Plato and Aristotle. Hutton's theory, too, was cyclic, and so was Lyell's. But in the nineteenth century, with the wrinkled apple theory of terrestrial cooling and with the rise of evolutionary theory, directionalist ideas began to command greater attention (although the ideas of Davis, discussed in Chapter 7, did not fall into this pattern). In the twentieth century, there was renewed interest in cycles, among biologists and geographers as much as geologists (e.g. Matthew, 1915; Taylor, 1919; Huntington, 1925). The cyclic way of thinking, which always tends to have explanatory appeal, came into its own again, and what we find in geologists' thoughts about the earth was, I suggest, possibly part of some broader intellectual movement, although there is insufficient space here to analyse all the cross-currents and interconnections.[32]

There were, then, numerous other workers in the first half of the twentieth century, particularly in the United States, who looked for cycles of sedimentary deposition. These seemed to be evident in the coal-bearing regions such as Iowa, Kansas, Illinois, Virginia and Pennsylvania, where regular sequences of coal, shale, limestone, sandstone and clay seemed to be evident, as first recorded by the Illinois geologist Johan August Udden (1912). Further work was undertaken by J. Marvin Weller (1930, 1931) of the Illinois Survey and Harold Wanless

(1931) of the University of Illinois. In a joint paper (Wanless and Weller, 1932) these two workers introduced the term 'cyclothem' to refer to the beds deposited within a single sedimentary cycle. Raymond C. Moore (1931), Director of the Kansas Survey, did similar work.

There is no need to consider the details of these investigations here and the arguments about which part of a cyclothem should be regarded as its 'base' or starting-point.[33] The interesting question for our present purposes is the debate about whether the cycles were controlled by movements of the earth's crust (diastrophic control) or rises and fall of sea level (glacio-eustatic control). Like Grabau, Weller (1930) and Moore (1935, 1936) favoured the role of earth movements. But Wanless and F.R. Shepard of the University of Illinois (Wanless and Shepard, 1936) argued that changes in sea level, caused by glacial comings and goings, provided a better explanation.

The debate is interesting because it provides a classic case of under-determination of theories by data. Either theory, or indeed a combination of the two (which was potentially the most difficult case to deal with), could account for the data. So, as far as the disputants of the 1930s were concerned, the debate just petered out. But it has been revived in the postwar period, with the help of the new data available from drilling logs (some of them deep-sea drilling logs) and computers to store, compare and contrast the vast mass of stratigraphic data that is now available. Also, the new results from 'seismic stratigraphy'[34] have proved invaluable in helping to sort out the theoretical and empirical mess. The microfossils in the drill cores facilitate correlation, and so too do reversals in the magnetic polarities of the rocks at different levels in the cores. Absolute radiometric datings can be obtained for some of the rocks from the cores.

Such work was stimulated chiefly by economic considerations – the search for oil. But the data obtained have also allowed newer and grander theories about the earth to be generated, and some resolution of the pre-war debates about eustacy. The question was: were there worldwide patterns of sedimentation? If there were, this would tend to favour the theory of glacio-eustatic control, especially if the data could be linked to the cycles of the Milankovich theory (see Chapter 7).

The leader in this work has been Peter Vail of the Exxon Production Research Corporation and Rice University, Houston, Texas. With a team of co-workers (the job is far too big to be undertaken single-handed), Vail has produced diagrams, now known as Vail curves, which show quite sharp changes in general sea level continuing throughout the stratigraphic column since the Precambrian (Vail, 1975; Vail et al., 1977; Haq et al., 1987). And, *mirabile dictu*, these can be correlated with the old subdivisions of the stratigraphic column worked out by the classic methods of nineteenth-century stratigraphy, chiefly in Europe, using palaeontological evidence and the search for unconformities (see Figure 8.5).

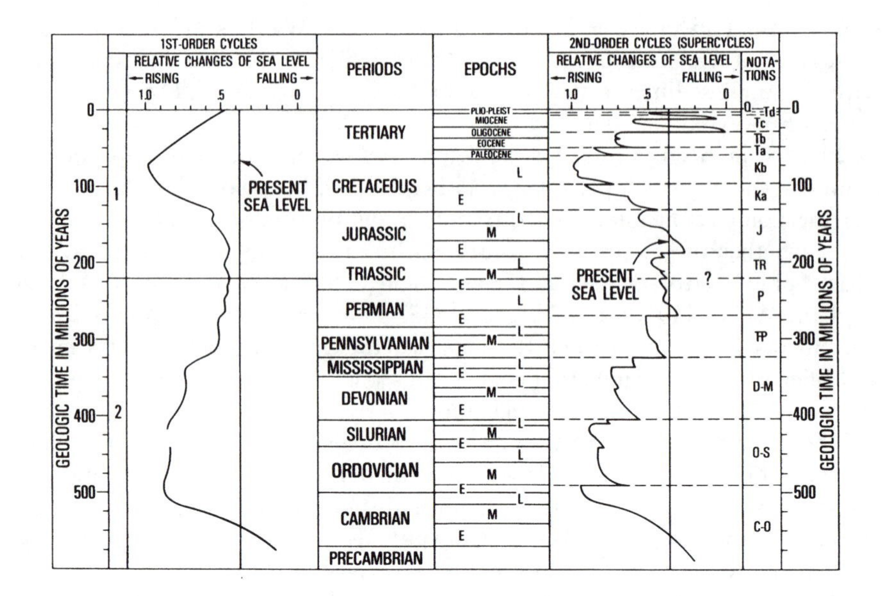

Figure 8.5 Example of a (low resolution) 'Vail curve' (Vail et al., 1977: 84) (by permission of the American Association of Petroleum Geologists)

The sea-level changes can also be correlated, for the upper parts of the stratigraphic column at least, with the Milankovich cycles (Fischer, 1981).

Thus some order in ways of thinking about the earth seems to be emerging. In the nineteenth century, following the work of Cuvier, there were semi-catastrophist accounts by French geologists of the different layers of the stratigraphic column (e.g. Coquand, 1843), each layer supposedly corresponding to some radical disturbance. It appeared from Coquand's work on Cretaceous rocks near Aix en Provence that old shorelines could be discerned and also the places where streams had entered the ancient seas, so there was evidence for ancient changes in sea level with concomitant changes in the sediments.

Today, many such changes in sedimentary type (from greensand to chalk, or whatever) seem to be explicable in terms of climatic changes, perhaps induced by astronomical causes such as were envisaged long ago by Croll and Milankovich. The palaeontological changes can be understood too. If the sea level stands high, this represents a warm global climate with stagnant oceans depleted of oxygen and poor in nutrients, as little material is brought in by sedimentation. In contrast, in cold periods the sea stands low but is circulating, well aerated and replete with nutrients. Such changes should surely be a motor for evolutionary change.

However, the ideas of the new 'seismic sequence stratigraphy' exemplified by the likes of Vail are by no means universally accepted, and there have been complaints (e.g. Sloss, 1991) that the new thinking is a kind of resurgent Neptunism. On the other hand, some regard the work of Vail and his co-workers as a development in ways of thinking about the earth that is every bit as important as plate tectonics – albeit less dramatic, of course. But no one really holds that glacio-eustatic control can be the whole story. There are undoubtedly the great Caledonian, Hercynian and Alpine orogenies that need to be set beside the gentler and briefer changes wrought by alterations in climate. Even so, the recent developments in sedimentology are, I believe, of the greatest interest, and we shall revisit them in Chapter 12, where ideas about the global 'see-sawing' of the elements of the earth's crust are discussed in relation to the 'controlling' activities of living organisms. Moreover, as we shall see, the prospect is now emerging of a unified view of theories of mountain building, etc., linked with ideas about sedimentology and eustacy.

CHAPTER 9
Thinking About Rocks and their Formation: Magma, Migma and All That Stuff

If you go into the field with a botanist you can expect to be told the names of the plants you encounter. Similarly, you can expect an entomologist to be able to identify the insects that are seen, or an astronomer will know the names of the constellations and the different kinds of heavenly objects that may be observed through a telescope. Likewise, geologists are expected to be able to identify different kinds of rocks and discuss how they have been formed. And in my experience they are more than equal to this task. Yet, so far as historians of science are concerned, remarkably little has been written on what might seem to be the most fundamental aspects of the geologist's task: petrography – the identification, naming and classification of the different kinds of rocks – and petrology – petrography plus ideas about how the different kinds of rocks have been formed (petrogenesis), their present conditions, their alteration and their decay.[1] Petrology links up with theories about mountain formation, and ultimately with the most fundamental ideas concerning the formation of the earth and its geological history.

With a few exceptions, such as glasses, rocks are made of mineral grains, and there is, of course, a separate branch of science, mineralogy, that deals with the systematic study of minerals. The study of mineralogy is older than that of petrology.[2] Sometimes minerals may be in the form of large and beautiful crystals, and such specimens were natural objects for collection by the 'virtuosi' of the seventeenth century, at the beginning of the modern scientific movement, or even earlier. They continue to be collected to the present day by amateurs, and are well known as objects for display in museums, where the number on view is usually mind-boggling to the visitor, who will notice that sometimes specimens that look quite different from one another may be labelled as the same type.

This raises the question of the basis of mineral classification. Should it be accomplished by consideration of chemical composition or external crystalline form? And what is a mineral species? Or even a mineral individual?

In the early days of natural history, it was usual to try to treat the mineral kingdom in the same way as the animal and plant kingdoms – animals, vegetables and minerals being the three great kingdoms of nature envisaged by the notable eighteenth-century Swedish systematizer, Carolus Linnaeus (1707–78). What is or is not an individual in plants and animals is usually reasonably clear,[3] and the recognition of species is a relatively straightforward matter[4] by virtue of the fact that organisms of the same species look similar and can reproduce with one another to give fertile offspring.[5] Also, species have (according to Darwinian theory) originated from a branching, diverging, evolutionary process. Minerals are quite a different matter. Historically, the fundamental question was whether external appearance (particularly crystalline form) or internal chemical composition should be the determining feature – or some combination of the two. The problem was exacerbated by the fact that substances of the same chemical composition may have different crystalline forms (e.g. calcite and aragonite are different forms of calcium carbonate), and substances with similar crystalline form may have completely different chemical compositions (there are, for example, many cubic crystals of different chemical composition). Moreover, there may be small quantities of 'impurities' present in minerals, so there may be reason to name a mineral specimen according to the presence of some rare or precious, but subordinate, component such as gold or silver. Or some materials may show a range of chemical composition. Such problems do not plague botanists and zoologists.

The characteristic classificatory method for the eighteenth-century botanist was to examine a restricted range of the external features of plants (for example, the number, proportion, figure and position of the sexual organs of flowers) and use this information for constructing a tabular arrangement of the species according to the similarities and differences. This arrangement could then be used to build a hierarchy of similarities and differences, and the naming of the plants could be undertaken. Type specimens could be collected, preserved and described, and other specimens could subsequently be identified by comparison with the determined type specimens. Every time a new type was discovered, it too would be described, named and a new pigeon-hole built for it; and thus the number of known species would be increased. The procedure was underpinned for Linnaeus by the metaphysical assumption that the different species had all been divinely created at 'the beginning' – in the Garden of Eden, according to Christian mythography.

Linnaeus attempted to deploy a system for the description and naming or minerals analogous to that used for plants. Using a procedure similar to that earlier deployed by Steno (1669; see Figure 3.1), he imagined a crystal's surfaces folded out into a plane. The resulting figure could then be described and named simply by counting the number of plane geometrical figures obtained. For example, a quartz crystal might be imagined folded out so that all its faces lay on a flat surface (Linnaeus, 1770, Tab. I, Fig. 1 and Tab. II, Fig. 1). It could then placed in the classificatory pigeon hole reserved for the 'eighteen faces, six rectangles, twelve triangles' kind.[6]

This procedure meshed with Linnaeus's 'externalist' or phenomenological scientific method. But for the kinds of reasons given above, relating to the difficulties of identifying and classifying mineral kinds, the procedure was far from satisfactory and had rather few followers among mineralogists, even though it meshed with the epistemic principles of eighteenth-century natural history. Yet, with chemical analysis being poorly developed until the end of the eighteenth century, better alternatives were not readily available, although many different systems of analysis and classsification were suggested. In Germany, Werner (1774, 1962) made a valiant effort to provide a satisfactory system of mineral classification based on the external characteristics of minerals such as texture, colour and shape, but he and his followers were led into a desperate system of description, as the following passage from Archibald Geikie's *Founders of Geology* shows in an amusing and yet accurate fashion:

> [Werner] employed his mother tongue, and devised a terminology which, though artificial and cumbrous, was undoubtedly of great service for a time. [However, while merely u]ncouth in German, it became almost barbarous when translated into other languages. What would the modern English-speaking student think of a teacher who taught him, as definite characters, that a mineral could be distinguished as 'hard, or semi-hard,' 'soft or very soft,' as 'very cold, cold, pretty cold, or rather cold,' as 'fortification-wise bent,' as 'indeterminate curved lamellar,' as 'common angulo-granular,' or as 'not particularly difficultly frangible'? (Geikie, [1905] 1962: 210).[7]

Such a passage obviously creates an unfavourable impression of Wernerian mineralogy. It should be said, however, that Geikie was anachronistic in his historiography, in so far as he failed to mention the peculiar difficulties of the mineralogist of Werner's day, before adequate methods for the chemical analysis of mineral substances had been developed and before a satisfactory theory of crystals had been devised. If minerals are to be described and named according to their external characteristics, then one will inevitably end up with a system like that of Werner.

In the second half of the eighteenth century, two French workers in particular, J.B.L. Romé de l'Isle (1736–90) and René-Juste Haüy (1743–1822), made important advances in crystallography. Or perhaps one should say that they founded the science. Initially following older chemical ideas, Romé de l'Isle had the idea that there might be some special saline 'principle' present in crystalline substances, responsible for their crystalline character; but this 'principle' could not be discovered by chemical analysis and the idea was of no practical use. So, like Werner and other eighteenth-century naturalists, Romé de l'Isle (1784) tried to deploy a cluster of external characteristics – crystalline form, hardness and specific gravity – as determinants of mineral species.

Romé de l'Isle's most significant procedure, however, was to imagine that all crystals were modifications of a few fundamental forms (Romé de l'Isle, 1772). By various mental 'bevellings' and 'truncations' of these basic forms, the great range of crystal shapes could be conceptually organized. The difficulty, however, was that there was no physical basis for the conceptual bevellings, and in effect the system was a kind of perpetual extension of *ad hoc* hypotheses. In any case, the method was not really new: it had been used in a simple way by Steno.[8]

Perhaps taking this objection into account, Haüy sought to develop a theory which linked external form to internal physical structure.[9] A possibly apocryphal story has it that one day he dropped a crystal of Iceland spar (calcite: calcium carbonate) on the floor and noticed that the fragments were similar in shape to the crystal that he had dropped. Then, when one of the smaller fragments was broken it too yielded a mass of still smaller pieces of the same general shape. Whether or not this anecdote is true, it does give a good idea of the way Haüy conceptualized crystals. He hypothesized that the process of division of calcite might be continued (mentally) until one reached a small portion of the crystal, still of the same shape as that with which one started, but incapable of further subdivision.[10] This he called the *molécule constituante* or *molécule intégrante*. Haüy then sought to show how different crystallographic forms of a substance might be produced by arranging the *molécules intégrantes* in different ways according to different rules of decrement.[11] For example, for calcite he stated (Haüy, 1784: 222) that no less than 1,019 kinds of calcareous spar might exist in principle, according to his proposed laws of decrement, even if only a limited number of these were known to naturalists.[12] Thus a general theory, linked to empirical findings, was established for the study of crystals. It achieved an integration of external form and supposed internal composition. The system provided a useful way of classifying crystals, according to a supposed small number of basic shapes for the *molécules intégrantes* or primitive forms and certain assumed laws of decrement. The application of Haüy's method is illustrated in Figure 9.1. (It can be seen how Haüy imagined two differently shaped crystals might be built up on the basis of the same cubical 'building blocks'.)

By the eighteenth century, there was a well-established distinction between crystalline rocks and rocks of mechanical origin, which corresponds approximately with our division between igneous and sedimentary rocks. However, in

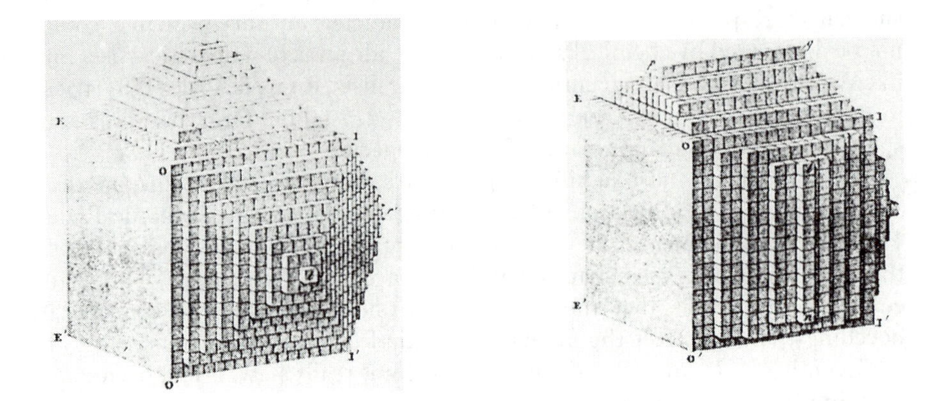

Figure 9.1 Structures of two different crystalline forms, built up hypothetically from the same cubical '*molécules*', according to different laws of decrement (Haüy, 1801, vol. 5, Plate 2)

the eighteenth century the crystalline rocks might be accounted for either in terms of Vulcanist/Plutonist/Huttonian theory or by Neptunist/Wernerian theory. The sedimentary rocks presented no major conceptual problem then or later (except for the glacial deposits and some special materials like the Chinese wind-blown loess) and I shall not discuss them further here,[13] although we shall have something more to say about sedimentary rocks in Chapter 12. The category of metamorphic rocks was only introduced in Lyell's *Principles* (1830–3) and will be discussed later. For the present, our problem is the igneous or crystalline rocks.

It might seem (to English-language readers) that Wernerism was decisively defeated by Hutton and British geology in general. However, while the rise of organized field mapping, described in Chapter 5, was a notable expression of 'English-language geology', the greater part of the basic work in petrography in the first half of the nineteenth century was carried out on the Continent, particularly in France and Germany. Thus the competent training offered by the Freiberg Academy and similar institutions bore fruit, although it must not be forgotten that igneous rocks as such played but a small part in the original Wernerian system, being thought of as the products of stratified rocks melted by the burning of coal beneath volcanic districts.

Some of the most important terms in petrography go back to antiquity. They include words such as syenite, porphyry, basalt and obsidian, or they are names, such as granite or gabbro, used by stonemasons. Syenite, for example, gets its name from Syene in Egypt, where it was used in ancient times for making monuments and obelisks (Pliny, 1962, vol. 10: 51). The term was given some precision by Werner (1788: 824), who used it for a rock near Dresden that was chiefly hornblende and feldspar, with subsidiary quartz. The term basalt (mentioned by Pliny (1962, vol. 10: 45) as 'basanites') was used by Agricola (1546: 7) in reference to a dark, fine-grained volcanic rock or lava. Porphyry was a term used by the ancients (*porphyrion*) for a purplish rock spotted with light crystals of feldspar. In the eighteenth century, it began to be used for a rock with large crystals in a fine-grained groundmass and the colour aspect was abandoned. The word granite – an ancient term, probably of Italian origin – was apparently first used in print by Andreas Cesalpinus (1596, vol. 2, ch. 11) for granular rocks in general, and in a more restricted manner by Wallerius (1747: 147) to refer to a rock chiefly made of quartz, feldspar and mica.[14] The term trachyte was used by Haüy before 1813 and by Brongniart (1813: 43) to refer to a rock of the Drachenfels found near the Rhine, of rough appearance, medium-grained, with alkali feldspars and minerals such as biotite and augite. Previously, it had been called trap porphyry by Werner. The term dolerite, so-called because it was difficult to identify in hand specimens (enough to make one weep), was used by Haüy (1822, vol. 4: 540, 573) to refer to a medium-grained equivalent of basalt composed chiefly of feldspar and pyroxene minerals. Gabbro is an old Tuscan name for dark, coarse-grained rock, chiefly made of pyroxenes and feldspars. It was first described in some detail by Giovanni Targioni-Tozetti (1768, vol. 2: 432–50).

Thus we can see that the common names for some of the crystalline rocks were of ancient usage, or they had made their appearance in the literature by the late eighteenth or early nineteenth centuries. There is not, of course, space here to attempt to cover the vast range of names used by petrographers.[15] Certainly a generally accepted system of nomenclature had not been established by the beginning of the nineteenth century.

The simplest kind of classification of crystalline rocks is based upon texture and mineral composition, the chief consideration being the proportion of silica (quartz) present. Rocks rich in silica[16] are generally light-coloured. As the proportion of silica decreases, one finds more of the darker minerals, rich in magnesium and iron ('mafic') such as amphiboles and pyroxenes. The type of feldspar found changes too, with greater calcium content and less sodium and potassium. Thus a very simple (and old-fashioned) scheme might look as in Table 9.1.

According to Huttonian-style theory, the fine-grained igneous rocks have been formed by fairly rapid cooling at or near the surface and thus often form lavas. The medium-grained rocks have been formed at intermediate depths, from magma that has cooled moderately slowly. They are said to be 'hypabyssal'. The coarse-grained rocks have been produced by slow cooling of magma at depth (plutonic rocks). If the cooling is very rapid (as, for example, at the margin of

Table 9.1 *Simple classification of crystalline rocks*

Texture	Acid (felsic)	Intermediate	Basic (mafic)	Ultrabasic
Fine	Rhyolites	Andesite	Basalts	Ultrabasic basalts
Medium	Microgranite	Microdiorite or microsyenite	Dolerite	
Coarse	Granites	Diorite or syenite	Gabbro	Peridotite
Lighter ←				→ Darker

an intrusion, where hot magma has been rapidly chilled by the surrounding rock), a glass may be formed (e.g. obsidian). Indeed, the glassy margins of basaltic 'dykes'[17] were an indication to Hutton that they had been formed by the intrusion of molten rock, and he found such indications at Salisbury Crag, overlooking Edinburgh. According to Huttonian theory, the colour and chemical composition of the rock were determined by the character of the magma that produced the intrusion.

As set out in summary form in Table 9.1, we have a view of the relation between what might be found in the field and what might be expected according to Vulcanist/Plutonist theory. But such a 'Vulcanist' classification[18] was by no means the only one used by geologists at that time. Nor indeed was the question of mineral classification according to chemical, crystallographic or a number of 'external' characters settled.

With the help of the atomic theory, chemistry in the early nineteenth century was at last beginning to make rapid theoretical progress. One of the more important factors in the development of chemical theory and practice was the introduction of the techniques of electrolysis for the decomposition of substances, either molten or in solution. In line with the developments in electrochemistry, the notable Swedish chemist and mineralogist, Jons Jacob Berzelius (1779–1848) devised a theory of chemical composition according to which substances were regarded as combinations of electropositive and electronegative components, and each component might have electrical subcomponents, until the ultimate chemical constituents were reached (Berzelius, 1811). On this basis, he regarded silica as the electronegative ingredient of minerals, functioning as an acid in their composition. Then, according to the proportion of silica in the minerals (determinable by the tedious methods of chemical analysis of the day), one might have 'siliciates', 'bisiliciates', 'trisiliciates'; 'subsiliciates' or 'double siliciates', etc.[19] Every mineral had one or more electropositive and one or more electronegative constituent. And, if these were known, a chemical basis for mineral classification was available (Berzelius, 1814).

Berzelius was not a thoroughgoing Vulcanist. He well knew that many crystals were formed in the earth from solutions and, being an 'electrician', he proposed (Berzelius, 1814: 19) that many minerals known to be insoluble under ordinary circumstances might crystallize from solutions within the earth under the influence of various (hypothetical) electrical currents. Such ideas were not, however, especially important. What was significant was that Berzelius offered a conceptual framework and a set of practical techniques for undertaking a chemical classification of the mineral kingdom. In fact, his interest in minerals was primarily chemical. He did not claim to be a geologist, although he did do some geological fieldwork (Bernhard, 1985).

Back in Germany, the system of mineral classification according to external characteristics was carried on by Werner's successor at Freiberg, Friedrich Mohs (1773–1839), remembered for the well-known Mohs' hardness scale for minerals.[20] His *Grundriss der Mineralogie* (1822–4) (*Treatise on Mineralogy* (Mohs, 1825)) deployed a classification based on externals, and so too did Breithaupt's *Handbuch der Mineralogie* (1836–41). Mohs met with Jameson in Edinburgh in 1818, and the two agreed on the basic principles of how the natural history system of mineralogy was to be organized. So, when a small student handbook for mineral classification was issued in Edinburgh by Mohs (1820), it offered a hierarchical system intended to be analogous to that used for plants and animals. It was a 'natural history' of the mineral kingdom organized along quasi-Linnaean lines. This text, therefore, is a counter-example to the claim made by Foucault (1966, 1970) that 'external' classifications characterized the eighteenth century (or 'Classical' period), whereas 'internal' systems were the norm in the nineteenth century. Berzelius's system fits Foucault's scheme, however.

In America, Charles Shepard (1835) sought to use the Mohs system, but then abandoned it in favour of trivial names. In his *System of Mineralogy* (Dana, 1837), Dana adopted a Mohs-type natural history system, but argued that, provided the crystallization was sufficiently good, there should be an agreement between the chemical identification of a mineral species and one based on examination of external characteristics. Dana kept the natural history arrangement in his edition of 1844, but also introduced the 'Berzelius' chemical classification. Eventually, in the edition of 1850, the natural history approach was dropped, although an effort was made to couple the chemical system with crystallographic principles.

Even in Britain, the land of Hutton, Wernerian geological theory, as well as rock and mineral taxonomies, continued to reach a wide audience well into the nineteenth century, chiefly through the efforts of Robert Jameson in Edinburgh. For example, the article on geology in the *London Encyclopædia* (Anon., 1829)[21] taught that there was a standard sequence of crystalline rocks, their geometrical order corresponding to the order in which they (supposedly) crystallized out of the primeval ocean:

15 Newer serpentine
14 Sienite [syenite]
13 Newer porphyry
12 Older flint-slate
11 Gypsum
10 Quartz
9 Older serpentine
8 Primitive limestone
7 Primitive-trap[22]
6 Older porphyry
5 Clay-slate
4 Topaz-rock
3 Mica-slate
2 Gneiss
1 Granite

But this sequence was so often 'disobeyed' that the Wernerian geognosy (as a petrological/geometrical/genetic doctrine) gradually collapsed under the weight of its anomalies,[23] even though, as we shall see, elements of Neptunist theory were sustained until almost the end of the nineteenth century, and in a certain sense carry through to the present.

It will be seen from the list of the *London Encyclopædia* that Wernerian theory had granite, with its generally large, well-formed crystals, with a non-laminar or non-foliated structure, as the oldest, 'primitive' or primeval, rock. Then came the foliated gneisses and schists; then the laminated slates.

In this list, it was the schists and gneisses that were the most difficult to understand. A gneiss can have much the same crystalline composition as a granite (the chief minerals being quartz, feldspar and mica), but some of the material displays distinctive layers of minerals, with the plate-shaped micas lying parallel to one another; other parts of an area of gneiss are less obviously banded, and in these parts the rock may look much the same as a granite. In a schist, there is usually much more mica evident, often giving it a distinctive sheen, and the entire rock is layered or foliated.

It was the Huttonian geologist Charles Lyell, in his *Principles of Geology* (Lyell, 1830–3), who proposed a major new category to accommodate such rocks as gneisses and schists: metamorphic rocks. Lyell (1833: 374) used the word 'hypogene' ('nether-formed' or 'formed at depth') instead of 'primary' or 'primitive'; and the hypogene rocks were divided into those that were unstratified (plutonic, e.g. granite) and those that were 'stratified' (metamorphic, e.g. gneisses and schists). He denied that there was any regular order of geometrical or chronological succession among the hypogene rocks, such as was favoured by Wernerian geology. Such rocks could, in principle, be of any age; but because they were presumed to have been formed under the action of heat and/or pressure they would, for Lyell, be expected to occur at the bottom of any system: the bottom of the Carboniferous, the bottom of the Silurian, or whatever.

According to Lyell (ibid.: 375 – emphasis in original), the metamorphic rocks were those that were '*altered* stratified', and the alteration was thought to be due to the action of heat and pressure. But this left an ambiguity as to whether the rocks had to be stratified first (i.e. sedimentary) or whether non-stratified igneous or plutonic rocks could also be altered so as to produce a layered structure; and either way there was the problem of how heat and pressure might produce a kind of laminated structure.

A few years later, in describing his geological observations in South America, Charles Darwin (1846: 141 – emphasis in original) used the terms 'cleavage', 'foliation' and 'stratification as follows:

> [By] the term *cleavage*, I imply those planes of division which render a rock . . . fissile. By the term *foliation*, I refer to the layers or plates of different mineralogical nature of which most metamorphic schists are composed; there are, also, often included in such masses, alternating, homogeneous, fissile layers or folia, and in this case the rock is both foliated and has a cleavage. By *stratification* . . . I mean those alternate, parallel, large masses of different composition, which are themselves frequently either foliated or fissile – such as the alternating so-called strata of mica-slate, gneiss, glossy clay-slate and marble.

Darwin's term 'foliation' has survived into modern usage,[24] but we use the term 'stratification' to refer to sedimentary rocks rather than the crystalline layerings of gneisses and the like.

Darwin and some of his contemporaries, such as Scrope, held that if magma crystallizes under pressure, while still in motion, it may give rise to foliated rock (Scrope, 1859, 1862: 300). Thus foliated structures might be associated with igneous rocks. But, on Huttonian principles, one might also expect to find a region of rock around an igneous intrusion metamorphosed by heat rather than movement or pressure. So far as the British Survey was concerned, that was more or less good enough for metamorphic theory till the 1880s. In their map work the surveyors quite often marked a metamorphic aureole (as we would say) round what was taken to be an igneous intrusion. This was done, for example, in north Wales, Shropshire and the Malvern Hills. But, up to the 1880s, the more complex problems associated with the mapping of metamorphic rocks were not a major issue and were not given much theoretical discussion.[25]

On the Continent and in the United States, things were different. Dana (1843) denied that schistosity was the result of sedimentation, but had the idea that sea water, heated by the action of volcanoes, might metamorphose granites and produce schistosity, and that some granites might initially have been sedimentary rocks. Thus some granites at least might have had a metamorphic origin. In France, the Chief Engineer of the Mines Department, Gabriel Auguste Daubrée (1814–96) (Daubrée, 1857, 1859, 1860), subjected many substances to high temperatures and pressures, with or without the presence of water, and came to the conclusion that metamorphism could occur (with new minerals being formed) without wholesale melting of the material. In particular, he was able to produce foliated, schistose structures by means of his experimental techniques.

Daubrée's attitude was opposed to uniformitarianism. He argued that the *antésilurienne* rocks (Precambrian, as we would say) were substantially different from those found higher in the stratigraphical column. They showed, he thought, every indication of having been formed in the presence of water, and probably under conditions of high temperature and pressure, under a primeval atmosphere utterly different from what we have today. Using the Wernerian language of an *océan primitif*, he took a stand against Hutton, saying that the origin of the planet was not 'lost in the night of some indefinite past', with geological processes proceeding endlessly and cyclically. Rather, we could look back to a past when things were quite different from the present. However, Daubrée did hold that the foliations of metamorphic rocks indicated pressure at the time of their formation, due to the earth's contraction (Daubrée, 1860: 119–24). Thus petrology was aligned with the broader ways of thinking about the earth that were espoused in the mid-nineteenth century.

Not surprisingly, Neptunist ideas were also upheld in Germany. Johann Nepomuk von Fuchs (1839), Professor of Mining at Munich, plausibly maintained that if granite had been produced by cooling of a fused mass the quartz crystals would have separated first so that there would not be a complete intermingling of quartz, feldspar and mica. He suggested, therefore, that schists, gneisses, granites and porphyries separated out from a primeval aqueous or pasty mass. The whole was not in solution, as Werner had supposed. The siliceous component was in a pasty state, from which crystals gradually formed. Much calcareous and magnesian matter was in solution due to the solvent action of carbonic acid, and by subsequent precipitation gave rise to beds of limestone and dolomite. Such arguments were plausible, but Fuchs went further and suggested that some sands had been formed by precipitation rather than by mechanical attrition of other rocks, which is scarcely credible. He believed, however, that the primeval atmosphere would not have contained oxygen, which is the modern view.

Gustav Bischof (1790–1870), Professor of Chemistry and Technology at Bonn, placed great emphasis on the role of water in depositing and modifying mineral substances. He questioned whether large masses of granite were produced by the cooling of molten magma, for, as he pointed out, large drusy[26] crystals were usually found in granitic veins rather than in massive granite. This, he suggested, was not compatible with the notion of masses of granite having formed by the slow cooling of liquid melts, for were this the case one would expect to find the largest crystals in the largest masses of rock, not the small veins (Bischof, 1854–5, vol. 2: 475–6). Bernhard von Cotta (1809–82), a teacher at the Freiberg Mining Academy, thought that some granite might be 'primitive' – a residue of the earth's first crust – and that plutonic processes involved water (von Cotta, 1866: 388). For mineral veins and ore bodies, he supposed that some resulted from igneous fluid injection, some were emplaced by sublimation and some were precipitated from deep-seated igneous solutions. Ore bodies could have become zoned according to variations of pressure and temperature with depth.

In the English-speaking geological world, some interesting ideas were put forward by Thomas Sterry Hunt (1826–92),[27] and these provide a convenient view of the development of Neptunism in the second half of the nineteenth century. Hunt was a member of the Canadian Geological Survey, moving to teach at the Massachusetts Institute of Technology (MIT) in 1872 and retiring six years later to consultancy work and (perhaps not quite) historical obscurity. But, as we shall see (Chapter 12), some of the ideas that he was trying to develop are now coming into their own in modern geochemistry. Hunt published prolifically and was a man with a considerable knowledge of the history of his discipline, recognized by such honours as an FRS and an LL D from Cambridge. But many geologists, such as the influential Archibald Geikie, had a low opinion of Hunt's work, and had no wish to see a resurgence of Neptunism, which it seemed to signal.

Although chemistry progressed considerably in the later part of the nineteenth century, it was premature of Hunt to think that he could tell a plausible history of the globe, from its earliest times, according to the reactions that necessarily occurred by virtue of the chemical properties of the elements and mineral substances. Nevertheless, this is what he sought to do. Intellectually, his 'genetic' history of the earth involved a bolder programme than that of the fieldworking map-makers; but in its deployment of chemistry it had to rely on less secure foundations than the physics of Hunt's contemporaries, such as G.H. Darwin or Osmond Fisher, or twentieth-century geophysicists, such as Jeffreys or Gutenberg (see Chapter 10). Brock (1979) has called Hunt's work chemical geology as opposed to geochemistry.

Working in Canada on the ancient Laurentian rocks of the Canadian Shield, one of the facts that Hunt had to try to explain was the presence of foliated (metamorphic) rocks intercalated with sediments such as limestone. There was no possibility of sorting out the stratigraphy of those ancient rocks by the 'Smithian' use of fossils, for there were no fossils to do the job. So Hunt (1858, 1867) proceeded to tell what might be called a chemical 'just-so story'. He imagined a great mass of chemicals in the primaeval earth interacting with one another according to their chemical affinities under the conditions of intense heat. An atmosphere of gases such as steam, carbon dioxide, hydrogen chloride, sulphur dioxide, oxides of nitrogen and (probably) oxygen would be liberated, whereas the cooling crust would be made of slaggy silicates. The great density of the atmosphere would lead to condensation of acidic seas in hollows on the thin crust, and interactions between the hydrochloric and sulphuric acids in the seas with the crustal silicates would lead to the formation of chlorides and sulphates in the seas and the precipitation of silica as 'quartz rock' (often found in Archaean or Precambrian deposits). The atmosphere would have consisted chiefly of water vapour and carbon dioxide. With the exposure of parts of the primitive crust, erosion and chemical reaction would have occurred, with conversion of silicates into aluminosilicates or clays. Limestone and salt deposits would be formed as chemical precipitates. Hunt claimed that his theory was a reasonable compromise between Neptunist and Plutonist geology. He did not think that granite was the primitive rock or substratum for everything, as Werner had envisaged.

On the grand question of metamorphism, Hunt (1858, 1859) first offered a chemical account of the formation of such materials by precipitations, and then, after work in the Appalachians, offered the view that 'crystalline schists had resulted from the consolidation of previously formed sediments, partly chemical and partly mechanical in their origin' (Hunt, 1871: 500), an opinion that would have been entirely congenial to Werner, but was more probably derived from De la Beche (1834). This view was anathema to most English-language geologists of the 1870s. But Hunt emphasized that great limestone formations could be found apparently 'interstratified' with gneisses, quartzites and conglomerates in the Laurentian of Canada, along with great deposits of iron oxides, metallic sulphides and graphite. It seemed to him that such deposits could only be understood if there were primitive life forms present in what was then customarily called the 'Azoic Period'. So, as far back as 1858, Hunt made the prediction that evidence for life might be found in these very ancient rocks.[28]

The suggestion that the Archaean (or Azoic or Precambrian) rocks might show indications of life was interesting and exciting and stimulated considerable research. And, quickly (1858), what seemed to be a good candidate was found in the Grand Calumet limestone near Ottawa by one of the Canadian Survey's collectors. The Director, William Logan, accepted the material as genuinely organic and it was displayed by him at the meeting of the American Association for the Advancement of Science in 1859 and in Britain in 1862. Further material was found by Logan in the Grenville limestone near Ottawa, and the McGill geologist J.W. Dawson (1820–99) was also persuaded that vestiges of Archaean life had been found. The claimed organism was dubbed *Eozoön canadense*.

The announcement of *Eozoön* by the Canadian Survey set off a heated controversy among geologists for the next few years (O'Brien, 1970). The Canadians were supported in Germany, particularly by von Gümbel, Director of the Bavarian Survey and himself a neo-Wernerian. He thought he could see evidence for *Eozoön* in limestones associated with the old Hercynian gneisses of Bavaria (von Gümbel, 1866a, b, 1868). Other favourable reports came from Bohemia (Hoffmann, 1869). But eventually the claims collapsed with ignominy when what appeared to be *Eozoön* was discovered in igneous rocks. The *Eozoön* thesis was subjected to savage criticism by two Irish authors, King and Rowney (1881).

This collapse was, I suggest, an important contributory factor associated with the lack of success of Hunt's later career (although he also had some personal differences within the Canadian Survey). And with the downfall of *Eozoön* went any sort of support for the neo-Wernerian programme such as was pursued by Hunt and von Gümbel. Rather, as mentioned below, theories of the origin of schists and gneisses by the action of heat and pressure subsequent to their original formation gained favour. But in three ways the neo-Neptunists of the later nineteenth century had important things to say. First, they recognized that life almost certainly played a part in the shaping of the early earth (a topic for further discussion in Chapter 12). Second, they apprehended that metamorphic rocks were not all the products of heat and pressure. (So, in a sense, we may see the nineteenth-century neo-Neptunists as forerunners of the 'migmatists', to be discussed later in this chapter.) Third, they sought to tell interesting 'just-so stories' about the chemical reactions that might have taken place in the earliest periods of the earth's history. In this, their efforts were surely premature. But their work has been followed in the twentieth century, and attempts to elucidate the chemical reactions that may have occurred very early in the earth's history are now regarded as an important part of the work of geologists (or geochemists).

In the nineteenth century, difficulties were always great – too great – for the successful development of chemical accounts of earth history such as that offered by Hunt, and it is hardly surprising that most geologists thought it prudent to concern themselves with field mapping rather than such speculations. The study of rocks by the experimental methods used by workers such as Daubrée was expensive, time-consuming and sometimes dangerous, because of the high pressures and temperatures involved. However, the help of the polarizing microscope proved most effective with petrography and aided map work and thinking about the earth.

Work with the microscope was initiated by the Scottish naturalist William Nicol (1768?-1851), who devised a technique for examining slithers of fossil wood in thin section. He first ground a flat surface, then cemented a slice to a microscope slide with Canada balsam, and then ground down the exposed surface of the slice to a thin film – so thin that it was transparent to light. The specimen could also be protected by a cover slip of thin glass cemented to the top of the specimen.

Not only this, Nicol (1829) invented an invaluable method for producing polarized light, using a crystal of calcite.[29] With this aid, he was able to make valuable contributions to the study of fossil botany, comparing modern and fossilized plant specimens. But he did not extend the technique to the study of rocks and minerals in general. This was done in the 1850s by the Yorkshireman, Henry Clifton Sorby (1826–1908), although before him Alexander Bryson of Edinburgh took over Nicol's collections and instruments and made some thin sections of rocks and minerals himself.

Sorby visited Edinburgh, inspected Bryson's collection and was stimulated to begin his own studies in this field. At first (1851) Sorby worked with fossiliferous sedimentary rocks, but then ([1856] 1857) he turned his attention to metamorphic rocks. He observed water-containing cavities in the quartziferous parts of mica schists which suggested that the metamorphic changes, whatever they were, involved some kind of aqueous process. This could, of course, give comfort to geologists with Neptunist leanings, and was supported by the experimental work of the likes of Daubrée.

But other workers found evidence for the role of pressure and/or heat without the action of water, superheated or otherwise. The American geologist Henry D. Rogers (1858, vol. 2: 911–16) thought cleavage and foliation arose from the action of heat passing in waves through the earth's crust, arising from pulsations in the fluid supposedly underlying the crust. Karl Lossen (1869: 282) of the Prussian Survey wrote of 'dislocation metamorphism',[30] thinking of sediments being converted into schists by the action of pressure. Then Johannes Georg Lehmann (1851–1925), lecturer in mineralogy and geology at Bonn and earlier a member of the Geological Survey of Saxony, published a detailed account of an important metamorphic complex near Chemnitz in Saxony. Using the petrographic microscope, Lehmann saw evidence of the crystals of a rock called a granulite having been deformed by pressure and also recrystallizing, as if the material had yielded under pressure but without necessarily having been heated or acted upon by water. He could also see a groundmass of bits and pieces of broken or distorted crystals in which fragments of what appeared to be the original rock were embedded (Lehmann, 1884: 248). But some of the crystals were mica and the whole was foliated. Thus it appeared that the rock had been crushed and broken, with some recrystallization, but the schistosity had nothing to do with the original stratification or sedimentation. This was 'dynamic metamorphism'.

Lehmann's ideas were quickly taken up in Britain by Charles Lapworth (1842–1920), of Mason College, Birmingham, who interpreted the great areas of schists of northern Scotland as having been formed by lateral forces and 'thrust faulting',[31] the schists supposedly having been formed by pressure and movement at the time of their emplacement, rather than having been somehow deposited in layers like a sedimentary rock, or metamorphosed after they had been deposited as sedimentary rock (Lapworth, 1883–4). Ideas like Lapworth's were also adopted by the official Geological Survey, and they provided a key to understanding the complex geology of the north-west Highlands of Scotland (Oldroyd, 1990a).

But what, then, of granites? These may often be found in great masses, as in the Andes, in Yosemite National Park in California, on Dartmoor, England, underlying the northern part of the Lake District in Cumbria or forming the greater part of the north of the Isle of Arran, near Glasgow. A mass that is closer to my present home is near Bathurst, to the west of Sydney, Australia. Such masses present no problem to straightforward Huttonian theory. According to that theory, they are to be construed as being produced by huge quantities of magma that have been forced upwards into the crust of the earth. Cooling there, they have formed large, coarsely crystalline masses, called batholiths.[32] The bottoms of such intrusions cannot usually be discovered in the field, but their tops can be seen when the overlying rocks are stripped away by erosion, and the general shape of batholiths can be discovered these days with the assistance of geophysical methods.[33]

But such a theory runs into serious difficulties when one begins to think of the physics of the earth as a cooling, rotating body. The Cambridge mathematician/physicist/geologist, William Hopkins (1793–1866), for example, sought to show mathematically, by consideration of the physics of cooling bodies, the shape of the earth and the rate of precession of the equinoxes, that the earth's crust was much thicker than was generally supposed by geologists (Hopkins, 1838, 1839, 1840, 1842). He had a serious point. If – as Huttonian theory suggested, and as Cordier had apparently confirmed – the earth had a massive liquid interior and a relatively thin exterior crust, it would be subject to immense tidal forces from the sun and moon. In consequence, a Huttonian earth, as generally understood, would be inherently unstable. So Hopkins thought it unlikely that there was a direct connection between volcanoes and a massive fluid interior of the earth, as classic Huttonian theory supposed. In his view, it was unreasonable to extrapolate, as Cordier had done, from the observed temperature gradient in the earth's outer crust to the conclusion that the interior must be liquid. This extrapolation did not allow sufficiently for the role of increased pressure.

Hopkins proposed, therefore, that the large crystalline masses of (what we call) batholiths did not represent intrusions so much as solidified lava lakes, which remained as residues from the original molten mass of the planet (Hopkins, 1840, 1842).[34] However, in his *Face of the Earth* (1904–24), Suess pointed out that his observations in the *Erzgebirge* in 1893 suggested that the granites there cut through the strike and folds of the mountains, which suggested that they had been emplaced after the formation of the country rock. Suess (1909: 552) also saw evidence that suggested 'melting and absorbing of the adjacent rock'. This was Huttonian theory again.

The issue is in fact highly complex and takes us to the heart of petrological theory. The question of the origin of granite led to a major controversy among geologists in the later years of the nineteenth century, the first half of the twentieth century and, in some respects, through to the present. First, there is the fact that granites often shade off into gneisses, or some of them seem to show indications of bedding and/or foliation. Also, some exposures suggest that granitic matter has somehow suffused into pre-existing metamorphic rocks, by quiet diffusion rather than forcible intrusion. Thus the layerings may represent replacement structures. On simple Huttonian theory there would be no visible layering.[35]

A second major difficulty lay in the fact that nineteenth-century geologists found themselves unable to synthesize granite successfully from its constituents, either by direct fusion or with the help of the presence of water (Fouqué and Michel-Lévy, 1882: 19–20), whereas there was fair success for some other rocks, such as basalt.

The third great difficulty may be called the 'space problem'. Sometimes, this may not seem to be a worry. For example, mapping of the Isle of Arran shows that the schists which largely surround the granitic mass in the north of the island have been arched up into a great dome before being stripped away from the higher ground by erosion. But in other areas (such as that near Bathurst, New South Wales) there is no equivalent arching. So in such cases one would doubt, on geometric considerations alone, that molten magma simply intruded from below and pushed up the surrounding rocks. This suggests, then, that the granitizing matter, whatever it may have been, somehow converted the country rock into granite. According to this view, granite could in fact be a kind of metamorphic rock, and the process of its formation might be called 'granitization' (Sederholm, 1923: 5).[36]

I mentioned above the idea of metamorphism arising from pressure. On a grand scale, if new rocks are produced by the action of both heat and pressure, the process is commonly called 'regional metamorphism' – a term introduced by Daubrée (1860: 155). Nineteenth-century geologists found it difficult to agree about the origin(s) of such metamorphic rocks. Several French geologists, notably Auguste Michel-Lévy (1844–1911), a member of the Geological Survey of France and disciple of Elie de Beaumont, argued that the schistosity or gneissose structures sometimes seen at the margins of granites arose from chemical interaction of the magma and the country rock, in the presence of hot water under pressure (Michel-Lévy, 1893–4: 20). This position was also adopted by other French geologists, such as Alfred Lacroix (1863–1948) and, to some degree, Pierre Termier (1859–1930), Professor at the Ecole des Mines in Paris. However, rather like Scrope earlier in the century, German geologists, such as Heinrich (Harry) Rosenbusch (1836–1914), argued that the metamorphic structures (foliations) were due to movements at the margins of the intruding, cooling magma (dynamometamorphism) (Rosenbusch, 1896, vol. 2: 85).

Termier's ideas were more 'extreme' than those of Michel-Lévy. Repudiating the concept of dynamometamorphism to account for the general metamorphism of areas such as the Alps, Termier (1904: 585) envisaged the process of metamorphism as being analogous to the spreading of an oil spot – his so-called *tâche d'huile* mechanism. He thought of hot, fluid magma percolating into and transforming the country rock until the whole was metamorphosed – perhaps into granite itself, which might then look as if it had been formed simply by the slow cooling at depth of liquid magma, even if its origin was quite different. With this view, there was (arguably) not the space problem that faced classical Huttonian theory.

Such ideas received significant support from the north in the work of Scandinavian geologists such as Jakob Johannes Sederholm (1863–1934), of the Geological Survey of Finland. Here, it may be noted, local observations seem to have outweighed initial training, for although Sederholm qualified in Germany under Rosenbusch he did not adopt his teacher's views on metamorphism.[37] At places along the coast of the Gulf of Finland[38] and in other parts of the 'Baltic shield', amazing rock surfaces may be seen, scraped clean by former glaciers and etched by the action of the sea. The rocks look as if they have been folded, squeezed and pummelled – as if handled by a cook making flaky pastry. Yet one can also say that the rock looks like a schist that has been impregnated by granitic matter. But, were this the case, the magma must have been exceedingly fluid. Indeed, the rocks look as if granitic matter got into place like water soaking into the pages of a book. The structure of the rock may also be made more complex by folding, as in the example in Figure 9.2.[39]

Sederholm studied such rocks assiduously over many years in the course of his investigations of the Precambrian rocks of Scandinavia. He called them 'migmatites', from the Greek word *migma* meaning 'mixture' (Sederholm, 1907, 1923, 1926, 1934), and also introduced the term 'ptygmatic folding', meaning 'flow folding' (Sederholm, 1907: 110). The name 'migmatite' appropriately suggests the heterogeneous character of such rocks. But, whereas in his early paper of 1907 Sederholm was inclined to view the process of formation of migmatites in a somewhat Huttonian, cyclic, fashion, with the changes in the country rock brought about by the infiltration of granitic magma, in his later work he spoke of a percolating granitic 'juice' or 'ichor'[40] which showed 'gradations between an aqueous solution and a very diluted magma, eventually also a magma containing much water in a gaseous state' (Sederholm, 1926: 89).

Now, since many of the migmatites approximate granite in character and composition, this is not far from seeing granite as a kind of metamorphic rock – in which case we have moved a long way from the original Huttonian conception of granite as the product of the slow cooling of a large mass of fluid

Figure 9.2 Migmatite from the island of Söderskär, 30 km south-east of Helsinki (by permission of the Geological Survey of Finland)

Figure 9.3 Examples of changes by migmatization

magma. Such a change of theory seemed to be necessitated by the rocks of the Baltic shield, where every conceivable transition from schist to gneiss to granite is discoverable. In many cases ghost-like structures of the original schists can be seen, even though the rocks are partly converted to granite.

As modern textbooks (e.g. Holmes, 1965: 182) point out, by examining rocks in different stages of transition one can envisage a whole sequence of changes by migmatization, and it is remarkable that the end-product may be granite in each case, even though the starting materials are quite different. Examples are given in Figure 9.3.[41] Such sequential changes are comprehensible if one thinks of the agent of change not simply as molten magma, but magma associated with water containing material in solution or water in gaseous form. The aqueous matter could have been derived from the sediments when originally buried, and driven off with other more volatile and mobile materials by the great heat and pressure deep in a geosyncline. The seeming percolation or infusion of liquid, like water seeping into the pages of a book, then becomes intelligible. It produces a form of metasomatism.[42]

Sederholm's essential idea was that some kind of material was injected during the process of migmatization. But this was challenged by his fellow Scandinavian the Swedish geologist P.J. Holmquist (1916), who thought that the banded structures might be ascribed to chemical metamorphic differentiation – 'exudation from within', so to speak, arising under conditions of extreme temperature and pressure, where part, but not all, of the rock has melted and the whole is deformed. Although such ideas may seem less plausible than those entertained by Sederholm, it must be recalled that the processes were presumably taking place at depth under great pressure and at high temperatures, where one might not expect to encounter aqueous mineralizing vapours.

Much subsequent work has been done on migmatites. The Swiss geologist C.E. Wegmann (1935) liked the idea of migmatization so much that he extended it to the formation of essentially homogeneous plutonic rocks: they were migmatites because they were formed from the combination, sometimes by melting as well as by the action of solutions, of 'local' and 'imported' rock. This was all very well, but it left unanswered the question of whether some granites, at least, were produced simply by the cooling of large masses of magma. The work on migmatites suggested that granite could be produced by alterations of pre-existing solid rock. But is this true of all granites? Maybe there are granites and granites? Might there still be a place for the old Huttonian idea of a great mass of fluid magma forcing itself up from the depths and then cooling so as to yield granitic batholiths?

The question of whether magma or migma should be granted precedence in the formation of granites and gneisses became one of the most controversial subjects in geology in the mid-twentieth century. The new controversialists willingly adopted the sobriquets of magmatists and migmatists. Among the magmatists, the leader was perhaps Norman Bowen, some of whose ideas will be outlined below. The leading English-language migmatist was Harold Herbert Read (1889–1970), Professor at Imperial College, London.[43] The controversy rumbled on among petrologists for thirty years or more. It was, in a sense, a replay of the old debate between the Neptunists and Vulcanists of the nineteenth century. Certainly, the disputants themselves liked to see the contest in that light, as Figure 9.4, which formed the frontispiece of Read's *Granite Controversy* (1957), clearly shows.

But, to appreciate the debates, we must once again step back into the nineteenth century. We have left the question of the cooling and crystallization of magma, in a Huttonian style, in a fairly unproblematic state. In fact, the process was anything but simple or uncontroversial, even without invoking the idea of migmatization. For there are many different kinds of coarse-grained igneous rocks, all the way from basic gabbros to acidic granites. Why should this be so? Was there perhaps one kind of magma, which somehow self-differentiated as it cooled and crystallized? Or could there be two or more kinds of subterranean magma, which gave rise to different igneous masses by mixing in various proportions?

Darwin (1844: 117–24) had the idea that a single kind of magma might self-differentiate under gravity, for some crystals would presumably form first from the melt and would fall under their own weight. The residual liquid might drain off into other fissures or chambers and would therefore be of different

Figure 9.4 Cartoon from Harold Herbert Read's *Granite Controversy*, comparing the magmatist–migmatist debate with the Vulcanist–Neptunist controversy (Read, 1957, frontispiece)

composition from the first crop of crystals. Dana (1849: 372), thinking about his observations of volcanic rocks on Pacific islands during his work with the United States Exploring Expedition (1838–42), supposed that a differentiation of magma might begin before crystallization – a process known as 'liquation'.

The idea that there were fundamentally two kinds of subterranean magma was advanced by the chemist Robert Bunsen (1811–99) (of Bunsen burner fame) in 1851 and 1853. He visited Iceland in 1845 and collected many rock samples, taking them back to Germany for analysis. It was found that there were two main types, which he called 'trachytic' and 'pyroxenic': silica-rich and silica-poor. As a chemist, Bunsen was not, perhaps, too concerned with the geological problems involved in the interpretation of his analytical results. He simply assumed that there were two large magma chambers under Iceland and that the various types of rocks he discovered were produced by different mixtures of the two kinds of magma.

Another idea was propounded by W. Sartorius von Waltershausen (1809–76) (Sartorius von Waltershausen, 1853), who worked on volcanic rocks from Sicily as well as Iceland. He supposed that the subterranean magmas were disposed in zones of increasing density and that eruptive rocks came from the different zones at different geological epochs, the older, more siliceous and lighter ones being drawn from nearer the surface. They were erupted when the crust was not so thick as it later became through cooling of the globe, when more basic rock crystallized. This theory was compatible with the long-held Neptunist belief that composition was related to geological age. However, in this case, the theory was Vulcanist in character, not Neptunist.

So in the middle years of the nineteenth century, there were really three contending Vulcanist theories: that of local differentiation, such as was envisaged by Darwin and Dana; the 'synthetic' theory of Bunsen; and the theory of a primitively or primevally differentiated magma, as suggested by Sartorius von Waltershausen. An attempted synthesis of the theories was offered by Joseph Durocher (1817–60) (Durocher, 1857a, b), with the suggestion that there might be two magmas, but arranged in zones, as in Sartorius von Walterhausen's theory. However, the two zones supposedly both remained fluid, so that there was no specific relationship between rock type and age. Durocher supported the idea of liquation within magma, by analogy with what may be observed in the cooling of metallic alloys. He also had the idea that metalliferous lodes might be formed as a result of sublimation of two kinds: *émanations motrices* (metallic) and *émanations fixatrices* (sulphurous). This was almost taking ideas about the earth back to Aristotelian theory (see p. 17).

In an influential publication, Ferdinand von Richthofen (1868) concentrated attention on volcanic eruptives rather than plutonic rocks. He proposed, on the basis of exposures observed in the Hungarian Carpathians and the Sierra Nevada range of America, that the order of extrusion for Tertiary rocks was general and dependent on the prior vertical order of the magmas, namely propylite,[44] andesite, trachyte,[45] rhyolite, basalt. The idea was that in earlier periods the eruptive rocks must have been chiefly siliceous, but by the Tertiary the siliceous magma was largely depleted, so the vast outpourings of Tertiary volcanics (forming, for example, the famous Giant's Causeway in Ireland or similar rocks in the Czech Republic) were chiefly basaltic. The rhyolite was subordinate in quantity and produced (somehow) by secondary processes.

As the nineteenth century proceeded, the number of different igneous rocks recognized, partly by chemical analysis and partly by their examination with the microscope in thin section, grew at an alarming rate, so that well before the end of the century it seemed impossible to sustain the idea that there was a special kind of magma for rocks of each kind of chemical composition; and it seemed even less plausible that there was a particular kind of igneous rock peculiar to each geological epoch. Pursuing the chemical approach, Harry Rosenbusch (1890) supposed that there was a differentiation (*Spaltung*) of magma, as reactions occurred therein according to the chemical affinities of the elements present in the melts. The details need not concern us here, but it may be mentioned that Rosenbusch thought that five basic types of magma could be produced by chemical actions in the magma: 'foyaite',[46] granite, granodiorite, gabbro and peridotite. These fundamental types had five fundamental types of chemical molecules or 'kerns', which, however, could supposedly dissolve different oxides or metals to give different types of eruptive rocks. A primary differentiation was thought to give the different magmas at depth while a secondary differentiation gave rise to different rocks at the surface.

Yet another view was expressed by Waldemar Brøgger (1851–1940), Professor of Geology and Mineralogy at the University of Stockholm. In a massive memoir of 663 pages, Brøgger (1890[47]) gave an account of the eruptive rocks of the Oslo region (which had many rocks unique to the area), which had the appearance of forming a continuous merging series. There had, therefore, seemingly been a common magma chamber in the Oslo area and not a universal layer of magma within the earth's interior. To try to account for the observations, Brøgger invoked what has come to be called 'Soret's principle'.[48] Suppose two parts of a salt solution are held at different temperatures. By osmosis, solvent will pass from the cooler more dilute part to the hotter more concentrated part. If, then, the whole is frozen, the concentration of the salt will be found to be increased in the part of the system that was previously cooler. Brøgger envisaged that something like this had been occurring among the complex suite of rocks that he had examined in Norway. There would naturally be temperature differences in the different parts of the magma chamber, with cooler regions towards the margins, and so Soret's principle might apply.

It will be seen, then, that, although the general Vulcanist or Plutonist principles were widely accepted by the end of the nineteenth century, there was great difficulty in finding an agreed set of principles to account for what was going on in the earth's interior during the production of igneous rocks, either intrusive or extrusive.

Besides the chemical and physical problems, there was also a fundamental geometrical problem: how could there be space for the emplacement of magma? As mentioned above, not all plutons (major igneous bodies formed at depth) seemed to show a doming of the overlying strata, such as might be expected according to old-style Huttonian theory. This question was given particular attention by the Canadian geologist Reginald A. Daly (1871–1957), of Harvard University.

In his well-known text, *Igneous Rocks and their Origin*, Daly (1914: 194) recalled how in 1893, when a student at Harvard, he was allocated an area to map in Vermont. It contained a series of igneous intrusions, and Daly found himself hard put to explain his field observations. Eventually, in 1902, he came up with the idea that parts of the earth's crust could collapse downwards and be assimilated into the molten material beneath, allowing the upwelling of magma. He called this process 'magmatic stoping'.[49] This model dealt with the 'space problem' reasonably effectively (Daly, 1903).

Daly also envisaged that there might be some separation of the different liquid constituents of a magma according to their different densities ('liquation'). And there could be further differentiation by the process of fractional crystallization: some crystals might form earlier than others and fall under gravity, thus altering the composition of the upper parts of the melt. This was essentially the idea previously propounded by Darwin (1844).[50] Daly's suggestion was that basic (mafic) crystals tend to form first and sink in the magma chamber, so that the upper part of a batholith or stock becomes relatively acidic in character, i.e. granitic. The general idea is shown in Figure 9.5.[51] This gave an explanation of Rosenbusch's (1882) empirical generalization that the order of crystallization for igneous rocks was: (1) iron ores and accessory minerals such as zircon and sphene; (2) mafic silicates (olivines, pyroxenes, amphiboles, micas, etc.); (3) feldspars; (4) silica. Thus Daly supposed the 'fundamental' magma to be basic or basaltic in character, and granite is produced in a batholith from the loss downwards under gravity of the basic minerals such as augite and olivine, which crystallized first. However, for small intrusions, such as dykes, and for many relatively thin lava flows, the differentiation process cannot occur, and in consequence such structures are usually made of basic materials, yielding basalts.

The idea that magmas may separate into different components during the process of cooling is, in principle, susceptible to investigation in the laboratory, as well as in the field, and should follow the fundamental principles of physics and chemistry. Many investigators worked on this kind of problem in the late nineteenth and early twentieth centuries, and such work continues to the present. With the establishment of the Geophysical Laboratory at the Carnegie Institution,

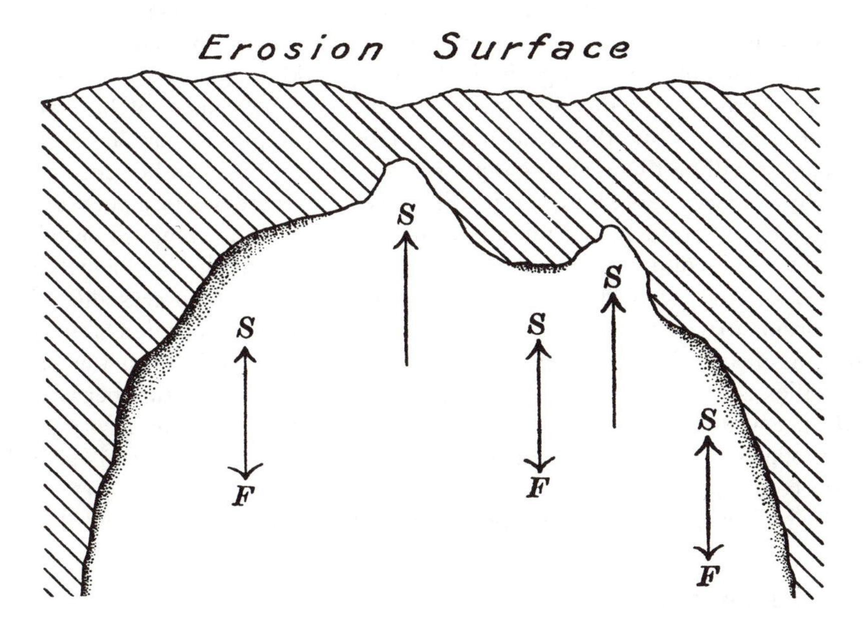

Figure 9.5 Differentiation of material in a batholith, according to Reginald Daly (1914: 244) (by permission of the Geological Society). S = 'salic'; F = 'femic'.

Washington, in 1906, where apparatus was devised for the examination of rock melts under conditions of very high temperature and pressure, real progress began to be made. As we have seen, the methods of chemical analysis, examination of rocks in thin section and field mapping did not really solve the problems of igneous and metamorphic petrology in the nineteenth century. All one got was a plethora of conflicting theories. The new 'high-tech' methods developed in Washington made significant progress possible. Theorizing continued, but could be controlled by experimental testing.

In the early years at Carnegie, the dominant figure was the Canadian/American, Norman L. Bowen (1887–1956), who over many years examined on the laboratory scale the processes that might occur in the cooling and crystallization of magmas. Various mixtures of minerals were melted together, and the products formed as they cooled and crystallized were carefully examined. One technique used was to quench a mixture rapidly and then identify its contents by examining under a microscope the crystals formed.[52] In this way, one could get an idea of the proportions of substances present in a melt before crystallization or after crystallization had been wholly or partly completed under cooling.

It is a commonplace of physical chemistry that an 'impurity' dissolved in a liquid lowers its melting-point. If one considers a two-component liquid (of constituents, say, *A* and *B*) over a range of composition, at one end of the range one may think of *A* as an impurity in *B* and at the other end one can think of *B* as an impurity in *A*. Thus the melting-points of each substance will be lowered by the presence of the other, and the more of the two 'impurities' that are present, the lower will be the melting-points. This state of affairs can be represented graphically (see Figure 9.6), and it will be seen that there will be a point on the graph (*E*) where there is a minimum melting-point. The composition of the mixture at this minimum is called the 'eutectic composition'. So, if a melt of *A* and *B* is slowly cooled, one would expect either *A* or *B* to crystallize first (according to which substance is in excess initially). The composition of the melt thus gradually changes until the eutectic point is reached. Then the liquid solidifies as a whole, as a eutectic mixture. This particular mixture behaves, in effect, as if it is a single substance. The ideas described here can, in principle, be applied to mixtures with three or more components, giving, of course, much more complicated diagrams.

Now, Bowen, in a series of papers (but most importantly in one published in

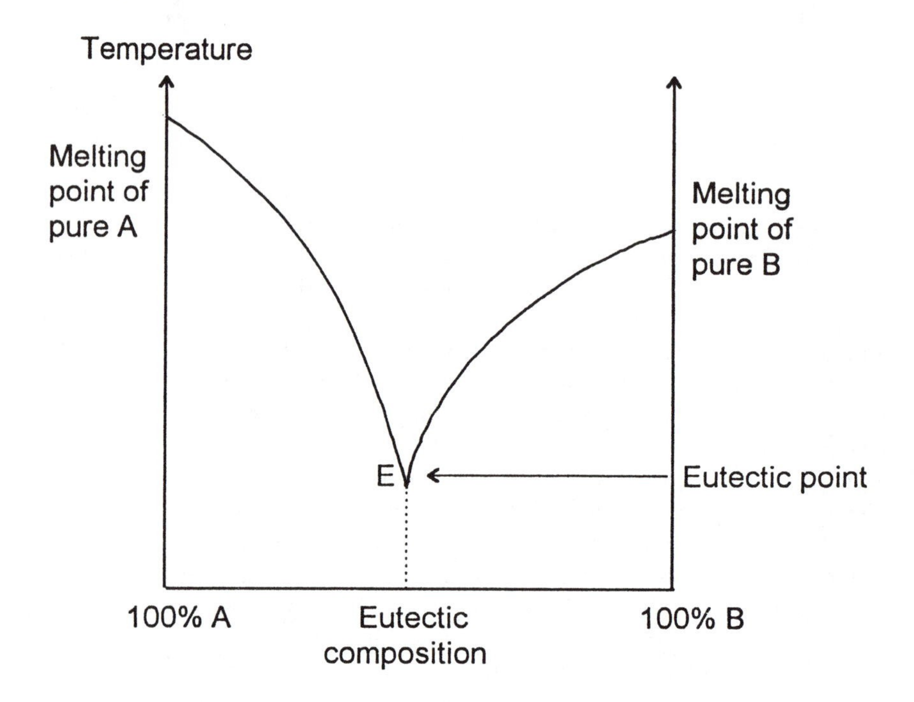

Figure 9.6 Variation of melting-points of a two-component mixture, illustrating the idea of eutectic composition

1922) and in his *Evolution of the Igneous Rocks* ([1928] 1956), developed a substantially different model from that of fractional crystallization, such as that deployed by Daly. He also rejected the idea that Soret's principle could provide an adequate explanation of the observable features of igneous rock masses. In a simple eutectic system such as that shown in Figure 9.6, once a crystal has separated from a melt, that would be it, so to speak. But Bowen envisaged that there could be reactions between the crystals formed early in the process of crystallization and the material of the still-liquid melt. For example, there could be ongoing reaction in a cooling feldspar melt, so that there might be a continuous series in the composition of the crystals of feldspars that are deposited, the materials forming a kind of 'solid solution'.[53]

Bowen also envisaged a situation such that a magma of basaltic composition (for example) might deposit crystals, which would then interact with the remaining melt; but the overall effect (allowing for the reactions between the first-formed crystals and the remaining melt) would be the deposition of crystals from a magma in a regular order. This has come to be known as 'Bowen's reaction series'. It had the backing of the experimental results generated in the Carnegie Laboratory, for melts could be quenched at different stages in their cooling and the results analysed. Thus there was a valuable experimental supplement provided to the thinking of geologists about what goes on in the processes of crystallization of magmas in the earth's crust. (It left aside, however, what was perhaps a more fundamental question, namely the source of the heat and hence the origin of the magma in the earth's crust.)

Besides the laboratory work and theoretical considerations, Bowen's argument about a 'reaction series' found support from field evidence, for example that in the earlier publication by the British petrologist, Alfred Harker (1909: 131). This showed both the appearance and the disappearance at different horizons of certain minerals at a site called Garabal Hill near Loch Lomond. They appeared in a certain order, as might be expected according to the principles of eutectic mixtures; but they also disappeared in a certain order, which was not to be expected according to eutectic principles. As Bowen ([1928] 1956: 59) pointed out, in a eutectic system no mineral ever disappears. The first-formed deposited mineral is joined by another and another and so on, until all the remaining materials appear together in a final eutectic product. But what Bowen found in practice (i.e. according to Harker's field evidence) was that there was a regular sequence of minerals produced successively in crystallization. Thus we have his famous 'reaction series' (Figure 9.7).

How, then, was Bowen's theory applied to the problem of granitic batholiths? His ideas were not dissimilar to those of Daly. Like Daly, he believed that in the first instance there is essentially one kind of magma: the basaltic or dark basic kind. If there were a large mass of magma of this kind cooling under the crust, the dark, basic ferromagnesian minerals would form first and would fall

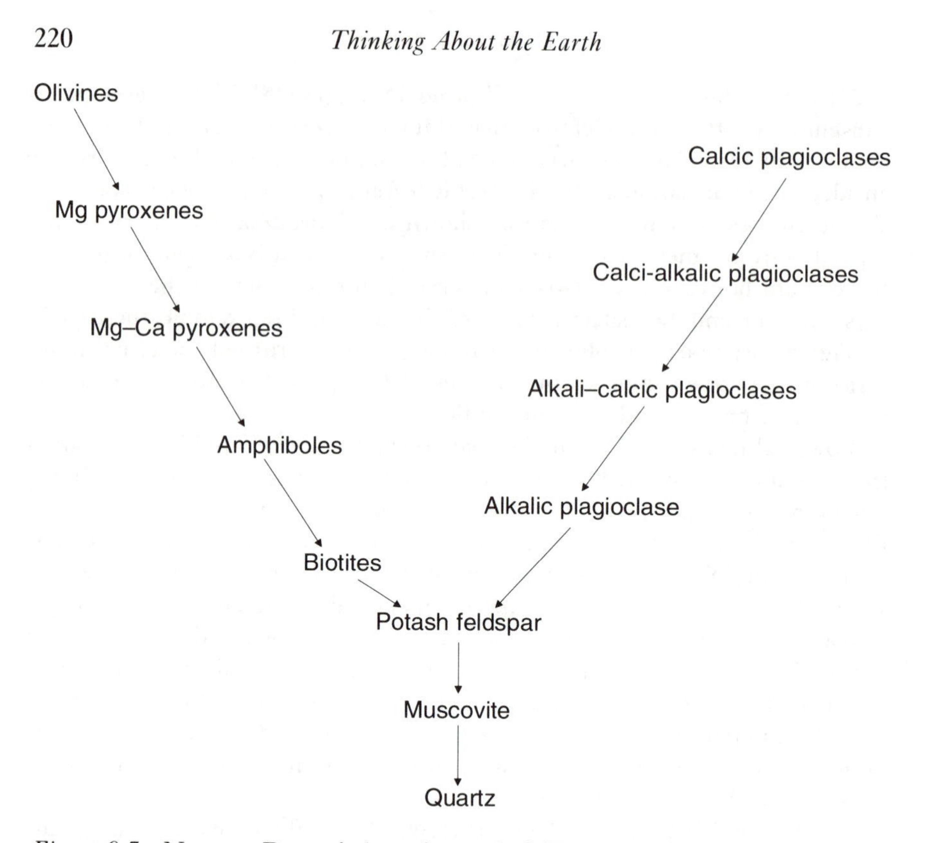

Figure 9.7 Norman Bowen's 'reaction series' (Bowen, 1922: 190)

out of sight (so to speak) by gravity – although they would react with the melt on the way, yielding the overall 'reaction series' referred to above. Thus, as in Daly's theory, a granitic batholith would represent the upper, light, acidic, residual material formed on cooling of a huge mass of magma. The dark, basic material would be out of sight and out of mind, deep down in the depths of the earth, having been first formed and having fallen away under gravity (Bowen, 1948: 87–8). As for the 'space problem', which rightly puzzled Read and the migmatists, Bowen simply argued that their solution to the problem was no better than his: material pushed up chemically occupied just as much room as material pushed up mechanically.

So the 'granite controversy' rumbled on for many years, and was one of the most important points of contention among geologists in the years before the advent of plate tectonics (see Chapter 10). It took in issues such as 'stoping'[54] and, of course, the long-standing problem of the origin of metamorphic rocks such as gneiss. There was much wordy debate about the interpretation of field evidence, as well as abstract theory. Both sides of the dispute made contributions that still stand. Bowen's notion of a 'reaction series' is still part of orthodox petrological theory; and the idea of migmatization, to account for phenomena such as those which Sederholm investigated in the Baltic region, is also accepted as appropriate to account for the formation of some rocks at least.

On the other hand, geologists today generally prefer the magmatists' story about the origin of batholithic granites. Here, they have found the work of Bowen and O.F. Tuttle of the Pennsylvania State University especially compelling (Tuttle and Bowen, 1958). These investigators carried out experiments on artificial melts with different proportions of water present and on naturally occurring substances, including granites; and they were able to obtain correspondence between the two. That is, it appeared that there was a minimum melting-point for the artificial systems containing only a small proportion of water, which corresponded with the compositions of the average granite or rhyolite. They concluded that granites could be formed from the cooling of silicate melts in the presence of water. Or, putting the matter another way for the present purposes, Tuttle and Bowen succeeded in doing what nineteenth-century investigators had failed to achieve: the production of 'artificial granite' from a cooling liquid. In their opinion, granites not formed at magmatic temperatures would be rare. They supposed that the melting occurred as sediments were carried down and buried in the depths of geosynclines.

However, it should be observed that Tuttle and Bowen (1958) maintained that the vapour that might be in equilibrium with a hydrous granitic liquid could actually carry away silica and feldspathic material from the liquid part of the system and transfer it to the cooler parts of the pressure vessel, sometimes leaving behind, in more concentrated form, the oxides of calcium, magnesium, iron and phosphorus. In one experiment, the vapour apparently removed nearly all the feldspar and quartz, leaving a more basic residue of garnet, pyroxene and apatite. Thus it appeared that there could be a chemical/physical process leading to the formation of basic zones within the melts. This provided a solution to the question of the differentiation of magmas that had concerned investigators in the nineteenth century, who at the same time had worried whether there was one, or more than one, kind of fundamental magma.

I suggest that the terms of the migmatist/magmatist debate were in a sense set by nineteenth-century concepts: the 'up-and-down' theory of the earth's crust of Lyell, rather than the idea of lateral forces and movements of plate tectonic or mobilist theory, and the old theories of 'fire and water'. However, with the establishment of plate tectonics, these older issues became subsumed under a new, wider and grander theory, which will be examined in Chapter 11. That is, the downward movement of silica-rich sediments of relatively low density into the earth's denser interior was accounted for in the grand new theory. When melted, such sediments might then rise up to form the great granitic batholiths on cooling.

But this does not mean that the granite problem is settled. On the contrary, it remains a source of contention among present-day geologists in their thinking about the earth. One theory concerning the emplacement of granite that has attracted both attention and favour is that of 'diapirism', about which something may be said here. The concept may be thought of in the light of the remarks of the previous paragraph. Obviously, not all rocks are of equal density, and although they appear solid to us they have their own (high) viscosity and can flow under heat and pressure. Generally, granite is less dense than other rocks, and so one might conceive of a 'blob' of granitic magma slowly rising through the earth's crust, rather as a bubble may rise to the surface of a jar of honey.

Back in 1945, Frank Grout (1880–1958) of the University of Minnesota sought to model such a process, using such homely substances as air or cylinder oil as analogues of magma and soft, wet clay or maize syrup as analogues for the country rock of the earth's crust (Grout, 1945). He could see a blob of oil form a dome and then rise up like a balloon through the clay. Some examples of his experimental results are shown in Figure 9.8. The bubbles were called 'diapirs' (Greek *diapeirein* = to pierce through), and the process is known as diapirism. It is a well-established phenomenon in relation to salt deposits, in which large masses of (low-density) salt apparently rise up through the earth's crust, as expected according to the theory of diapirism.[55]

Considerable efforts have been made by geologists to map the areas of granitic

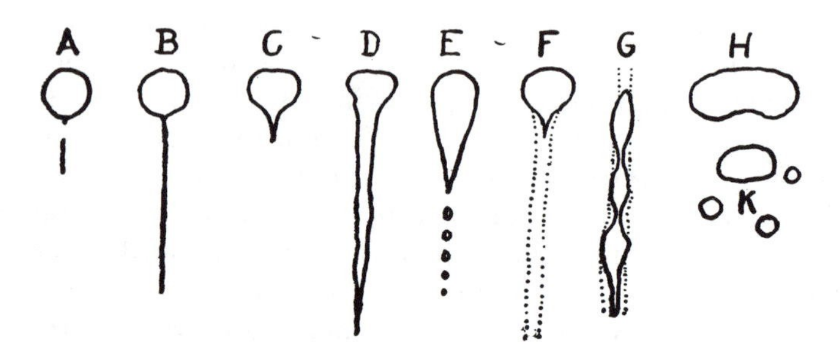

Fig. 1. Forms of fluid bodies rising in viscous matrix.

A and B. Cylinder oil in corn syrup, both at room temperature.
C. Cold oil in cold syrup.
D. Cold oil in warmer syrup.
E. Warm oil in cold syrup.
F. Water in syrup, leaving a trail of diluted syrup.
G. Water following the trail of an earlier globule like that in F.
H. Large air bubble in water or thin syrup.
K. Smaller air bubbles.

Figure 9.8 Frank Grout's experimental simulation of the emplacement of granitic magma (Grout, 1945: 263) (by permission of the Geological Society)

outcrops to see whether they agree with the shapes that might be expected according to the work of modellers such as Grout. There has been some success, but it has been limited. The diapiric concept has been modelled theoretically by Bruce Marsh (1982) of Johns Hopkins University, who considered that diapirism alone could hardly be sufficient to account for the rise of granitic magma, although he was sympathetic towards the diapiric model. To rise high enough from the depths without running out of steam (so to speak) – i.e. before cooling and becoming too viscous to move further – some easy line of passage seems to be needed and, according to Marsh's suggestion, faults would provide what is required.

Certain granites in Donegal have been a particular focus of interest in the granite controversy. They were seen by Read (Pitcher and Read, 1960) as evidence in favour of stoping. However, Donny Hutton of Trinity College, Dublin, suggested that the granite might indeed be diapiric and showed a 'shouldering aside of the country rocks' (Hutton, 1977). Thus, as so often in the history of geology, field evidence does not speak unambiguously and is constantly modulated by the theories to which the geologists subscribe and the controversies in which they are engaged. I cannot bring the story of ideas about the origin of granite to a satisfactory conclusion here: it is an ongoing saga. Geologists still have plenty to think about.

Chapter 10

Thinking with Instruments: Earthquakes, Early Seismology and the Earth's Hidden Interior

The nineteenth-century ideas concerning the earth's interior discussed in Chapter 9 were necessarily speculative by virtue of the kind of evidence available. It was certainly drawing a long bow if one sought to determine whether the earth's interior was liquid or solid on the basis of its external shape, and astronomical arguments relating to rates of rotation, precession and nutation.[1] As Bruce Bolt, (an Australian) Professor of Seismology at the University of California, has put it (Bolt, 1976: 59): 'Before seismology, most notions of the bowels of the earth were vague and fanciful'. But the situation began to change rapidly towards the end of the century and into the present century with the establishment of the science of seismology: the study of earthquakes.

While what we know today of the earth's deep interior – its bowels – is largely based on information gleaned from the records of earthquakes, the early results of seismology were insecure and open to innumerable interpretations. However, it soon became clear that the little wiggles on seismograms could provide information about important features of the earth's internal structure. It is through the seismograph and seismometer[2] that we have information that allows our thinking about the earth to be greatly refined and for profound theories about the planet to have an instrumental base, providing empirical support for the ideas suggested by the mapping of the earth's external crust.

Earthquakes were, of course, matters of great concern as far back as antiquity and, as was pointed out in Chapter 1, there were myths that may be construed as pertaining to volcanic and seismic phenomena; and there were some naturalistic theories. There are records in China of an instrument, dated AD 132, which determined the direction of seismic impulses (Needham and Wang Ling, 1959: 624–35).[3] In modern times, the great Lisbon earthquake on All Saints' Day, 1755, aroused immense interest and concern in Europe (Kendrick, 1956). The

English clergy thought it was the result of a divine intervention against the excessive Catholicism of the Portuguese capital. The Catholic survivors thought the trouble was due to the presence of Protestants in their midst, and these unfortunates were forcibly baptized into the Roman Catholic Church. In fact, the Lisbon earthquake, although it had a profound impact on European thought, was a lesser affair than earthquakes that have occurred in recent times in countries such as China. Even so, the Lisbon disaster, with its destructive 'tidal wave' and great fire, was bad enough. Portugal never really recovered from the disaster economically until quite recent years. The calamity threw doubt on the deistic assumptions of eighteenth-century philosophers, as satirized by Voltaire: the belief that we lived in the best of all possible worlds, divinely created in such a manner that God's existence was revealed by His created work, the earth. Kant, however, reckoned that earthquakes should be a warning to mankind against hubris (Reinhardt and Oldroyd, 1983).

There were limited scientific studies of earthquakes in the eighteenth century, notably by John Michell (1724–93), Woodwardian Professor of Geology at Cambridge (Michell, 1759–60). As the motive cause, he suggested the action of pent-up vapours generated when water penetrated to the earth's (presumed) subterranean fires. The effects were transmitted as waves, Michell thought, and he sought to explain the observed foldings of strata in this way. Michell also had the idea that, since the effects of an earthquake diminished outward from a focus, one could in principle draw lines backward to locate the position of the centre of the disturbance.

In the eighteenth century and well into the nineteenth century, the simple pendulum, variously modified, was the tool used to detect small seismic disturbances. For example, Andrea Bina (1751), a teacher of philosophy at several Benedictine monasteries in Italy, proposed suspending a heavy lead pendulum over fine sand and producing traces with a pointer attached to its bob. In keeping with the vogue of the time for explaining many phenomena as electrical manifestations, Bina thought that preliminary tremors were due to water conveying electrical fluid from the bowels of the earth into the air and hence disturbing the pendulum. The use of the pendulum was extended by several eighteenth-century Italian investigators (Dewey and Byerly, 1969).

Following some minor earthquakes in Scotland in 1839, James Forbes (1844) invented an instrument in which the pendulum was turned upside down, so to speak. That is, a rod transfixing a metal ball was supported vertically on a piece of stiff wire. The movements of the upper end of the rod could be recorded by a pencil marking a piece of paper, thus providing a permanent magnified record of the movement of the ball. The advantage of this arrangement, compared with an ordinary pendulum, was that it avoided the necessity of using a long pendulum to obtain long periods of swing. In Forbes's instrument, the ball remained more or less stationary, while the earth moved around it, so to speak. But friction between the pencil and paper limited the sensitivity of the device.

As the nineteenth century progressed, the systematic collection of records of seismic events was begun, particularly by K.E.A. von Hoff (1826–35), and by Alexis Perrey (1807–82), Professor of Mathematics at Dijon and Director of the municipal observatory. Perrey looked for periodic effects that might be related to the configurations of the members of the solar system at the times of earthquakes; but although he thought he could discern some law-like regularities, they were in fact illusory.

An uncommonly severe earthquake occurred in southern Italy in 1857, and the Irish engineer Robert Mallet (1810–81),[4] whose earlier career was linked to the armaments industry but who had also made some theoretical studies of earthquake phenomena (Mallet, 1848), obtained a commission and grant from the Royal Society, London, to travel to Naples in 1858 to make a detailed investigation of the effects of the quake (Mallet, 1862). It was, it seems, thought that an explosives expert might be well suited to studies of earthquake phenomena. In his report, Mallet gave a minute account of the devastation generated by the earth movements and recorded much of it photographically. He also examined the geology of the Naples region, to see if this might be related to the damage produced by the earthquake.

Mallet made a close check of the cracks in buildings and the way structures had been disturbed by the earth movements. Thereby he sought to determine the directions in which the disturbances had travelled, and by tracing these back to where they met he was able to determine approximately the focus of the earthquake. The cracks in the buildings also gave some indication of the angles at which the waves of disturbance had emerged from the ground. Again, by tracing these backwards, Mallet was able to estimate the depth of the point of origin of the disturbance. That the disturbances were due to waves seemed to be evidenced by the observation that for two adjacent pieces of similar statuary one had been twisted in a clockwise direction, the other anticlockwise. Moreover, it seemed to Mallet that the waves travelled most easily 'end on' to the strata, rather than transverse to them. The supposed subterranean cavity where the shocks originated he called the *centrum*, while the point on the earth's surface immediately above, where the shocks were presumed to arrive first and were of greatest intensity, was the *epicentrum*.

But Mallet went further. He prepared a large map (Figure 10.1) which showed what he called 'isoseismal lines' – that is, lines of equal seismic disturbance. Moreover, he extended his thinking beyond Italy to the eastern Mediterranean region to show in a general way – on the basis of historical records – those areas that were particularly subject to earthquakes and which areas were relatively safe. Indeed, he extended his thinking about earthquakes to the whole world, in a report to the British Association in 1858 (Mallet, 1859), which depicted the

Figure 10.1 Map showing isoseismal lines, as determined by Robert Mallet for the Naples earthquake, 1857 (Mallet, 1862, vol. 1, map in back pocket) (by permission of the President and Council of the Royal Society)

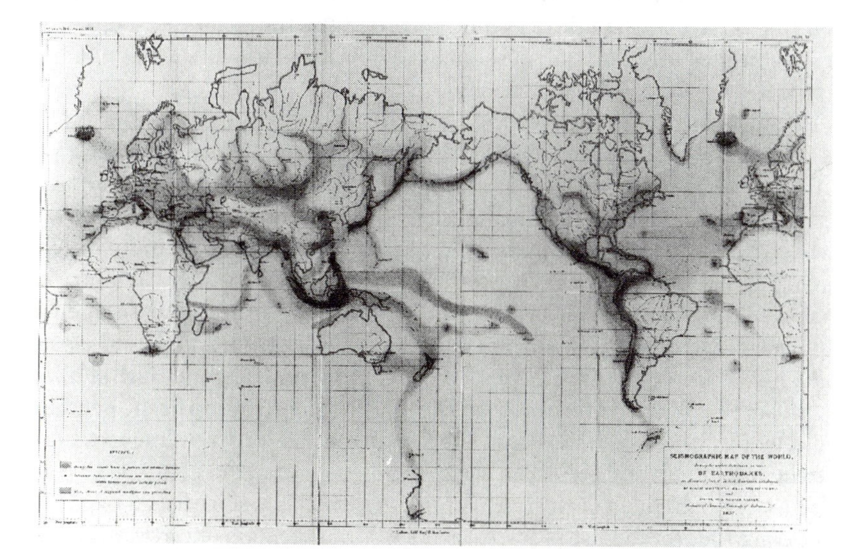

Figure 10.2 Worldwide distribution of earthquakes, as mapped by Robert Mallet (1859, Plate XI) (by permission of the Geological Society)

earth's major earthquake zones in remarkable detail (Figure 10.2). Mallet believed that earthquake waves consist chiefly of longitudinal motions, or compressions and dilations of the material of the earth; but this generalization was found to be mistaken when sensitive instruments capable of detecting transverse motions were subsequently constructed, and it appeared that both longitudinal and transverse vibrations were generated by earthquakes.

During the second half of the nineteenth century, a considerable number of earthquake recording devices were invented and largish 'databases' were developed to build up a reliable stock of information on seismic events (McConnell, 1986; Ferrari, 1992). Various recording instruments were devised. Chiefly pendulums were used. Their swings could be observed; or the pendulums could be designed so as to be dead-beat and the motion of the frames relative to the almost motionless pendulums could be recorded. A particularly elaborate instrument was constructed in 1856 by Luigi Palmieri (1871, 1872) at an observatory on Mount Vesuvius. It was called a *sismografo elettro-magnetico* (Figure 10.3). This splendid Heath-Robinson-like instrument (or rather set of instruments) used the spiral spring (*E*) to detect vertical motion,[5] while horizontal motions along different axes were detected, and recorded electrically, by the movement of mercury in the U-tubes. The closing of an electrical circuit activated electromagnets (*C* and *D*), which caused the clock to be stopped, thus recording the time of the arrival of the disturbances from an earthquake, and the rotation of the drum (*i*) was started, with paper being unwound and marked at *m*. But these elaborate devices only recorded the times and, to some extent, the magnitudes of seismic disturbances. They did not display the vibrations graphically.

As said, many instruments were devised and used, particularly in Italy, in the nineteenth century (Ferrari, 1990, 1992). Perhaps the simplest and most useful was that made by Father Filippo Cecchi (1822–87), Director of the Osservatorio Ximeniano in Florence, in collaboration with Father Giovanni Cavallere (1807–74). They constructed a series of seven pendulums of different weights and lengths, similar in principle to the old instrument of Bina, and capable of responding to vibrations of different intensities and frequencies. Movements were traced on to smoked glass or paper, wound round rotating drums. As a simple warning device, Cecchi used a reverse pendulum similar to that of Forbes. When disturbed, a small rod balanced at the top of the pendulum would be displaced and fall. It could be arranged that the movement activated an electrical circuit and set off an alarm bell.

Thus through much of the nineteenth century Italian scientists took the lead in the construction and use of instruments for the detection of earthquakes. The Italian interest was of course stimulated by the fact that Italy is a country that has frequent earthquakes, but it had the effect that many Catholic (especially Jesuit) institutions round the world played an active and important role in the

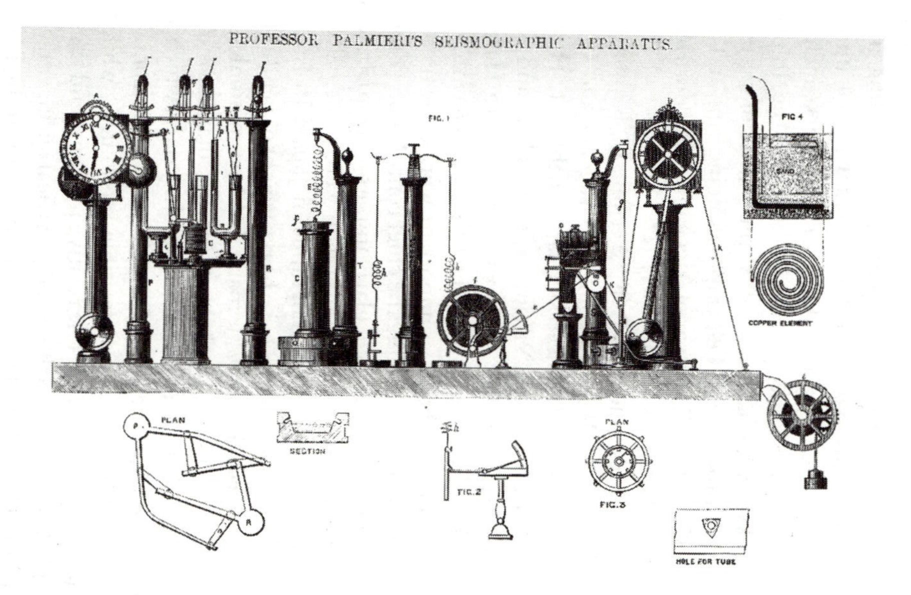

Figure 10.3 Seismograph invented by Luigi Palmieri in 1856 (Palmieri, 1872: 407) (by permission of the British Library)

twentieth century, as a global network of recording stations was established. Such a network of recording stations became necessary for the development of theoretical seismology and the elaboration of thoughts about the earth's interior, as well as for the purely practical purposes of earthquake recording and attempts to alleviate earthquake damage through forewarning of impending disasters.

The use of a horizontally suspended pendulum to detect horizontal earth movements was described by Friedrich Zöllner (1873). However, the most significant nineteenth-century developments in recording instruments were achieved initially by a group of English lecturers who took up posts in the new Japanese education system in the 1870s and 1880s. These included John Milne (1850–1913), Professor of Geology and Mining at the Imperial College of Engineering, Tokyo, who has been called the 'father of seismology'; Thomas Gray (1850–1908), Professor of Telegraphic Engineering at Tokyo University; James Ewing (1850–1935), Professor of Mechanical Engineering and Physics at Tokyo University, Cargill Knott (1856–1922), Professor of Physics at Tokyo; and some other figures, such as W.E. Ayrton, E. Kipping, E. Naumann, T. Mindenhall, J. Perry, W.S. Chaplin, G. Wagener and C.D. West. These men, together with some Japanese scholars, such as I. Hattori, F. Omori, K. Yamakawa, K. Furuichi, K. Nakamura, C. Kochike and K. Sekiyu, established the Seismological Society of Japan in 1880, with Hattori as President (Tuge, 1968: 105). The Society published the *Seismological Journal of Japan*, which contained much of the most important work done on seismology in the late nineteenth century, Milne being the chief contributor. (But, as said, Italians and their Jesuit outposts were particularly important contributors in the work.)

It was Gray (1881) and Ewing (1881) who were principally responsible for the technical improvement of seismographs in Japan. In Ewing's instrument (Figure 10.4), during an earthquake the heavy weight or pendulum bob, *a*, which pivoted about the vertical axis, *bb*, remained almost motionless, along with the rod, *d*, which had a stylus at its end able to mark a rotating smoked plate, *f*. The light-supporting bracket, *c*, was pivoted about the axis *ee*, which, however, would be shaken or vibrated by the earthquake. Thus the horizontal pendulum was designed so as to remain as far as possible motionless, while the rest of the apparatus moved with the earth. The relative motion could thus be recorded on the rotating plate, *f*. Gray's machine had a vertical spring attached to a pivoted bar, one end of which was designed to give a magnified indication of the movement by a stylus marking a horizontal plate.

As shown in Figure 10.4, two of Ewing's detectors could be arranged at right angles, so as to record the different components of a horizontal vibration; and it was not long before instrument makers were producing Ewing seismographs commercially (e.g. Milne, 1886: 39).

With instruments of increasing sensitivity becoming available, it occurred to Milne (1883) that one might, in principle, detect an earthquake anywhere in the

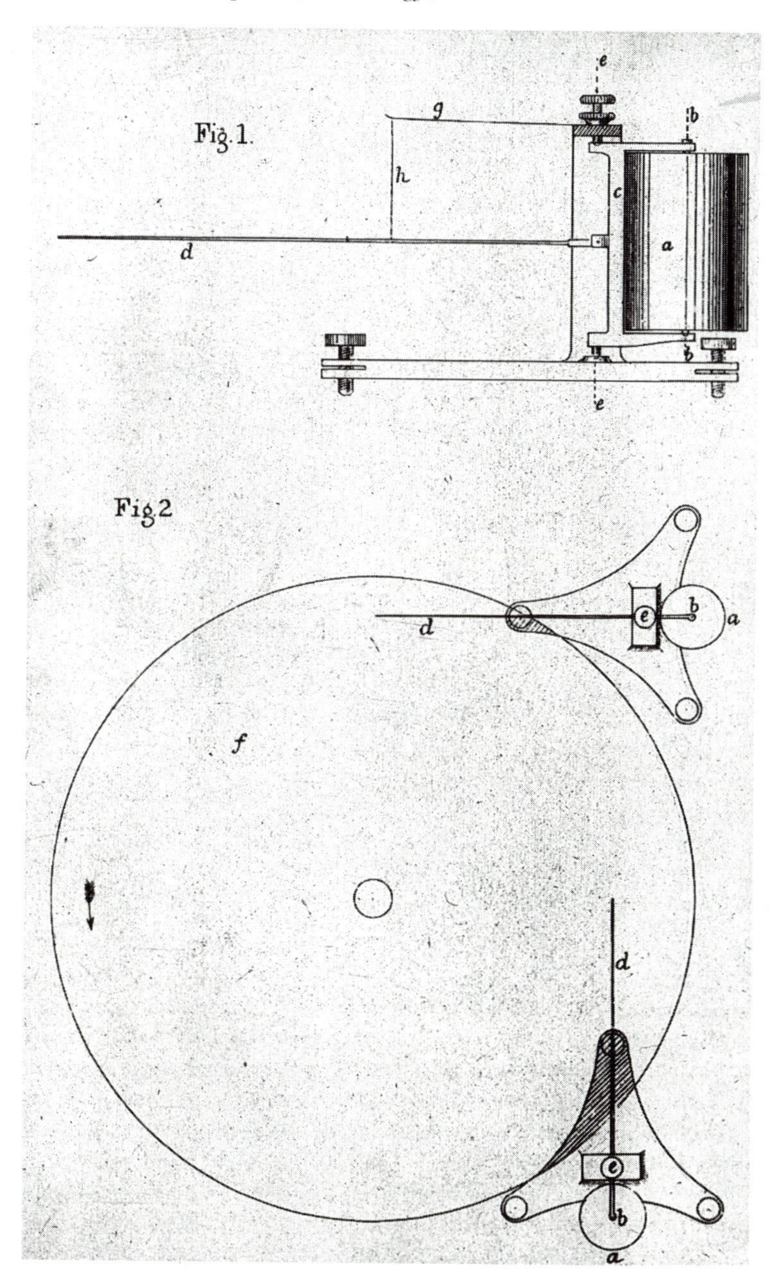

Figure 10.4 James Alfred Ewing's seismograph (Ewing, 1881: 441) (by permission of the Geological Society)

world if appropriate detectors were available. In 1889, this prediction was fulfilled when Ernst von Rebeur-Paschwitz (1889) noticed that seismometers at two stations in Germany – Potsdam and Wilhelmshaven (near Bremen) – responded at about the same time to a large earthquake that had occurred in Tokyo on the same day (18 April). It appeared, therefore, that the waves of disturbance had indeed passed right through the earth, and, allowing for the differences in time due to differences in longitude, it was estimated that the shock took 1 hour 4.3 minutes to travel through the body of the earth. One was thus in a position to gauge the velocity of transmission of the disturbance; and later seismologists could utilize such information to try to glean information about the material constitution and conditions of the earth's interior, to which the velocities of transmission were related. But here the mathematics began to get difficult.

From 1889 on, discoveries were rapid, and they led in a remarkable way to some understanding of the nature of the earth's deep interior with the help of scientific instruments, whereas previously understanding had been achieved largely by speculation or by elaborate inferences from astronomical theory, as, for example, in the work of Hopkins or Lord Kelvin.

By the beginning of the twentieth century, leadership in the construction of seismometers had moved decisively from Italy to Germany, and German scientists also began to move to the fore in theoretical seismology and theories of the inner constitution of the earth. Emil Wiechert (1861–1928), who became lecturer in physics at Königsberg University, travelled to Italy in 1899 to examine Italian instruments, and on his return began work on the production of an instrument, the principles of which became standard for many years in the twentieth century (Wiechert, 1903).

Wiechert's first fully operational machine was established at Göttingen, where it is still in working order. Again, a reverse pendulum was deployed: a large, heavy, cylindrical inertial mass. Its movements relative to a flat plate, firmly connected to the ground, could be detected. (Or one might say that any movement of the ground relative to the stationary inertial mass was being detected.) A small cylinder, attached to the mass, passed through a central hole in the plate. This small cylinder was attached to two thin rods, perpendicular to one another and pointing north–south and east–west. Their motion was conveyed via two levers to two other rods, parallel to and above the first pair. The ends of these upper rods were connected to two pistons in hollow tubes, which acted as air damping devices. The other ends connected to two other levers, which were joined to two light writing arms made of aluminium wire with platinum ends. These traced vibrations on to strips of smoked paper moving on two rotating drums. The motion of the drums was controlled in such a way that a continuous trace could be maintained for twenty-four hours on one piece of paper. The pens, controlled by an electrical relay, were lifted briefly every minute so that the time of any seismic motion during a day could be known.

Figure 10.5 A Wiechert seismograph, operated at the Osservatorio Collegio Alberoni, Piacenza (Ferrari, 1992: 139)

The construction of Weichert's instrument is rather difficult to describe in words, but the general form of its upper part can be gauged from Figure 10.5. The damping system incorporated into the Wiechert seismograph, clearly visible in the illustration, should be noted particularly. It allowed intense vibrations to be recorded but did not prevent the detection of slight movements as well. Such instruments were soon in use in various parts of the world, not long after their invention by Wiechert.[6]

Also in the early twentieth century, electrically operated instruments were being constructed (e.g. Galitzin, 1906). Briefly, a pendulum moved relative to a magnet, and the current generated was passed through a galvanometer. A beam of light was reflected off a mirror attached to the galvanometer suspension, and movements of the reflected beam could be recorded by means of a moving roll of photographic paper.

These technical details may seem to be taking us away from our theme of thoughts about the earth. But they are worth consideration, first because I shall presume in what follows that reasonably accurate and sensitive devices were available to seismologists for the registering of earthquakes from the beginning of the twentieth century, and second because I shall want to give some thought later in this chapter to the role of instruments in thinking about the earth. In the nineteenth century, the main tools mediating knowledge between the geologist and the earth, as maps were constructed, were the hammer, the hand lens and, in later years, the petrographic microscope. In the present century, less blunt instruments have come to be of ever-increasing importance. And the question arises about the truth of our ideas about the earth when they are based on readings generated by complex and sophisticated instruments, which necessarily become more and more like 'black boxes'. The observer can see by inspection how the Wiechert seismograph functions. The same cannot be said of modern instruments. And so, for the present purposes, I shall not attempt to give further details of these devices. The philosophical implications of this 'blackness' of modern geophysical instruments are interesting and will attract our further attention. But for the present we take up the account of the development of theory rather than instruments.

In 1899, the Director of the Geological Survey of India, Richard Oldham (1858–1936)[7] reported (Oldham, 1899) that a huge earthquake in Assam in 1897 seemed to have given rise to two distinct preliminary impulses when the waves arrived in Italy. He thought that the evidence suggested the existence of a solid iron core at the earth's centre, surrounded by solid, stony material. But the reasoning was shaky (Brush, 1980: 710). Oldham followed up this discovery with an elaborate theoretical paper (1900) and another major paper (1906) which gave further arguments for the existence of the earth's core.

It was shown that the preliminary oscillations generated by an earthquake were of two main kinds: 'condensational' and 'distortional', later called primary (or 'pressure') (*P*) and secondary (or 'shear') (*S*), and that the evidence provided by the two kinds of waves could be used to make inferences about the inner constitution of the earth. At any given observation station, the preliminary waves from an earthquake arrived in two batches, so to speak. The *P* waves, arriving first, evidently travelled faster than the *S* waves. The *P* waves seemed to emerge from the ground at a steep angle, giving a vertical shaking. The somewhat slower *S* waves, arriving a little later, caused lateral shaking and more damage. The existence of these two kinds of waves could be expected from the principles of wave theory, if the earth were treated as an elastic body (capable, in a sense, of vibrating like a musical instrument). But there were sceptics towards this view, such as Cargill Knott.

Besides finding indications of two distinct kinds of waves, Oldham (1906) also found, from the earthquake records in different parts of the globe, that the velocities of the waves increased with the distance travelled, which suggested that they travelled faster at greater depths in the earth, presumably where the temperatures and pressures were higher. Also, because the two kinds of waves apparently travelled at different speeds, the difference between their times of arrival would be proportional to the distance of the earthquake. Thus, by triangulation from three observatories, the position of any earthquake on the earth could in principle be calculated. (This suggestion was due to Milne, not Oldham.)[8]

In his paper of 1906, Oldham referred to his discovery, made on the basis of his examination of the records of several earthquakes, that if the angle of the earthquake[9] was more than 120° from the position of the recording instrument, then the arrival of the *S* waves was greatly delayed – by about ten minutes. The situation is displayed in Figure 10.6, the upper curve of which shows the times of arrival for the *S* waves and the lower one the times for the *P* waves.

There could be at least two reasons for the break in the graph for the *S* waves: the central part of the earth might for some reason be unable to transmit the *S* waves so that when they did arrive they had done so by being somehow refracted round the outer part of the earth; or for some reason the *S* waves might travel more slowly in the centre of the earth than in its outer part. At the time, Oldham preferred the second hypothesis. The observations would also be intelligible, however, if the inner part of the earth were liquid and thus unable to transmit the (transverse) *S* vibrations while being able to transmit the (longitudinal) *P* waves (or pressure waves). Either way, by simple geometry, Oldham was able to show that a core, of uncertain nature but of radius about 0.4 times that of the

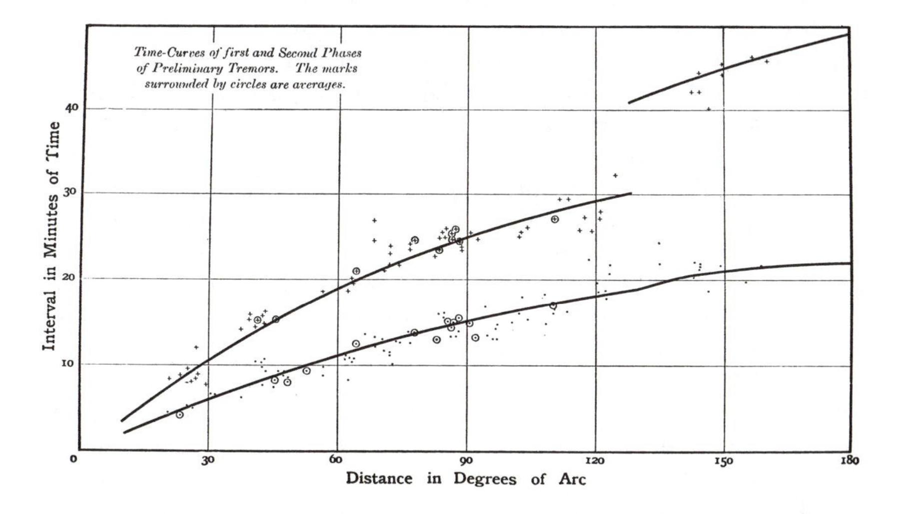

Figure 10.6 Richard Oldham's travel times for different kinds of seismic waves, according to distance from seismic disturbance (Oldham, 1906: 462) (by permission of the Geological Society)

earth, inhibited the arrival of *S* waves if the angular distance from the focus of the earthquake were more than 120°.[10] This figure was refined by the German (later American) seismologist Beno Gutenberg (1889–1960)[11] in 1914, when he obtained a figure of 2,900 km as the depth of the earth's core.

It should be noted that modern textbooks sometimes say that Oldham discovered that the earth has a liquid core because, being liquid and unable to transmit waves of transverse vibrations, it casts a shadow for the *S* waves. This is not strictly correct. Oldham thought that the core might be fluid, but he did not at first firmly espouse this view. In letters to *Nature* in 1913 and 1914 Oldham did begin to move towards the idea of a fluid core causing the 'shadow' for the *S* waves. But he was preceded in this by a Russian scientist, Leonid Leybenzon (1879–1951) in 1911 (Brush, 1980: 713). Unfortunately, as it was written in Russian, Leybenzon's work attracted little attention, and the community of geophysicists as a whole did not at that time accept the notion of a liquid core.

In fact, a considerable number of speculative hypotheses might be generated to explain the seismic data. For example, in a complex and heavily mathematical paper, Wiechert and Zoeppritz (1907) envisaged (as did most of the German community of seismologists of that time) a fairly homogeneous interior for the earth. They thought it possible, however, that when the seismic waves were reflected at some surface they divided into two sets, so that at distant receiving stations the arrival of the direct waves was complicated by the arrival of the reflected waves, which had travelled part of their journey as *P* waves and part as *S*. The critical distance for the onset of this confusion was supposedly 120° from the origin. What Oldham had thought were delayed *S* waves might be due to the arrival of reflected *S* waves that had travelled a greater distance, rather than being slowed down in the core. Many other explanations might be, and were, given. The situation was akin to the classic one beloved by philosophers of science: empirical data can in principle be explained by an infinity of hypotheses. While the philosophers' infinity was hardly approached, it was the case, none the less, that there was such a quantity of confusing seismic information, excessively difficult to sort out, that it was compatible with a considerable number of conflicting explanations. It could not be said that the earth's interior structure was immediately evident on the basis of the countless little wiggles on the seismographs at different stations all round the world.

In 1925, Gutenberg published an important monograph, which among other things set out the lines of evidence being used by geophysicists to gain information about the rigidity of the earth's interior. These were: ocean tides, the displacement of the 'vertical' (plumb-line) by lunar and solar attraction, the effect of the earth's deformations on gravitational attraction at different parts of the globe, the period of nutation of the pole, seismic waves, and the supposed dependence of the earth's stability on a rigid centre. Of these, only seismic waves seemed to point towards the existence of a liquid core. But, a year later, one of the world's leading geophysicists, the influential Sir Harold Jeffreys (1891–1989), Plumian Professor of Astronomy at Cambridge, came round to the view that the idea of a liquid core was acceptable (Jeffreys, 1928), and the scientific community endorsed his arguments, although Gutenberg held off from stating in so many words that the core was liquid until as late as 1957.

What was involved in Jeffreys' work was another elaborate exercise in mathematical modelling, such that the various factors listed by Gutenberg might be reconciled with one another. The details of Jeffreys' calculations lie beyond the scope of this book. The simple point to be made here is that the arguments now rested on the more refined information available from twentieth-century seismographs and the elaborate mathematical tools of the astronomers. What was offered was the best available theoretical model to account for the mass of evidence, not all of which pointed in the same direction. Thinking about the earth was getting far away from what might be achieved by the geologist in the field with his maps and hammer. And so things have remained to this day so far as the deep interior of the earth is concerned.

The picture of a liquid core to the earth held the field until 1936, when the Danish seismologist Inge Lehmann (1888–1993) tentatively proposed that there might be a solid core at the centre of the earth (within the liquid core). She used the evidence of the 1929 Buller earthquake in New Zealand, which was recorded in many seismological laboratories in Europe.[12] What she did was analyse the mass of data relating to the times of arrival of the waves of different kinds for observatories at different distances from the epicentre on the other side of the world. Making a number of simplifying assumptions, notably constant velocities for the *P* waves in the mantle (see note 13) and the core, she then showed that a good approximation to the observed data could be obtained using a model of a three-shell earth. Her idea was that some *P* waves were bouncing off the edge of a small solid inner core, so as to be deflected to a different observation point from that which would have been the recording site if the central core of the earth had been entirely homogeneous and liquid (Lehmann, 1936[13]). Her suggestion has since received general acceptance.

Earlier in this book, I have used the notion of a crust of the earth rather freely, as it was a common (though by no means universally accepted) hypothetical entity in the nineteenth century: a thin solid crust enclosing a huge fluid interior. In a less well-developed form, the idea goes back at least to the time of Leibniz and Descartes. But we have just been thinking of the earth's interior as solid, as far down as the (outer) core, and capable of transmitting both *S* and *P* vibrations. So does the crust extend uniformly down to the liquid core at 2,900 km depth?

Further light was thrown on this question in 1909 by the work of Andrija Mohorovicic (1857–1936), a professor at the University of Zagreb in Croatia, studying the seismic waves produced by an earthquake that occurred in October that year in the Kulpa Valley, thirty miles south of Zagreb (Mohorovicic, 1910[14]).

He noticed that seismic stations less than about 800 kilometres from the earthquake recorded the arrival of two sets of *P* and *S* waves. (It was Mohorovicic who introduced this terminology.) It might have been the case that there were two shocks, but it appeared that the time difference between the two sets of waves decreased according to the distance from the centre of the shock, so two shocks could hardly have been responsible. To account for the observations, Mohorovicic proposed that there was some kind of boundary within the earth at a depth of about 45 kilometres, with the waves travelling faster below the boundary than above it. The slower-moving waves travelled directly to the observer through the outer layer. The others apparently went down into the lower layer, which then transmitted vibrations more rapidly, and they were then (somehow) refracted up again to reach the observer.[15] The boundary between the two layers is today called the Mohorovicic discontinuity, being so named by Macelwane and Sohon (1936, vol. 1: 204), and it is held to mark the boundary between the crust and the mantle – that is, the material of the earth's interior down as far as its core.[16]

Finally, for the present purposes, mention should be made of the so-called asthenosphere, the existence of which was first proposed by the American geologist Joseph Barrell (1869–1919) in 1914. It seemed to Barrell (1914: 659) that the theory of isostasy required that there be a layer of the earth under the crust or lithosphere which would be able to yield to strains of limited magnitude. This supposed 'sphere of weakness' he called the 'asthenosphere', from the Greek word for 'weak'. Some evidence for the existence of the asthenosphere, which manifests itself as a zone of diminished velocity for the transmission of seismic waves, was provided by Gutenberg (1926). He was investigating the question of how far down the earth's interior was crystalline. This was still the issue of the day in that the nature of the core – solid or fluid? – was not settled to everyone's satisfaction. In the course of his enquiry, Gutenberg observed that there were indications of a decrease in the wave amplitude and velocity for the outer portion of the mantle for depths between about 70 and 100 km. This hinted that the mantle was more plastic at this depth than at greater depths where the pressure was greater. In 1928, Gutenberg suggested that at a depth of about 80 km the melting-point of the rock was reached.

It cannot be said that Gutenberg 'discovered' the asthenosphere back in 1926, and his paper of that year did not really concentrate on the external layer of the mantle. As said, he was interested in the earth's core at that time. However, for many years, he was an advocate of the idea of plasticity of the upper part of the mantle, which might therefore be the possible locus for movements of portions of the earth's crust relative to the mantle, as required by the mobilist theory of continental drift (see Chapter 11) (e.g. Gutenberg and Richter, 1931, 1939; Gutenberg, 1948,[17] 1954, 1955). In that sense his 1926 paper was important. But it appears that it did not carry much weight in people's minds at the time as concrete evidence for the existence of a mobile or plastic asthenosphere.

The existence of the asthenosphere – a plastic layer of the mantle with diminished velocity of transmission for seismic waves – was only confirmed to everyone's satisfaction some thirty years after 1926 by the examination of world-wide records of atomic blasts. These provided the opportunity for 'controlled' large-scale seismic experiments, such that Gutenberg's 'layer' could be examined more closely.[18] But, earlier, the influential Jeffreys (1929: 193, 231) had accepted the general idea of the existence of the asthenosphere. And with the support of one such as Jeffreys the idea proved acceptable, albeit not empirically proved. Its great theoretical significance for geology was not, however, evident at the time.

So the asthenosphere, now thought to extend from the Mohorovicic discontinuity down to about 600 km, appears to be more fluid and yielding than the region below and hence is less able to transmit seismic waves; and minor reflections can be discerned from its lower surface for *P* waves that travel from a seismic disturbance right through the earth and back again. The reason for the existence of the asthenosphere is easily understood. Rock melts when heated, and temperature evidently increases with depth. But melting becomes more difficult with increased pressure. In the upper part of the mantle, the temperature effect is dominant, so the material is more fluid there, but at greater depths the pressure increases to the extent that the mantle is solid. At greater depths still, the temperature wins and we have the fluid core. But right at the centre the pressure wins once again, and so the inner core is solid. Needless to say, there must be changes in composition as well as temperature and pressure.

Evidence for actual fluidity at the level of the asthenosphere was reported by the Russian seismologist G.S. Gorshkov (1958), as a result of his study of seismic vibrations arriving in Kamchatka from the Aleutian Arc and Japan. It appeared that, underlying a volcanic region at a depth of about 50–70 km in southern Japan, there was a convex lens of material able to transmit *P* but not *S* waves. This suggested the existence of liquid material at about the level of the boundary between the earth's crust and the mantle. The argument was similar to that previously deployed in relation to the fluidity of the earth's core. However, the molten matter at the level of the asthenosphere appeared to be quite localized.

We thus have the following picture of the earth's interior, largely based upon the information that can be gleaned from seismograms.[19] There is an outer crust called the lithosphere. Below this there is the Mohorovicic discontinuity, which leads to the somewhat plastic, or occasionally fluid, asthenosphere, which is itself the upper part of the generally solid mantle of the earth. Deep down, below this, there is the liquid core, and right at the centre of everything is, not hell with Satan and the world's three most notorious sinners frozen in a block of ice, as

Dante imagined, but a modest-sized, solid inner core. Modern seismologists think there may be other subsidiary shells (Bolt, 1982), but further refinement of the model is not necessary for our present purposes.

Further important information from East Asia relating to the earth's interior was obtained by the study of earthquakes in the 1930s by the Japanese seismologist Kiyoo Wadati (1934–5), of the Central Meteorological Observatory of Japan, following in the tradition of seismological research established in that country the previous century. He studied the distribution of deep-focus earthquakes in his region and found that they followed three principal lines: (1) from Vladivostok to the north of Hokkaido; (2) from the island of Kyushu down the archipelago to the south-west of Japan; and (3) from Vladivostok, through central Honshu and onwards to the island of Ogasawara. He was also able to draw contour lines for the occurrence of earthquakes at different depths, from which it appeared that there was a 'weak surface' that 'exist[ed] slopewise in the crust near the Japanese islands having some connection with volcanic activity' (ibid.: 324). He suggested that this might provide evidence for Wegener's theory of continental drift (see Chapter 11), for the 'weak surface' might be a 'trace' of continental displacement. But this prescient suggestion attracted little interest at the time for geologists.

Wadati's ideas were 'rediscovered' in the 1950s by Hugo Benioff (1954) of the California Institute of Technology (Caltech). Studying the records of many earthquakes in seismically active regions, he found evidence of steeply inclined 'reverse' fault planes (see Figure 11.14b), which were of somewhat different form in oceanic and 'marginal/continental' regions. The oceanic faults appeared to extend down to about 700 km, while the ones adjacent to or dipping towards and under continental margins changed their inclination at about 300 km and then extended down to 700 km. Interestingly, a plane of fault in the area of Japan itself appeared to have been ruptured and rotated, so that it was now in two parts that converged on Vladivostok. Hence the peculiar geometry of Wadati's fault planes was explained.[20] Benioff mentioned the idea that the frictional heat developed by the faulting might be responsible for melting some of the rock near the fault zone, which would account for the typical association of the earthquakes, the faulting they implied, and volcanic activity. The idea of 'Benioff zones' (or better 'Benioff–Wadati zones') of faulting eventually came to be a major component in the development of the plate tectonic paradigm (see Chapter 11), although in the 1950s the seismic information did not, in itself, seem sufficient to warrant acceptance of continental drift/mobilist theory.

We may now reflect on the character of the knowledge of the earth's interior provided by seismology and its instruments. Interpretation of the blips and wiggles produced by seismographs was described by Inge Lehmann in some unpublished autobiographical notes as 'a black art for which no amount of computerizing is likely to be a complete substitute' (quoted in Brush, 1980: 716).

And surely there was, at the time that the main features of the earth's interior were delineated, much truth in this statement.

There was, I think, implicit in Lehmann's statement the idea that there was a strong 'tacit' component to the work of the seismologist (Polanyi, 1958, 1966). This may be compared to the way in which a cardiologist discerns diagnostic details on an electrocardiogram that are of no significance whatsoever to the lay person. To interpret such instrument readings – or seismograms – one needs to be deeply embedded in the culture of the science (or black art) concerned. The skill can only be accomplished through years of experience in the relevant community. This means, however, that the knowledge vouchsafed by the seismologist has to be taken on trust by the lay person, so far as the details of the earth's interior are concerned.[21] Our thinking about the earth's interior is necessarily refracted through the minds of the seismologists, and their thinking is refracted through their instruments.

The comparison between the seismologist and the cardiographer is, I believe, just. Indeed, seismologists use the term 'tomography', common in medical parlance, to refer to the processes whereby they form three-dimensional 'images' of the earth's interior by means of their instruments. They look for regions in the interior which appear to be 'geologically fast' or 'geologically slow' in the transmission of waves. In this way, indications have been found of convective movements within the mantle, so important for the theory of plate tectonics (see Chapter 11). And peculiarities in the transmission of surface waves can give information about special structures in the outer parts of the earth. It is as if the seismograph is used as a kind of X-ray machine for building up a shadowy picture of the internal structure of the earth.

What, then, does this tell us about the reality of the layers of the earth's interior, as propounded by seismology? This question is clearly one where philosophy walks on to the stage. Are the shells of the earth's interior 'convenient fictions', like the epicycles of the old Ptolemaic astronomers, which successfully 'save the appearances', or are they (literally) part of the real world on which we make our habitation? According to the philosophy of 'instrumental realism' of Ian Hacking (1983), for example, theoretical entities are real if they are experimentally manipulable. His motto is: electrons are real if you can spray them (as in a television set, for example) and make them do what you want them to do. This pragmatic approach has, I believe, a good deal to recommend it.[22] But it can hardly find application in the case of the earth's deep interior. We cannot manipulate that, and I doubt that we ever shall, unless we think of waves generated by atomic (or smaller) explosions as falling under that head, or unless the Moho project (see note 16) is revived.

According to Don Ihde (1979, 1990), who takes a phenomenological approach derived from the philosopher Martin Heidegger, but which is in some ways not so very far away from Hacking's pragmatism, instruments serve as marvellous extensions of our senses. One can 'feel' one's way around with them. To illustrate: I happen to be adept in the dense traffic of Sydney in squeezing through small gaps that other drivers (no doubt politer than I am) seem to prefer not to attempt to enter. It is as if, through much experience of driving, my car is an extension of my own body, like a cat's pair of whiskers. So it is, suggests Ihde, with many forms of technology – computers perhaps most obviously, I suppose, although the hammer as an extension of the hand is a simpler example to contemplate.[23]

It might seem that, for the seismologist (and for those working in other areas of geophysics), their instruments also act as extensions of their very senses. But it is, of course, a 'black art' compared with driving a car, for few have the requisite tacit knowledge to use seismographs to peer into the earth's interior. For the rest of us, we have to take their findings on trust; and so it is with most science. Thus, although modern atlases and school textbooks commonly show diagrams of the earth's interior with an onion-like internal structure, I do not think that such knowledge is part of the person in the street's everyday thinking about the earth, when in an unreflective mode, as Barfield described this person's common conception of the cosmos in the Middle Ages and the Renaissance (see p. 42). The medieval persons in the street had to take on trust what they were told by the intellectual élite of their day: mostly the clergy. If they would differ with them, they would have to acquire the skills of scholastic disputation and spend a lifetime in the study of the Scriptures. In our day, if we would take issue with the received view of the structure of the earth we would need to construct better seismic instruments, obtain a better database of information about earthquakes and learn the mathematical techniques of geophysicists. Not being in a position to do this, we must necessarily form our view of the earth's interior by taking on trust what the 'authorities' tell us. But it is a more esoteric view than that of the medieval world picture. It is not one that most people think about, either in their reflective or unreflective moods. In the modern era of science, thoughts about the earth's interior are chiefly confined to geologists and geophysicists.

This said, the role of the instrument in (geologists') thoughts about the earth can be analysed more closely. Instruments may have a different role from merely being an extension of the senses. It is clear that scientific instruments have a role of mediating between the world and the observer. Ihde (1990) represents this by the simple 'formula':

Human–Instrument–World

In the cases of the car acting as my 'whiskers', the vehicle effectively becomes part of my sensory apparatus, at least when it is functioning as it should and is not, in itself, an object of my concern. We might choose to say, then, that there is an 'embodiment' relationship between myself and my car. This state of affairs may be indicated with the help of parentheses, thus:

(1) (Human–Instrument) → World

A more obvious example than the car might be a microscope. We observe a thin section of a rock through the microscope. The instrument serves as an extension of the geologist's senses, amplifying some features of the direct sensory experience but diminishing others, such as the rock's hardness or 'feel'. In both examples, we can regard the instrument as being 'embodied', or as being in an 'embodiment relationship' to the human observer.

But in other cases it is the instruments themselves that are the direct object of examination, and the readings on their dials, printouts or whatever have to be interpreted. Thus we are engaged in a kind of hermeneutic activity. This may be represented as follows:

(2) Human → (Instrument–World)

Ihde (1990) gives the example of an engineer observing the dials of the instruments in the control room of a nuclear reactor (serving as an example of 'the world'). All attention is focused on the instrument panel, not the actual interior of the reactor, which a human is not physically able to enter. The engineer has to interpret the instrument readings in order to gauge the internal state of the reactor. It is, therefore, an exercise in hermeneutics of a kind. It is an important question, of course, whether the instruments furnish a correct 'picture' of the reactor's interior. This will depend on the design of the instruments, the adequacy of the theories deployed in their design, their state of repair, and the skill with which they are used and interpreted. The 'instrument–world' relationship is by no means transparent. But the intention of the designers and operators of the instruments is that they will 'refer' to the conditions of the reactor's inaccessible interior.

Cases 1 and 2 above are represented as distinct from one another. But one should, rather, think of there being a kind of continuum, for different kinds of instruments. And so it is with seismic instruments. If a seismic instrument is simply some device for enhancing one's capacity to detect a weak earth tremor – say a pendulum of some kind, perhaps even hand-held – then one would be justified in calling it an embodied instrument. But clearly that is not usually the case. Rather, we have instruments that detect and record vibrations, representing them visually, not as sound, and with a good deal of associated electronic gadgetry today. The relationship between the earth's vibrations and their depiction in the form of a seismogram is in principle reasonably straightforward, however. It is easy to see how the early seismographs were constructed and how they worked. Today the instruments are more complex and may well be treated by their operators more or less as 'black boxes'. Nevertheless, they have been manufactured by humans and are not inordinately complicated by the standards of modern scientific instruments, and their workings can be understood by the operators if they wish to take the trouble. The relationship between the seismogram and the movements of the earth at the site of the seismic observatory is relatively straightforward or 'transparent'.

Nevertheless, the reading of the seismogram is a problem in hermeneutics: the seismologist has to recognize which is the wiggle for the *S* wave, which for the *P* wave, which for the *P'* wave (see below), and so on. This may in itself be a 'black art', but it is only the beginning of the problem. A single seismic observatory is only of limited use when it comes to elucidating the internal structure and constitution of the earth. A whole network of observatories is required, and the records of earthquakes have to be exchanged, correlated and analysed. As we have seen, such work was begun in the early days of seismology, by men like Milne or Mohorovicic or by the network of Jesuit geophysicists. Such research programmes have, by their very nature, to be social in character.

With data, the theorizing can begin – the thinking about the earth. Models of its internal structure are constructed. At first they are simple and then gradually refined. In his early astronomical work on the moon, Newton first supposed its orbit to be circular and the earth and moon to be point masses. Having obtained a reasonably satisfactory result with these gross simplifications, he then refined his model, giving the bodies volume and trying out non-circular orbits. So too with Oldham. The data from different observatories gave him the idea that some part of the earth's interior was impermeable to *S* waves. He hypothesized therefore that there was a sphere of material within the earth which produced a 'shadow zone' for this kind of wave, and he did a rough calculation to determine the radius of this sphere. He made the simplifying assumption that the waves travelled in straight lines, even though he well knew that they would have been transmitted along curved pathways because of the variation of pressure and density within the earth causing refraction. With success for the preliminary calculations, geophysicists could then proceed to more complex models.

As time went on, more and more possible wave paths between a seismic disturbance and a recording station were contemplated. There could be *P* waves in the mantle; *S* waves in the mantle, waves passing through the core (*K* waves); reflections at the core, designated by *c*; waves passing through the mantle and the core (Lehmann's *PKP* or *P'* waves); waves reflected at the earth's surface (e.g. *PP*); waves reflected at the antipodes (*PKPPKP* or *P'P'*), etc., etc. A *P* wave might be converted to an *S* wave (or vice versa) on reflection at the core, adding further complications. Over the years, the travel times for different distances for all these kinds of waves, and many more, were worked out and reduced to a single diagram by Jeffreys (1937–40 [1939]), with the help of the New Zealander Keith Bullen. The computational labour in this work was immense, but the resulting diagram (Figure 10.7) was a godsend to seismologists. It provided a 'ready reckoner' to help convert the wiggles of the seismograms to physical events supposedly taking place within the earth. With the help of

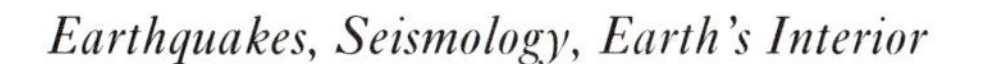

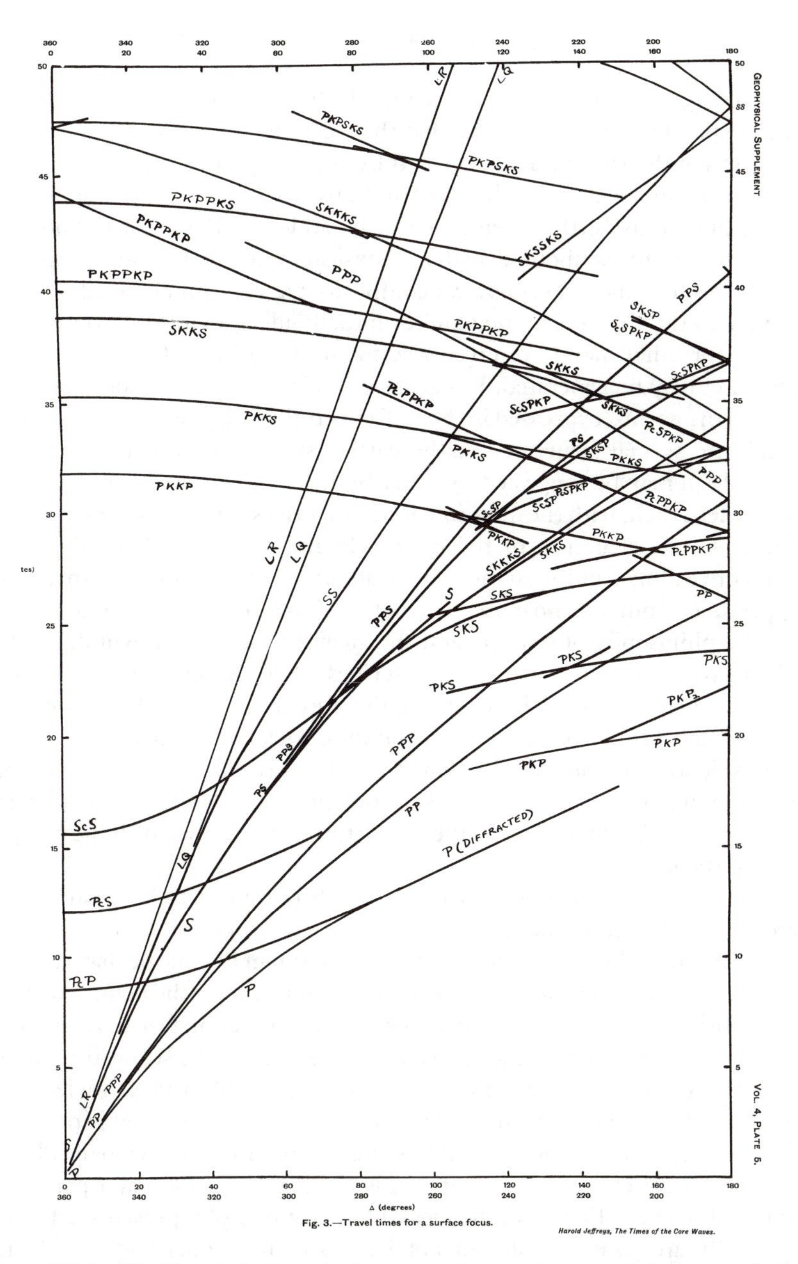

Figure 10.7 Sir Harold Jeffreys' diagram representing travel times for different kinds of seismic waves (Jeffreys, 1937–40 [1939], Plate 5, facing p. 604) (by permission of the Geological Society)

the Jeffreys–Bullen diagram, the 'black art' of seismology was somewhat whitened.

Such a diagram certainly helps to render the 'instrument–world' relationship more transparent. But it is evident that the (seeming) transparency is one that involves much theory; and it is achieved by the comparison and correlation of instrument readings at many different stations. Even so, with the help of the diagram, numerous further elements can be added to the theories about the earth's interior. Ideas about densities, physical states and elastic properties, temperatures, chemical/petrological compositions, etc. can all be added more effectively to the theoretical mix; and eventually ideas about the events of the earth's history that may have given rise to an object of such a structure and composition may be generated. In particular, seismologists can look for deviations from the norms, expressed in the Jeffreys–Bullen diagram. Such deviations may indicate special structures of the earth's such as subduction zones (see p. 267), which may be suggested by other branches of geological theory.

We are left, then, with the question of whether the seismic instruments 'refer' truthfully to the structure of the earth. Or are we just looking at a lot of instrument readings and weaving a story about the earth's interior that 'saves the appearances' but does no more than that? This is a deep and complex question within the philosophy of science, and the answer that is given will depend on the theory of truth to which one subscribes. The geologist/seismologist is looking for correspondence between the theories and the way things are in the world. But this cannot be checked by standing back and comparing the world and the scientific picture of the world, in the way that one can compare a photograph of a person with that person. For the picture of the earth's interior that we have is only obtainable via the scientific instruments and all the complex associated theory.

My own view on this question is that truth is gauged by consistency and coherence. All the bits of the story have to fit together like the pieces of a jigsaw, and the better they do this the more certain or confident can one be that a truthful account has been accomplished. But total certainty cannot be achieved. What counts as fitting, consistency and coherence is not something that is transcendentally obvious. It is something that is fought out within the scientific community, as the battle of science goes on, in endless publications, reviews and conferences. In this process, the manner in which the information is presented is of signal importance, and it is interesting to note the way geologists like to reduce their theories to maps and diagrams.[24] Diagrams are a major part of the 'rhetoric' of science. They give the pleasant appearance of representing the way things really are. But one should not be too easily seduced by them. The instruments that enabled them to be produced have been swept under the carpet in the drawing of the diagrams. Yet their role is inescapable. And their 'transparency' may be an illusion. Were it not so, there would be much less controversy in science, much of which has to do with the efficacy of instruments (Collins, 1981). By the time we reach the end of Chapter 12 of this book, I hope I shall have given the reader an overall sketch of what geologists think about the earth and how they have come to think about it the way they do. Much of the story fits together remarkably well. But it would be a delusion to suppose that, because of this pleasing fit, all is secure, correct and truthful. Even thinking about the results of seismology alone, it will be apparent that a vast number of models can be devised to fit the data. It is a vastly 'underdetermined' field.[25] So, as will be seen in Chapter 11, there can be and are 'respectable' theorists who offer ways of thinking about the earth that are quite at variance with the received views. Matters remain open. The 'instrument–world' relationship remains clouded. For this reason, I have thought it worthwhile to make a detour through a corner of geophysics, first to raise the question of the role of instruments in the ways that geologists think about the world in science, and second because some discussion of the geophysical ideas is needful as a prolegomenon to the discussions of theories of continental drift and plate tectonics in Chapter 11.

Chapter 11
Movement of Poles and Continents; or Getting Bigger?

Most modern geologists believe they have a general theory that provides a satisfying way of thinking about the earth. This, of course, is the theory of plate tectonics, which grew from a concept propounded earlier in the twentieth century called continental drift. Today, there are hardly any outposts of opposition to the theory of plate tectonics, and students of the earth sciences are frequently unaware that there have been, or could be, plausible alternatives. The theory, together with its associated techniques for assessment and application, may well be called a paradigmatic paradigm (in Kuhn's (1962) sense of the word paradigm), providing a 'world view' appropriate for thinking about the earth. Thus several authors have seen the processes that led to the establishment of the plate-tectonics theory as exemplary illustrations of the way in which scientific change takes place, and some have seen it particularly through Kuhnian eyes.[1] In the present chapter, I shall outline the development of the theory of plate tectonics and relate it to our earlier discussions about such matters as isostasy, petrology, radioactive materials in the earth and the origin of granite.

As seen, the dominant view of the earth in the nineteenth century was one of stability, modulated by a process of contraction due to the earth's cooling. For the thoroughgoing steady-statists such as Lyell, different parts of the earth's surface rose and fell, yo-yo-like, in an irregular fashion. But this view depended on the idea of the earth as being (or at least having) fluid in its interior, with only a relatively thin crust. For Dana, in contrast, the continents and oceans were primordial features, produced when the earth first solidified from a molten state.[2] This 'permanentist' view, though subject to difficulties, was widely held in the United States for many years (e.g. Dana, 1890; Willis, 1910).

As we have seen, Suess, like Dana, was a contractionist, but he did not think the collapses of the oceanic basins were primordial events. The process was ongoing, which explained how marine sediments could be found on inland continental regions or in folded mountain areas. Suess's theory offered roughly the same result as Lyell's doctrine, but it did not have the same gradualist character. It could make some sense of the phenomena of biogeography in that similar species on distant continents[3] might be accounted for by supposing that they had all formerly been part of one supercontinent – Gondwanaland – parts of which had then subsequently foundered. But given the discovery of the generation of heat by radioactive minerals, the earth might not be cooling. Indeed, it might be getting hotter (e.g. Joly, 1909). So Suessian 'contractionism' became less plausible in the twentieth century.

While Dana's theory had little success in making sense of the global distribution of plants and animals, it was not wholly incompatible with the evidence for huge crustal shortening suggested by the Alpine theories of Continental geologists such as Heim and Bertrand in the 1880s. Suess's theory did not fit comfortably with the apparent magnitude of the crustal shortening and was incompatible with the theory of isostasy.

But the theory of isostasy had its own difficulties in that the work of physicists such as Hopkins and Thomson suggested that the earth's interior must be largely solid, with only occasional pockets of fluid (Hopkins' lava lakes). Thus, according to the physicists, there was little liquid in which, or on which, the continents might 'float'. According to the theory of Fisher (1881), the earth's interior was largely solid, with only a thin annular ring of liquid matter. This would, I suppose, have been agreeable to the idea of isostasy, but Fisher's theory was uncomfortably *ad hoc* and did not attract much support. The idea of a somewhat plastic asthenosphere was, as we have seen (p. 238), intended as an aid to isostasy theory.

As said, there was also the problem of crustal shortening, brought to the fore by the work of Heim and Bertrand. There seemed to be too much shortening for the 'wrinkled-apple' theory of global shrinkage. To try to deal with the problem, Otto Ampferer, later Director of the Austrian Geological Survey, suggested (Ampferer, 1906) that there must be an 'underflow' of material in 'swallowing-up zones' to make possible the foldings observed in the Alps.[4] But this suggestion failed to find acceptance, the prevailing idea being that the earth's interior was solid and thus unable to accommodate the entry of downward-moving material.

In short, in the early years of the twentieth century there were fundamental problems in basic ways of thinking about the earth – even though there had been such successes in working out the history of its outer skin by stratigraphy and mapping; even though there were such advances in mineralogy and petrology, and despite the fact that so much more was known about palaeontology, with the fossil record becoming more intelligible in the light of evolutionary theory.

Of course, a number of people, especially von Humboldt, had noticed the congruence of the coastlines of South America and Africa and the possible jigsaw fit of other continents and islands, and this led to the proposal of a few speculative theories about the possible original connection of continents (e.g. Snider-Pellegrini, 1858; Taylor, 1910). Frank Bursley Taylor (1860–1938), who worked under Chamberlin at the US Geological Survey, developed and radically modified the ideas of Suess. Taylor drew particular attention to the area between Greenland and Canada, the outline of whose land masses suggested fractures of the crust and movement of Canada and its northern islands away from Greenland in a south-westerly direction, so that Baffin Bay, the Davis Strait and the Labrador Sea formed a rift valley. Generalizing, then, he proposed large-scale crustal creep, leading to the dispersion of the continents from the polar regions. But the cause of such movements was weakly attributed to 'some form of tidal force'.[5]

A radical theory that set out to deal with some of the theoretical problems mentioned above was developed by the German meteorologist and geologist/geophysicist Alfred Wegener (1880–1930) and first made public at a meeting of the Geological Association at Frankfurt-on-Main in January 1912 (Wegener, 1912a, b). The (English) title of the address (shortened in the printed version) was 'The geophysical basis of the evolution of the large-scale features of the earth's crust (continents and oceans)'.

According to Wegener's own recollections, he first arrived at his ideas in 1910 by meditating on the apparent jigsaw fit of the continents.[6] But he also sought to explain the biogeographical evidence that had partly motivated Suess's proposal of an original supercontinent, Gondwana. Also, Wegener took into consideration important structural features that seemed common to distant continents, such as parallel fold lines in South Africa and South America.

But Wegener's theory also took into account the 'American' idea of isostasy. He agreed that continents are formed of less dense material than the ocean basins, and he postulated that the material of the ocean floor underlies the continents, so that the latter 'floated', so to speak, on the former. Also, Wegener presumed that the substrate for the continents could slowly move, the underlying material behaving as an extremely viscous fluid that could yield under stress. An additional suggestion was that the earth had originally been covered by a thin layer of continental material, which had subsequently broken apart and the continents had 'drifted' on the underlying material. Mountain ranges were formed by the ridging up of the leading edges of the slowly moving islands of continental material. Wegener supposed that, coincidentally, in the Mesozoic all the continental crustal material was concentrated in one supercontinent, which he called 'Pangäa' ('Pangaea') in the second and third editions of his book, *Die Entstehung der Kontinente und Ozeane* (Wegener, 1915). Subsequently, Pangaea had slowly split up, the fragments drifting apart yielding the earth's present continental configurations. This process was represented in diagrams that have been reproduced in innumerable publications, but are also shown in Figure 11.1.

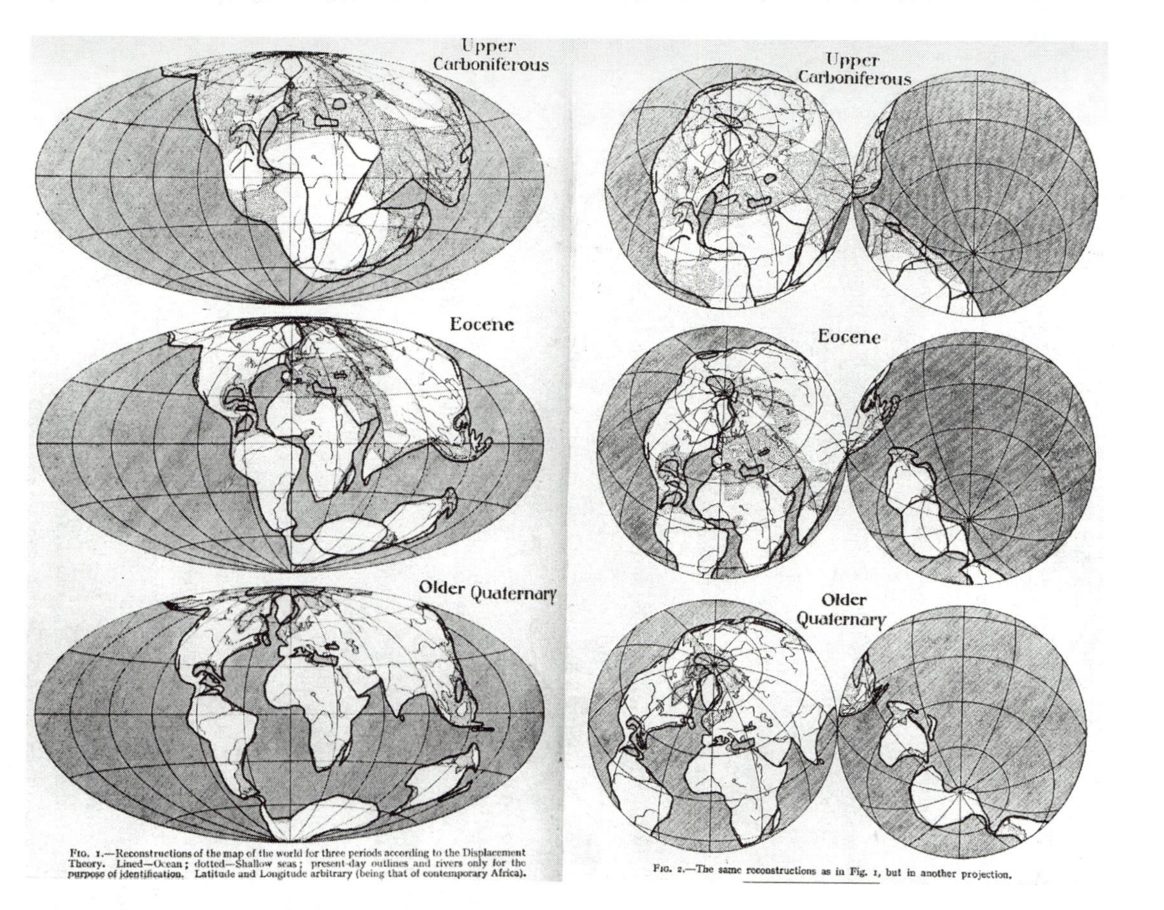

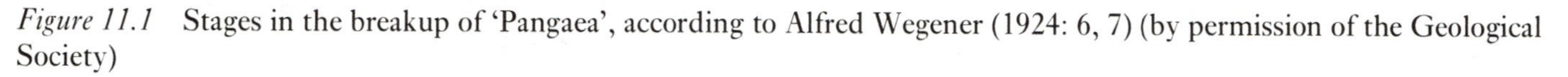

Figure 11.1 Stages in the breakup of 'Pangaea', according to Alfred Wegener (1924: 6, 7) (by permission of the Geological Society)

A great virtue of Wegener's idea was that it allowed for the occurrence of inland mountain ranges, such as that of the Urals or the Himalayas. The former was supposedly the remains of some long-past intercontinental collision, such that the effects of mountain building had to a large extent been destroyed by erosion. The Himalayas, where tectonic activity continued, were the result of a relatively recent collision. In contrast with Dana's old theory of mountain formation, Wegener's 'drift' hypothesis thus accounted for both inland and 'coastal' mountain ranges. Dana could only explain the formation of ranges, such as the Appalachians, on the edge of a continent. As a cause of drift, Wegener suggested that it might be due to tidal friction and differential gravitational effects arising from the earth's oblate shape.[7]

However, as is well known, Wegener's theory initially found rather few supporters. It has sometimes been stated by historians of science that the problem was that geologists could not see how or why the light sialic continents could 'plough' through the denser sima substratum. What could be the force driving the process? Wegener's proposed source was unequal to its task, and he ignored the possibilities of movement due to convection. According to Oreskes (1988), however, the 'ploughing problem' wasn't really the main issue, which chiefly had to do with the American geologists' liking for isostasy, which seemed to evince sound engineering principles. Even so, the 'ploughing problem' was certainly one that the influential Sir Harold Jeffreys raised in his book *The Earth* (1924: 261). And this surely had something to do with the largely negative response that Wegener received. (However, according to Oreskes,[8] Jeffreys was not particularly influential in the United States in the 1920s, as he was somewhat disliked personally in America and was thought to have plagiarized some of Chamberlin's ideas.)

How, then, did matters stand with respect to the acceptance of Wegener's theory? Although mobilism was not finally endorsed by the geological community as a whole until the 1960s, there was more support for drift theory than is sometimes realized. 'Drift' seemed to be compelled by the evidence of the jigsaw fit of coastlines (or rather the continental shelves), biogeographical data and the structural and petrological similarities of distant continents; and there were geologists willing to try to find suitable causal mechanisms.

Among the early supporters of drift theory, perhaps the most significant initially was the Swiss geologist, Emile Argand (1879–1940), Professor of Geology at Neuchâtel. Argand's work came straight out of the Swiss tradition of tectonic studies of Escher von der Linth, Heim, Hans Schardt and Maurice Lugeon (Argand's teacher). According to accounts of Argand (e.g. Lugeon, 1940; Wegmann, 1970; Carozzi, 1977), he had a remarkable capacity for thinking in three dimensions and for depicting three-dimensional structures in pictorial form. Such a gift was an urgent necessity for anyone wishing to unravel the mysteries of the structures of the Swiss Alps, and Argand did just this. He began to think of folds, thrusts, etc. as developing through time, and sought to reconstruct the many stages of development of the Alps until they reached the condition in which we see them today. It was as if the mountains unfolded like an embryo; so, in a lecture to the Swiss Geological Society in 1915, Argand (1916) coined the term 'embryotectonics' to describe his procedure.

Although Argand's work up to this stage was still conceived within the 'fixist' (but also 'contractionist') paradigm of Suess, he converted to Wegener's views when these were published in 1915, and then began an even greater project, namely to try to unravel the tectonic history of Asia. This he did, not by personal fieldwork in Asia, but by developing his own work in Europe and utilizing reports from other geologists. It was a mammoth undertaking, the results of which were presented to the Thirteenth International Geological Congress in Brussels in 1922 (Argand, 1924, 1977). An accompanying map, dated 1928,[9] shows folded areas and 'tabular areas' for all of Eurasia (and parts of Africa in the Atlas Mountains and Red Sea areas), only parts of Tibet, Yunnan, north-east Siberia and Borneo being left blank.

Argand sought to show how the plastically deformable surface of the earth had, over millions of years, been shaped into mountain ranges by the effects of continental drift. In particular, he sought to unravel the immensely complicated structures of the Himalayas, which, following mobilist ideas, he thought had been produced by a 'collision' between the Indian subcontinent (Gondwana) and Eurasia – a 'duel between Indo-Africa and Eurasia'. In this giant, glacially slow collision, which had closed the Tethys Sea (taken from Suess's theory), India had supposedly been thrust under the Eurasian continent. The resultant mountain range – the Himalayas – contained squeezed, folded and metamorphosed sediments from the previously intervening Tethys Sea, and also materials caught up from the underlying sima, so great was the magnitude of the collision (Argand, 1977: 182–3). The idea of the inclusion of sima material was especially important. Argand had for long given attention to the folding of the basement rocks in Europe, as well as the more obvious overlying rocks, the structures of which had been understood in general terms in the nineteenth century. The model for the Himalayan situation was, therefore, the tectonic structures which Argand had earlier developed for the European Alps (ibid.: 187–8).

Argand envisaged a 'traction process' operating on the continents, leading to their drift. This gave rise to 'ripped buttonholelike depressions' in which the sima would appear; but his suggestion that the origin of the traction could be found in 'variations of the basal or frontal adherence between sial and sima' (ibid.: 134) was obscure. For Argand, 'fixism', with its notion of collapsing continents as earth contraction occurred, was incompatible with isostasy (ibid.: 125). However, he wanted to retain the idea of geosynclines in his theory, which he did by assuming that traction caused stretchings in the raft of sial on the sima.

According to Carozzi (1977), Argand's tectonic story was too complex for geologists to take in at the Brussels Conference, and even afterwards when the memoir and map were published, although some European geologists borrowed his ideas without always giving him due credit. In the period following the Second World War, Argand's ideas were more often mentioned than read.

Whereas Argand had difficulty with combining fixism and isostasy, it being hard to see how light sial might founder into denser sima, perhaps producing the oceanic deeps, this was precisely what Reginald Daly thought might happen. According to his theory (which was closely related to his thinking about magmatic stoping (see Chapter 9), the old cooling and contracting (wrinkled apple) theory of the earth, notions of isostasy, geosyncline theory and the geographical distributions of the ancient 'shield' areas of the earth's surface), the early earth might have had an irregular shape, with bulges near the poles and equator. It continually receives heat on its outer surface from the sun, and also loses it. But, for the interior, it is all loss, so to speak, the crust being a poor conductor. So the interior is supposedly cooling more than the external crust. This cooling causes further collapse of the oceanic basins and accompanying

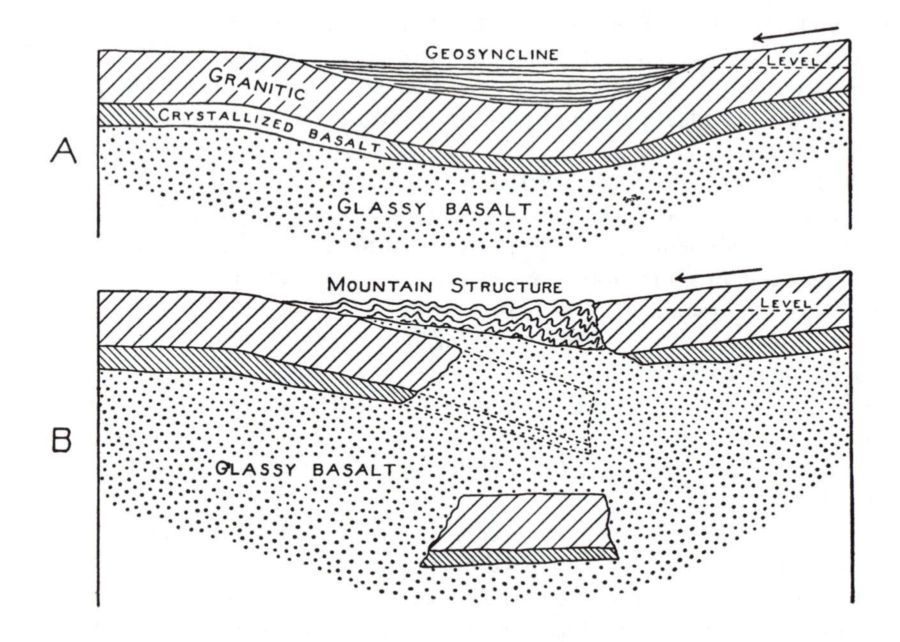

Figure 11.2 Diagram illustrating the crumpling of geosynclinal prisms of layered rocks by the sliding of a large block of the earth's crust, according to Reginald Daly (1926: 269)

doming of the continents (Daly, 1926: 263–7). Such changes then begin to lead to a 'sliding' of continental regions towards the oceanic basins, eventually giving rise to large-scale 'stoping' and crumpling of geosynclinal sediments, as shown in Figure 11.2.

The further development of the proposed process is shown in Figure 11.3. Despite the physical peculiarities of Daly's theory, it is interesting to see how similar his diagram is to that suggested according to modern plate-tectonic theory. For example, one can see in Daly's diagram something akin to the Benioff–Wadati fault zone of modern theory (see p. 240). It should be observed that Daly's theory allowed for a 'drift' (or 'slide') of continents, compatible with his petrological ideas and with the principles of physical theory (gravitation). His theory was substantially different from Wegener's, but was certainly mobilist.

Daly did not invoke the ideas of John Joly to account for the plasticity of the crustal substratum, although he might well have done so. Joly (1925), who was

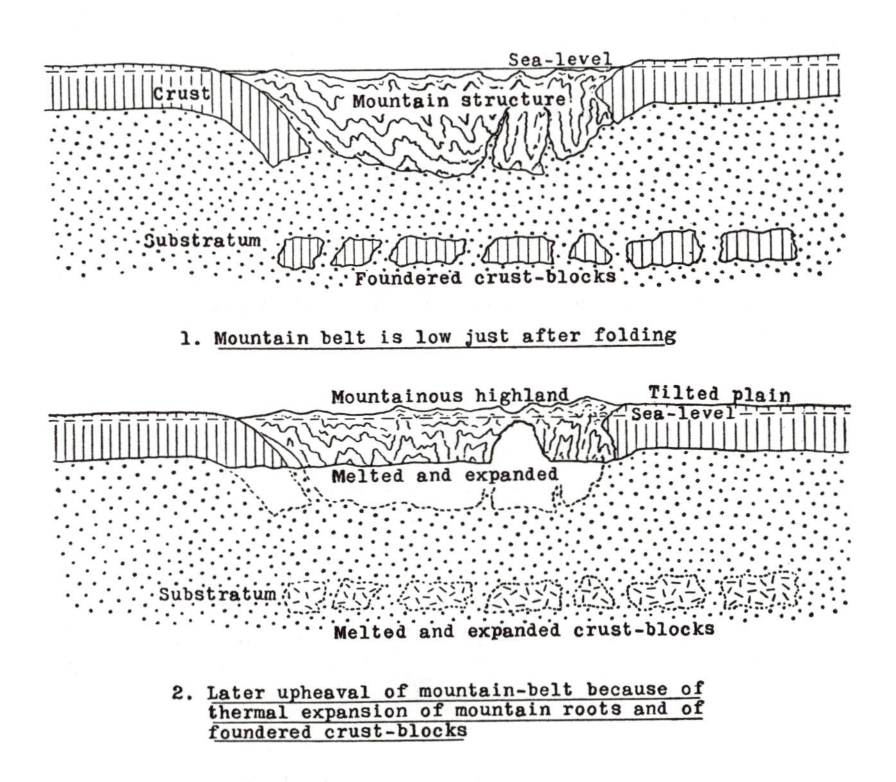

Figure 11.3 Diagrams illustrating the formation of mountains, according to Reginald Daly (1926: 276)

so interested in the role of radioactive substances within the earth, thought that the heat they generated might get trapped under continental masses to the extent that the underlying rock might become fluid, and thus mobility of the continents might be possible from time to time.

Earlier, at the University of Graz in Austria, Robert Schwinner had developed the idea of slow convection currents in a 'tectonosphere', with crustal plates riding on the backs of a convection stream (Schwinner, 1915, 1920), much as in modern plate tectonic theory. Likewise, Arthur Holmes (1928, 1928–31, 1929) proposed that the movements of the continents might be attributed to vast, slow-moving convection currents within the earth's interior (Figure 11.4). He pointed out that Jeffreys himself (1927) had shown that the viscosity of the mantle was insufficient to prevent convection. Holmes (1944/1965) made his ideas widely known in his influential textbook on physical geology.[10]

Wegener's major supporter was the distinguished South African geologist Alexander du Toit (1878–1948), of the University of Johannesburg. In 1923, he visited South America specifically in order to compare rocks there with ones with which he was familiar in South Africa and found many striking similarities. This helped make him a devotee of Wegener's ideas and 'mobilist' theory in general. Du Toit's book *Our Wandering Continents* (1937) gave a fine exposition of the drift theory, although his version differed from that of Wegener in that two original supercontinents (a northern Laurasia[11] and a southern Gondwana) were envisaged rather than Wegener's single Pangaea (Figure 11.5).[12] Du Toit deployed Suess's name, 'Tethys', for the intervening sea. He also accepted Barrell's concept of a plastic asthenosphere. Again, with du Toit as with Wegener, most geologists read the mobilist theory but did not believe in it.

Thus many of the leading elements of the modern geological synthesis were

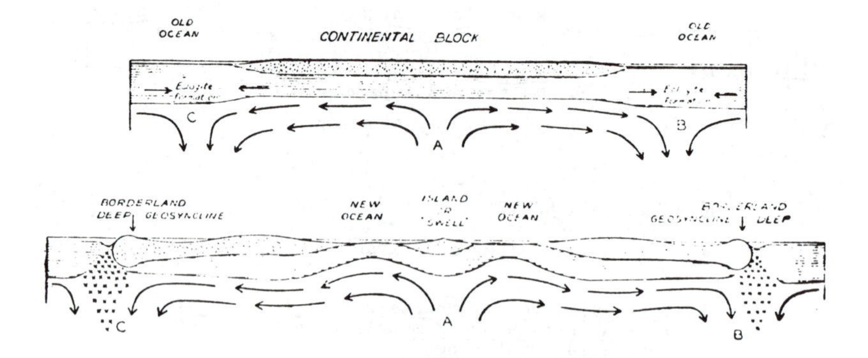

Figure 11.4 Production of the surface features of the earth by internal convection currents, according to Arthur Holmes (1928–31: 579) (by permission of the Geological Society)

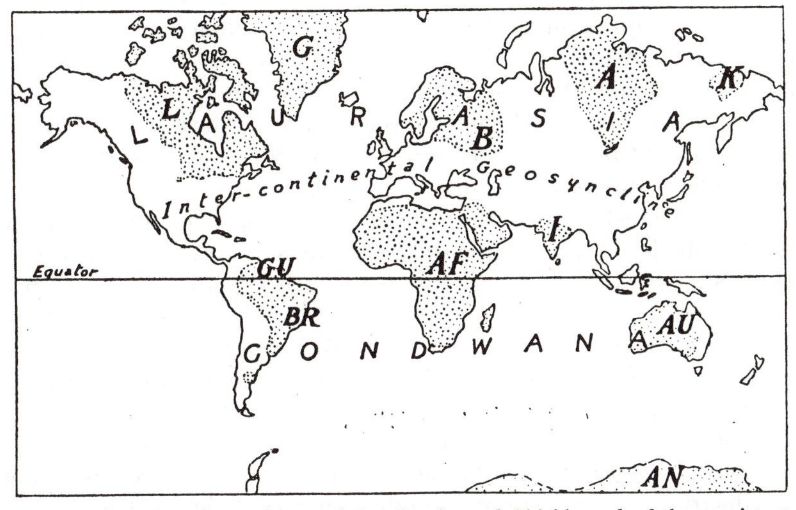

FIG. 6.—Showing the positions of the Continental Shields and of the persistent Geosyncline parting Laurasia and Gondwana: A, Angara; AF, Africa; AN, Antarctica; AU, Australia; B, Baltica; BR, Brazil; G, Greenland; GU, Guiana; I, India; K, Kolyma; L, Laurentia. (The Mercator's Projection used exaggerates areas in high latitudes).

Figure 11.5 Alexander du Toit's representation of Laurasia and Gondwana (du Toit, 1937: 39)

available for decades, but still did not succeed in gaining acceptance. There was plenty of geological research to do without accepting the grand theory of drift. The reasons for the rejection have exercised historians considerably. As has been shown by Howell (1991), there were several 'convectionists' writing in the first half of the twentieth century; and it was convection theory that eventually came to be accepted as the 'motor' for drift.[13] But the early convection theorists (including Holmes) believed that the tops of the rising hypothetical convection currents were located under the continents, not the oceans. This was because, following Joly, it was supposed that a significant proportion of the earth's heat was generated by radioactivity, and the heat being generated could supposedly dissipate from the oceanic regions, whereas it got trapped under the continental masses. Thus the early convectionists had their thermal currents moving in the opposite direction to that proposed by later theorists such as Harry Hess (see p. 267). In any case, it must be recalled that in Holmes's day the mantle was thought to be solid (as evidenced by seismology and Jeffreys' complex equations), so for most geologists it was difficult to conceive of convection occurring in the mantle.

As in all scientific controversies, the rejection of Wegener's drift theory involved a social decision, and as is often the case the feeling generated in a conference was significant. In the case under consideration, a meeting convened by the American Association of Petroleum Geologists in 1926 (van der Gracht, 1928) 'decided against' Wegener, with reasons which in retrospect do not seem especially compelling.[14] In Britain, a meeting of the British Association held at Hull in 1922 had been somewhat more sympathetic, but was by no means overwhelmingly supportive (Evans et al., 1923).

It was, then, chiefly American geologists, beginning to become world leaders in the twentieth century, who played the decisive role in the rejection of Wegener's model. Several broad reasons for their rejection can be suggested. There was the tradition of Dana's 'permanentism'. There was the tradition of the 'American' theory of isostasy,[15] which really had nothing to say about lateral movements. And there was the remarkable Panama Isthmus, which geographical feature was of such profound importance for the American nations. It was, then, a theory of 'isthmian links' that came to hold the field in American geology, and for much of the rest of the world, until the 1960s. This theory was, in a way, Lyell's doctrine revisited. It proposed the rising and falling of land masses, as required to allow the migration of animals and plants and thus explain the otherwise inexplicable features of biogeography (such as lemurs found in Madagascar somewhat similar to those in India).

There were several ideas propounded in the nineteenth century that were interesting precursors of the idea of isthmian links or of former supercontinents.

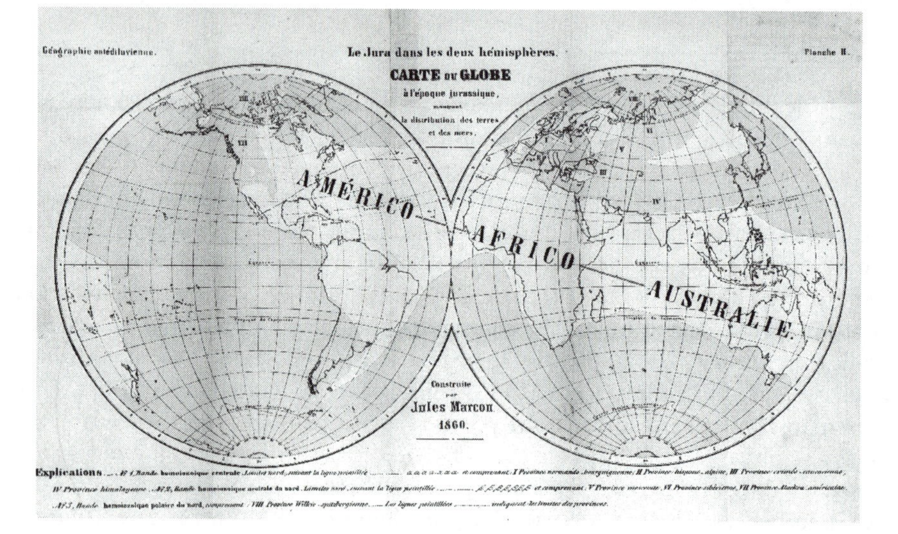

Figure 11.6 The 'antediluvian geography' of the Jurassic era, according to Jules Marcou (1857–60, Plate II) (by permission of the library of the Sedgwick Museum)

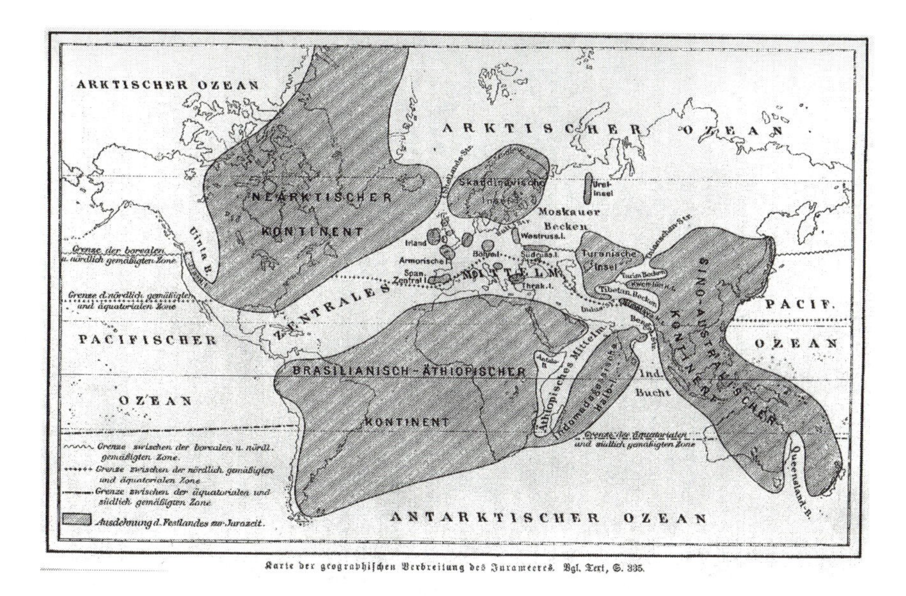

Figure 11.7 Distribution of continents in the Jurassic era, according to Melchior Neumayr (1887, vol. 2: 336) (by permission of the library of the Sedgwick Museum)

The earliest of these known to me is that of the French geologist, Jules Marcou (1824–98), in his *Lettres sur les roches de Jura* (1857–60), which proposed the existence of a grand continent spanning America, Africa and Australia in the Jurassic (Figure 11.6). In his *Erdgeschichte* (1886–7), the German geologist Melchior Neumayr (1845–90) offered a very different disposition of the continents for the same era, but one which also provided for land bridges (not necessarily isthmuses) to account for the quirks of biogeography (Figure 11.7). Such maps were necessarily speculative and were not based on satisfactory databases, but they are interesting in that they reveal an awareness in the nineteenth century of the need to account for some of the phenomena that were later addressed by the theory of isthmian links, continental drift, plate tectonics or the expanding earth.

In fact, the theory of isthmian links could deal with almost any problem in biogeography by the invocation of a sufficient number of *ad hoc* appearances and disappearances of land bridges. And this is precisely what was done in the version of the theory proposed by the influential Bailey Willis in a paper of 1932 that became paradigmatic for the next thirty years or so, even allowing for objections from authors such as du Toit and Wegener himself, who pointed out that the theory of the rising and falling of land bridges to accommodate

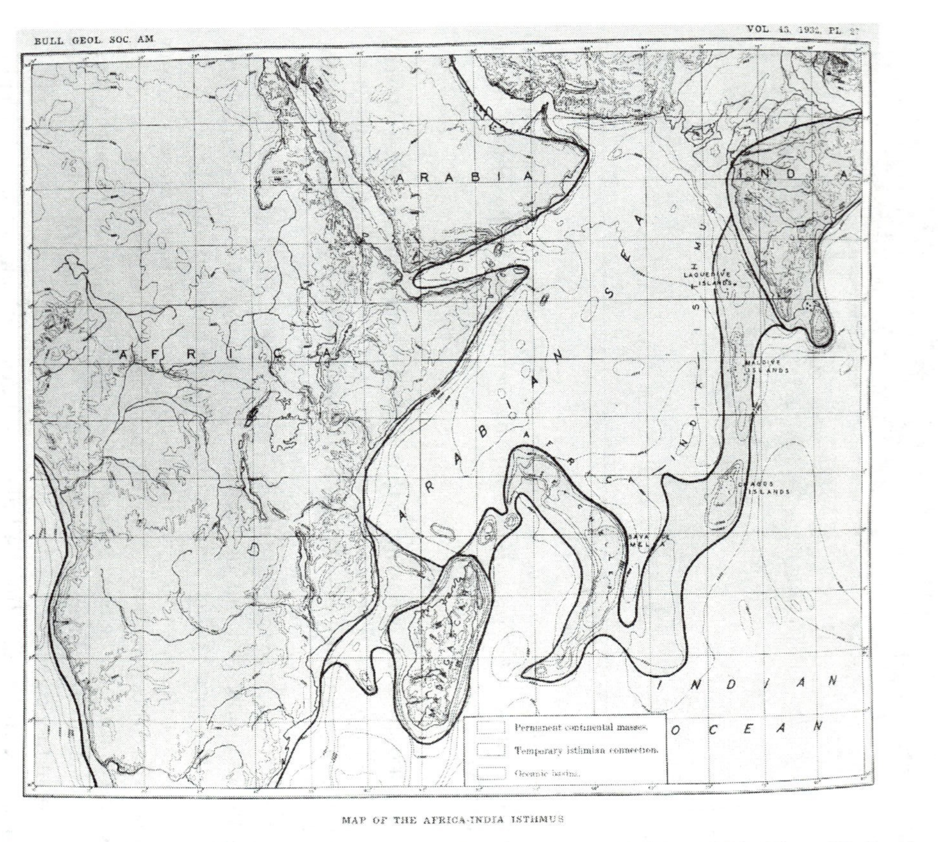

Figure 11.8 Illustration of theory of isthmian links, according to Bailey Willis (1932, Plate 27, facing p. 939) (coloured in original)

biogeographical data was incompatible with the 'permanentism' of Dana-tradition American geology (Wegener, [1929] 1966: 17). The nature of Willis's theory, which was developed in conjunction with another influential American geologist, Charles Schuchert (1932), is evident from Figure 11.8.

But it was not a satisfactory solution to the problem. Willis's theory gave little regard to the principles of isostasy, could not deal satisfactorily with animals or plants that had to cross different climatic regions by means of the proposed isthmuses, and seemed to find only limited support from studies of bathymetry – the survey of the ocean depths by echo-sounding or other means. For example, a map of the floor of the Atlantic Ocean, published by Taylor back in 1910 (p. 216), showed the mid-Atlantic ridge, and was quite incompatible with Willis's proposed transatlantic bridge.[16]

This brings us, then, to consideration of the question of oceanography in the further development of ways of thinking about the earth.[17] In the 1920s and 1930s, the Dutch geophysicist Felix Vening Meinesz (1932, 1934) made gravity surveys from submarines (to minimize the effect of the rolling of the vessels on the pendulum apparatus) and sought to link his findings to the idea of convection currents within the earth (Vening Meinesz *et al.*, 1934).[18] The chief object of the investigations was to gather data to help determine the figure of the earth, but the results suggested ideas of wider significance. It was found that there were gravity deviations or anomalies that could not be explained by the figure of the earth. In general, the ocean regions showed positive anomalies, and all the areas of major tectonic activity revealed significant gravity anomalies. Particularly remarkable were the long narrow belts discovered adjacent to island arcs (for example to the south of the main Indonesian archipelago), where there were marked gravity deficiencies. This suggested an excess of light sialic rock at such points, and seemed to be compatible with the theory of internal convection currents favoured by Vening Meinesz. Later, such regions became associated with the concept of 'subduction zones' (see p. 267).

However, major developments in oceanography did not occur until the Second World War, and more intensively in the following two decades, when systematic surveys of the ocean floors were undertaken, the research being funded to a large degree, especially in the United States, for military reasons, but also with an eye to petroleum exploration. Some of the leaders in this work were Harry Hess, a US naval officer who later became a professor at Princeton; Maurice Ewing, Director of the Lamont-Doherty Geological Observatory at Columbia University, with his co-worker Bruce Heezen; the Frenchman Xavier le Pichon, subsequently professor at the Collège de France, but working at the Lamont unit at the critical period of the late 1950s and 1960s; Bill Menard of the Scripps Institution; and Kenneth Hsü, who sailed on the famous survey and drilling vessel *Glomar Challenger*. Through this period there were further gravimetric surveys, as well as magnetic, seismic and bathymetric surveys and surveys of the flow of heat from the ocean floors (Revelle and Maxwell, 1952).[19] Many samples of ocean-floor sediments were collected and examined, and, by setting off explosions on the surface of the ocean and examining the reflections of the blasts from below, information could be gleaned about the rock layers underlying the oceans. Basaltic boulders dredged from the ocean floor were dated radiometrically.

The results were startling and largely unexpected. A system of midocean ridges circling the globe was discovered,[20] which correlated with regions of earthquake activity, such as had been well known and mapped since the time of Mallet (see Figure 10.2). Also, the ridges seemed to have 'rift valleys' running along their axes.[21] There were submarine canyons running down the sides of the continental shelves to the abyssal plains. And the deepest waters (ocean trenches) were found quite close to land, as, for example, parallel with the Chilean coast, or close to arc-like archipelagos such as the Aleutian Islands. Then some peculiar flat-topped submarine mountains were discovered, rather like flat-topped volcanoes in form. These were named 'guyots' by Hess (1946), who had conducted bathymetric surveys in the Pacific for the US Navy during the war. Such submarine features are illustrated in any good atlas today.[22]

Gravity investigations suggested that the ocean floors were generally in a state of isostatic balance, but there were negative anomalies over the midoceanic ridges and the deep trenches (as previously noted by Vening Meinesz). Also, there was a greater than average output of heat from the regions of the ridges and the trenches (Bullard *et al.*, 1956). Seismic studies and deep-sea dredging suggested that the layers of sediment out on the abyssal planes of the oceans were much thinner than expected, and that the underlying rock (down to the Mohorovicic discontinuity) was only about five kilometres thick. It appeared to be basaltic in composition (sima).

The results of magnetometry investigations of the ocean floor were perhaps the most remarkable, but before considering them we should notice the palaeomagnetic work on continental rocks conducted by Stanley Keith Runcorn, who became professor at King's College, Newcastle upon Tyne, and was subsequently at Imperial College, London, until his recent tragic murder in the US. He was interested in the problem of the origin of the earth's magnetic field, for which the physicist P.M.S. Blackett was endeavouring to develop theoretical models (the details of which need not concern us here). Runcorn, like Blackett, was also interested in the question of palaeomagnetism – a topic that he was pursuing until his untimely death, extending it to studies of the magnetism of the moon, other planets, asteroids and even nebulae.

In the 1950s, these workers and others began a systematic survey of the 'remanent' magnetism of the rocks of the British Isles, the assumption being that the direction of the magnetism was caused by, and hence parallel to, the direction of the earth's magnetic field at the time that the rock was formed. It appeared that the further back in time one looked (i.e. the older the rocks whose magnetism was being investigated) the more did the direction of the remanent magnetism differ from that of recently formed rocks. For Blackett, the results seemed to suggest that Britain had somehow 'rotated' and moved northwards. Runcorn and his group, on the other hand, suggested a movement of the poles through time, rather than the movement of land masses over the whole surface of the earth (Runcorn, 1955).

A group newly established at the Australian National University in Canberra also entered the field. Edward Irving (1956) established what was required to determine whether the poles were shifting, or whether the continents were moving relative to one another. One had to trace '(magnetic) pole pathways' through time in different countries. This was done by Irving, Blackett and Runcorn, the latter coming to the drift hypothesis with some reluctance. But his tracking of pole movement through time in both Britain and America strongly suggested that the poles did not coincide until geologically fairly recent times. The earlier divergences (Figure 11.9) had to be explained, then, by assuming that the continents had, through long ages, been slowly moving relative to one another (Runcorn, 1956, 1962). Thus more evidence was gathered in favour of mobilism; but still it did not convince the majority of geologists. The basic theory of geomagnetism was uncertain, and the sequence of geomagnetic pole reversals was not securely established, nor was the cause of the reversals understood.

Returning, then, to consideration of investigations of the ocean floors, geomagnetic studies were producing fascinating and unexpected results there as well. It appeared that there were 'stripes' of rock of different polarities on the ocean floor, the stripes running roughly parallel with the mid-ocean ridges. An example of this is shown in Figure 11.10. As may be seen, the regular magnetic pattern is disturbed in some places by what might appear to be 'faults', and bathymetric surveys also revealed a corresponding pattern of faulting on the ocean floor.

In the 1950s it was still possible to fit the information newly emerging from oceanographic studies into the traditional ways of thinking (except perhaps for the theory of isthmus links) or to ignore it. George Lees, President of the Geological Society of London, was disturbed that the new results from geophysics had 'unduly influenced geological opinion' (Lees, 1953: 217), and suggested that the real business of geologists was still stratigraphy. More usefully, Kenneth Landes, speaking in favour of the contractionist hypothesis to the American Association for the Advancement of Science in 1951, saw oceanic ridges as mountain ranges similar to those on land. The deep-sea trenches might be the geosynclines which geologists had postulated for so many years (Landes, 1952). According to the geosyncline theory, as understood in the 1950s, there were ancient stable areas of the earth's crust (cratons) such as the 'shield' areas of Canada, Scandinavia or Western Australia. There were geosynclines at their

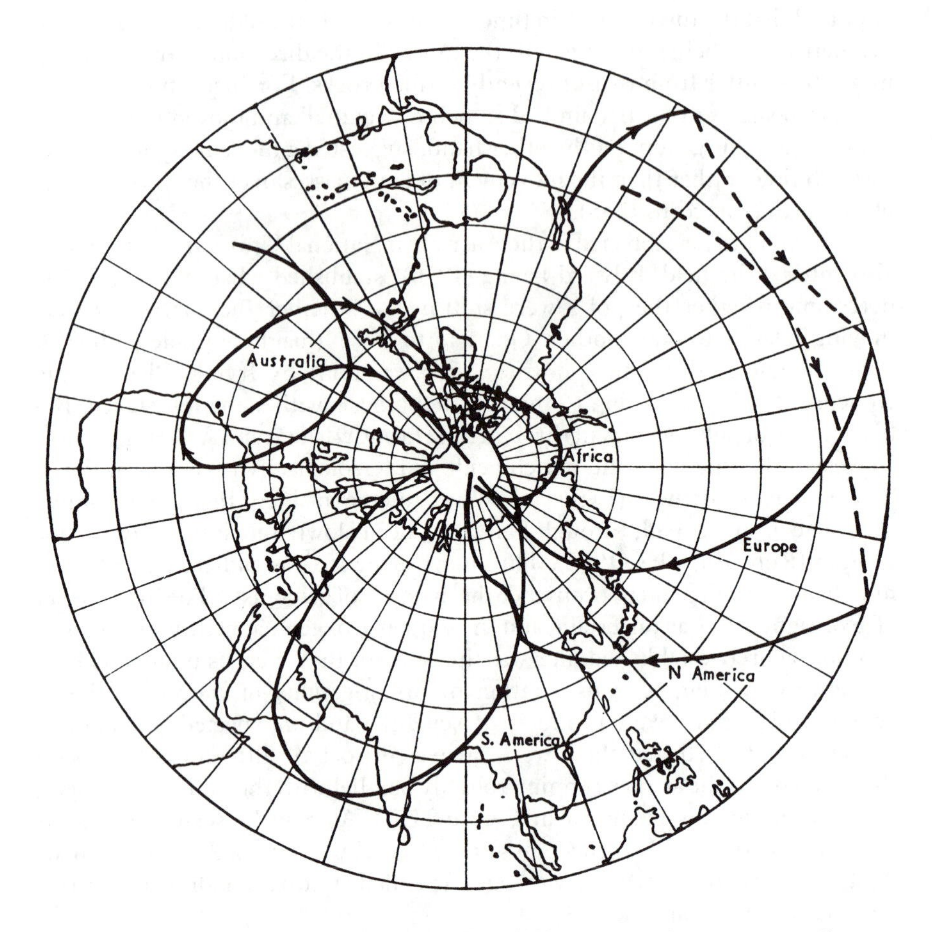

Figure 11.9 Pole-wandering curves for different continents, according to Stanley Keith Runcorn (1962: 24) (by permission of Academic Press and S.K. Runcorn)

margins which filled with sediments, later to be compressed by lateral forces as the earth contracted. New mountain ranges were formed, and these were added to the shield zones, the continents thus gradually increasing in size by the addition of the products of successive geosynclines (Dott, 1974, 1978). This way of thinking was by no means new. Charles Schuchert (1916) had conceived of Australia as growing in this fashion by the accretion of the eastern half of the continent – formerly the Tasman geosyncline, so named by Schuchert – to the western craton. He showed the idea pictorially for the North American continent in 1923 (Figure 11.11). The geosyncline theory of accretion and mountain

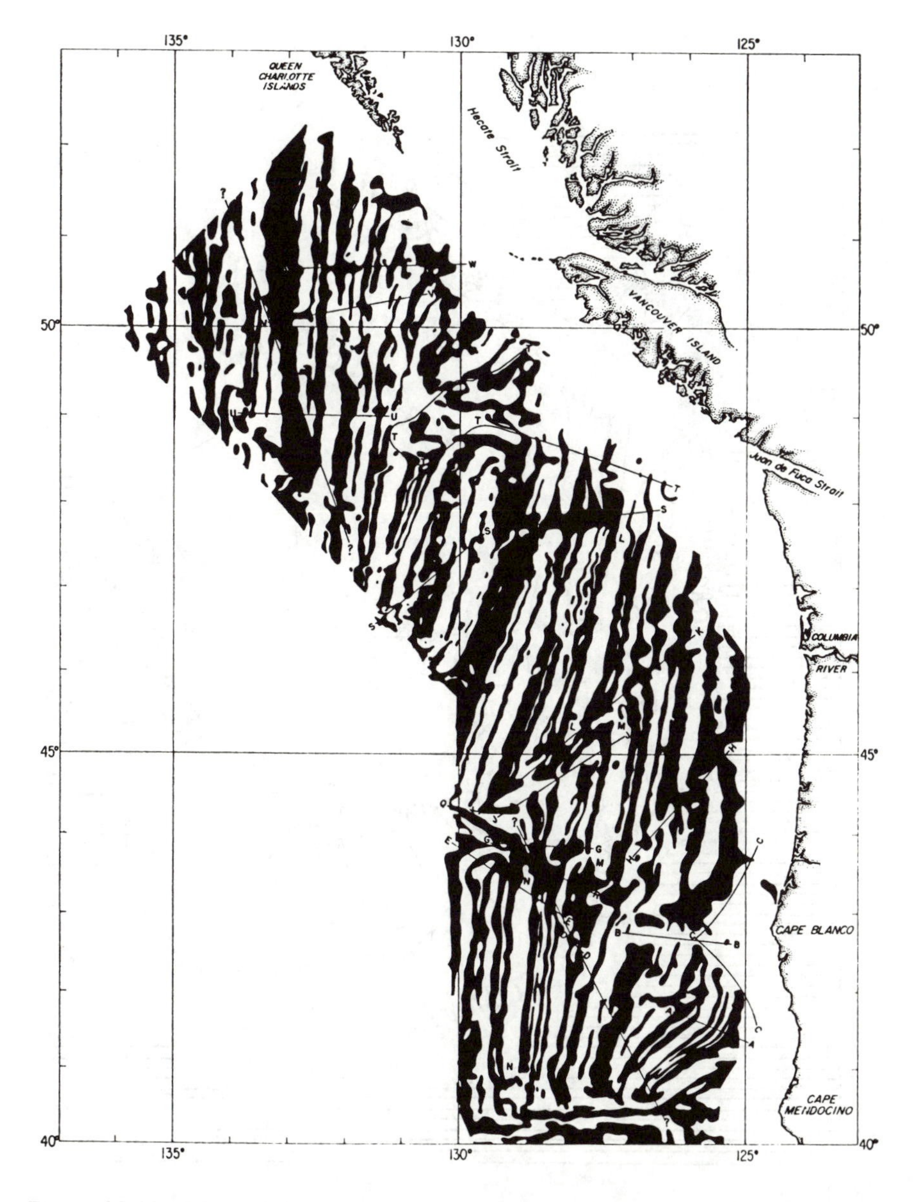

Figure 11.10 Map of the magnetic anomalies for the rocks of the ocean floor off Vancouver Island, Canada, according to Ronald Mason and Arthur Raff (areas of positive anomaly shown in black) (Mason and Raff, 1961: 1268)

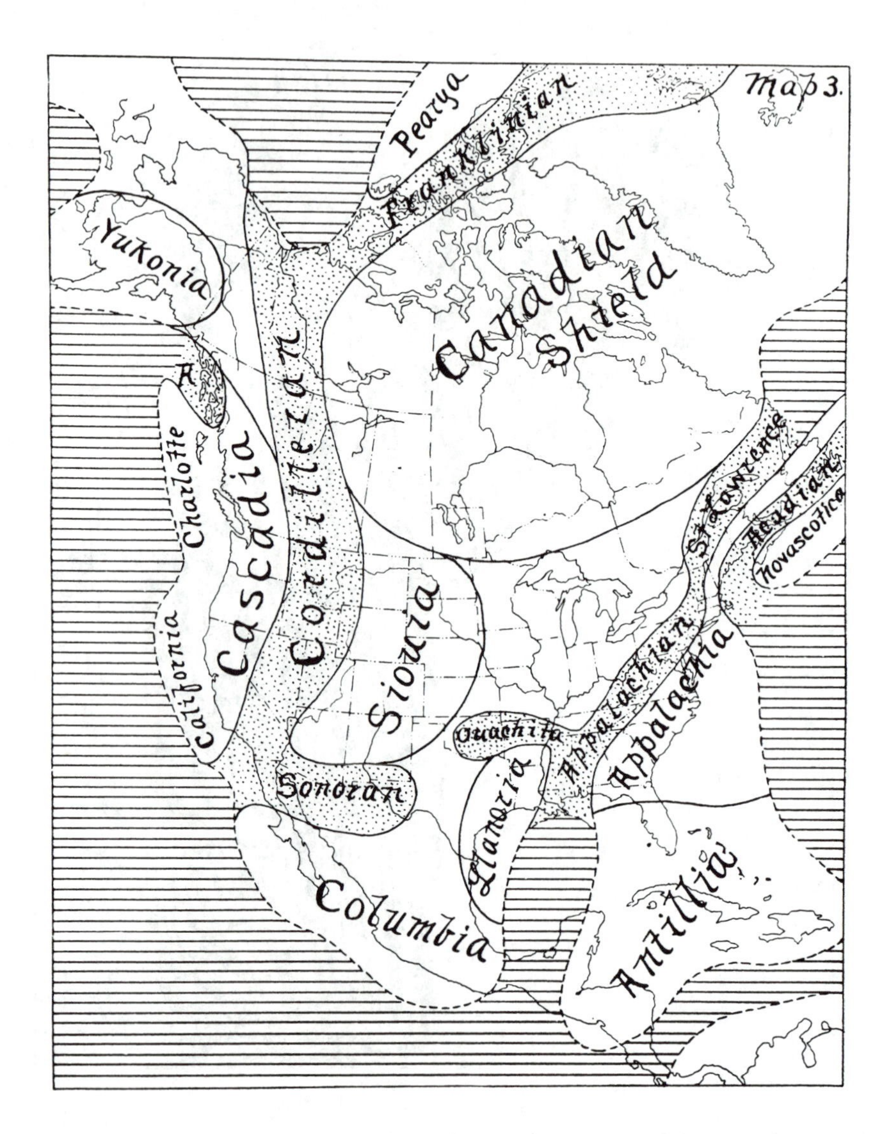

Figure 11.11 Growth of North American continent by successive accretions, according to Charles Schuchert's version of geosynclinal theory ('borderlands' white, geosynclines and embayments stippled) (Schuchert, 1923: 215)

formation, traceable to Dana, was still struggling along in the 1950s, although the cause of uplift after geosynclinal sedimentation remained uncertain and controversial, despite the work of Griggs (1939).

The Canadian J. Tuzo Wilson, Professor of Geophysics at Toronto University and later to become one of the key figures in the 'plate-tectonics revolution', liked the idea of continental accretion (Wilson, 1951, 1954; Jacobs *et al.*, 1959). But he was also endeavouring to take into consideration the new results of oceanography and geophysics, rather than insisting on the purity of older continental geological research.

The 1960s saw the independent introduction of what is now well known as the hypothesis of sea-floor spreading by Hess and Robert Dietz, a marine geologist of the US Navy Electronics Laboratory at San Diego. It appears (Frankel, 1980) that Hess formulated the idea first, made it known at a scientific meeting in 1959 and wrote of it with significant modifications in a preprint in 1960. But Dietz (1961) was the first to publish in the general scientific literature. Hess formally published in 1962. However, questions of priority need not concern us here.

Perhaps trying to fend off potential accusations of being too speculative – apparently a heinous geological sin in the 1950s – Hess (1962: 599) famously (and quite unnecessarily, in my view) described his paper as 'an essay in geopoetry'. He had for a good many years, along with Vening Meinesz, been developing ideas about convection currents within the earth and had spoken in 1959 about the idea of mid-oceanic ridges being at the tops of rising convection currents. In his paper of 1962 (which regrettably made no reference to Holmes or other early 'convectionists', although this could have been a tactical manoeuvre), Hess envisioned that there had been a major convection current in the primordial earth which had separated the lighter sialic matter at the surface to form a single supercontinent. Subsequently, convection currents had been at work in the mantle, basaltic matter rising at the ocean ridges and descending into the trenches, the process subsequently known as 'subduction'.[23] This basaltic material could carry the 'light' continents on its back like a conveyor belt. So there was no need to suppose, as did Wegener, that the continents somehow ploughed through the heavy basaltic sima of the ocean floors – the idea that Jeffreys found so repugnant. As Hess put it ('poetically'?): 'The continents do not plow through oceanic crust impelled by unknown forces; rather, they ride passively on mantle material as it comes to the surface of the crest of the ridge and then moves laterally away from it' (ibid.: 609).

Dietz's article (1961) was briefer than that of Hess, but it introduced the useful term 'sea-floor spreading', and it suggested that the moving surface was at a greater depth than Hess envisaged. Dietz maintained that the old division between the crust and the mantle (marked by the Mohorovicic discontinuity) was best given up in favour of a distinction between lithosphere and asthenosphere, for it was between these 'spheres' that the lateral movement supposedly occurred. The movement, for Dietz, was down in the ultrabasic interior of the upper earth, not just below the sialic matter of the continents. His theory was welcomed in the pages of *Nature* by the distinguished J.D. Bernal and by Tuzo Wilson.

What also made 1962 an important year was the finding by the survey ships from Lamont and Woods Hole that the mid-Atlantic ridge was in some places offset by fracture zones, sometimes several hundred miles in length. Such zones had previously been found by Menard in the Pacific (Menard, 1986) but had not been associated with mid-oceanic ridges. They were also observed by Russian survey vessels. The idea of sea-floor spreading also allowed one to make sense of the observations of guyots. These curious flat-topped submarine cones had every appearance of being old volcanoes that had had their tops planed off by marine erosion. But some of them had tops well below sea level. This could be understood if some parts of the sea floor were now descending, carrying the guyots with them, towards subduction zones.

The factor that most people today find particularly convincing in relation to the sea-floor spreading hypothesis has to do with palaeomagnetism. As can be seen from Figure 11.10, the pattern of palaeomagnetism for the rocks of the sea floor, as represented diagrammatically, have something of the appearance of the coat of a zebra. In 1963, two Cambridge scientists, Frederick Vine and Drummond Matthews,[24] of Sir Edward Bullard's geophysics unit, proposed that the results of palaeomagnetic research for the sea floor might be combined with the idea of sea-floor spreading to give a unified hypothesis. The idea was that basalt upwelled from the mid-oceanic ridges and spilled sideways, so to speak. It slowly spread out across the ocean floor, and was magnetized in a certain direction according to the direction of the earth's magnetic field at the time of the rocks' crystallization. Since the formation of basalt and its spreading might be expected to be approximately the same on both sides of the ridge, one might understand how there might be a symmetry between the patterns of remanent magnetism in the rocks on either side of a ridge (Vine and Matthews, 1963). Their hypothesis was developed on the basis of data from the area of a ridge in the Indian Ocean. Essentially the same hypothesis was developed, and earlier, by the Canadian Lawrence Morley, but was refused publication by *Nature* and the *Journal of Geophysical Research* – a blot on the escutcheons of both these journals, as it appears with hindsight.

An important further piece of evidence was provided by Sir Edward Bullard and his co-workers at Cambridge (Bullard *et al.*, 1965). According to Bullard's own account (1975), he became interested in the efforts of the Tasmanian geologist S.W. Carey (see p. 275) to fit pieces of the continental jigsaw together, and sought to perform such exercises himself, establishing in advance criteria for goodness of fit. Using the fit of the continental shelves, rather than present coastlines, Bullard obtained excellent results, particularly so far as South America and Africa were concerned. They were presented at a Royal Society symposium in 1964 – another social event where scientists could meet to battle out whether mobilist doctrine was to be accepted or rejected yet again. At the meeting, Bullard and his co-workers[25] concentrated solely on the geometry of the fit(s), and refused to be side-tracked into other questions by the anti-mobilists. This was, no doubt, tactically advantageous. But still the geological community was not convinced. M.G. Rutten, for example, objected (not unreasonably, one might think) that Bullard had arbitrarily omitted Iceland, whereas he included Rockall Bank, which conveniently fitted into the pattern.[26] All the various components of what was to become the paradigm (or even the dogma) of the theory of plate tectonics were not yet on the table.

Nevertheless, the new ideas were developed by Tuzo Wilson, who soon became the major player in the emerging field that was to become plate tectonics. He did collaborative work with Vine and Matthews, paying special attention to the fault systems found on the floors of the oceans. If the earth were flat, one might conceive of a smooth production of magma along an oceanic ridge, its movement laterally and its eventual disappearance into an oceanic trench (subduction). All this might be 'powered' by slowly moving subterranean convection currents. But, given that the earth is a sphere, the patterns must necessarily be much more complex, and to accommodate the production and movement of the ocean floor on a spherical surface the earth's crust must be 'torn' to adjust to the movements. Moreover, given that there are several such convection systems at work, they must somehow accommodate themselves to one another. Wilson (1965) went a long way towards showing how this might be possible. He did this by introducing the idea of a new kind of fault to the repertoire of structural geologists: 'transform faults', which allowed 'transformation' of the spreading of the ridges into the production of a new sea floor.

Before the 1960s, the types of faults known to geologists were 'normal' (where the forces are tensional), 'reverse' (where the forces are compressional), 'thrust' (where the compressive forces are extreme and special, low-angle reverse faulting occurs), and 'tear' or 'transcurrent (strike-slip)' (where two parts of the earth move laterally with respect to one another). Wilson's proposed 'transform fault' is depicted in Figure 11.12, along with the 'traditional' types of geological fault.

In the case of a transcurrent fault, it will be understood that the maximum rupture must be near the central region of the fault and that it dies out towards its ends. One must expect to find indications of stresses and strains on both sides of the fault for most of its length. In Wilson's suggested transform faults, however, things are somewhat 'easier' because of the production of new material at the surface. Indeed, it will be seen from Figure 11.12e that there is only crustal rupture along the lines of the transform faults themselves and there is no problem as to how this resolves itself at the end of the fault, as is the case for the ordinary

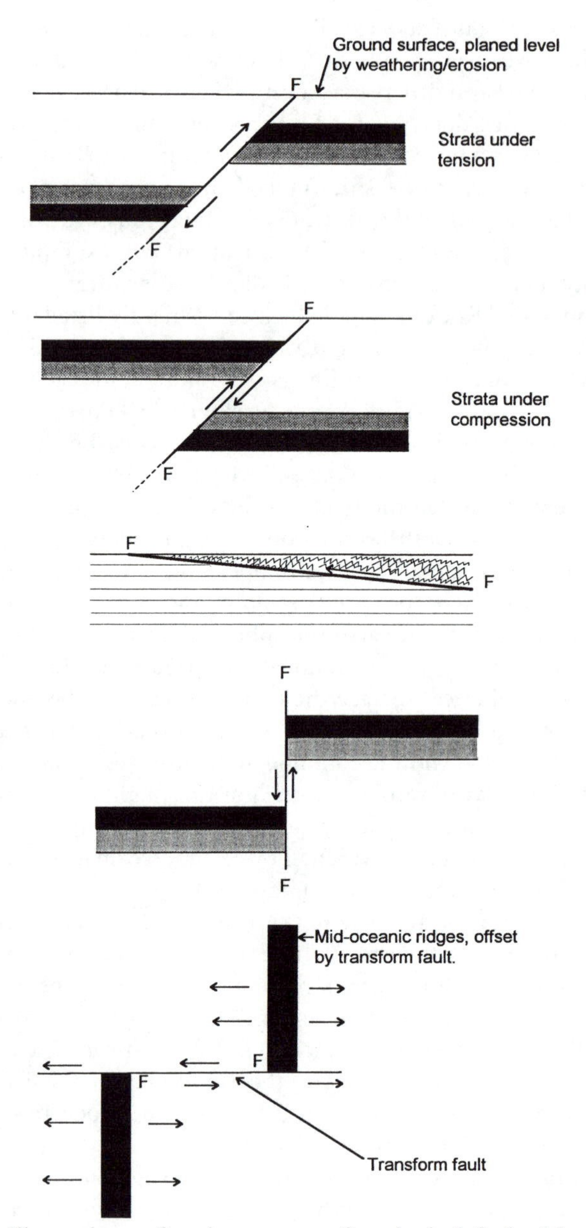

Figure 11.12 Illustrations of various types of geological fault: (a) normal fault (represented in section); (b) reverse fault (represented in section); (c) thrust fault (represented in section); (d) tear, transcurrent or 'strike-slip' fault (viewed in plan); (e) transform fault (viewed in plan). (For the transform fault, new material is thought to be formed at the mid-oceanic ridges, spilling 'sideways' on to the ocean floor.)

transcurrent fault. Moreover, by the step-like arrangement of transform faults on the sea floor, the (approximately) spherical surface of the earth is able to 'accommodate itself' to all manner of subterranean convection movements.

So Wilson began to advance the idea of the surface of the earth as consisting of a number of 'plates' which moved around relative to one another, but also were generated at the mid-oceanic ridges and swallowed up, so to speak, in the 'subduction zones' of the deep oceanic trenches.

Given the hypotheses of Morley (Morley and Larochelle, 1964), Vine and Matthews (1963) and Wilson (1965), several ways of testing the new ideas presented themselves. First, one might expect the basalts more distant from the ridges to be older than those newly formed and adjacent to the ridges. Second, one might hope to find a symmetry in the pattern of palaeomagnetism – the zebra stripes – on either side of a ridge. Such work was now possible because of the recent work on establishment of the global pattern of geomagnetic reversals (see Cox et al., 1963;[27] Glen, 1982). Vine, together with Wilson, worked on the development of a model as to what might be expected for the pattern of geomagnetic reversals for the so-called 'Juan de Fuca Ridge', off Vancouver Island, Canada. They were also able to compare their model with actual empirical results, and obtained a reasonable concordance (Vine and Wilson, 1965). Shortly thereafter, the data being collected by the oceanographic vessel *Eltanin*, which was trawling its magnetometer over the Juan de Fuca Ridge, were analysed by Walter Pitman.

With the more refined palaeomagnetic reversal scale recently to hand, the earlier sea-floor spreading hypothesis, which had been regarded with scepticism by the Lamont group under Ewing, began to seem more plausible – or better able to be corroborated by empirical information. Pitman was delighted to find that one of the traverses of the ridge – the famous '*Eltanin*-19' profile – fitted almost perfectly with the predictions made by Vine and Wilson (1965). The investigations were announced at a meeting of the American Geophysical Union in April 1966, and publication rapidly followed (Pitman and Heirtzler, 1966). Thus began the final stage of the 'plate-tectonics revolution'. The Lamont laboratory, which up till then had been generally sceptical under the leadership of Ewing, caved in. And so too did the vast majority of the world's geologists, until we reach the present situation where the theory seems 'necessarily' right. It is even taught in some primary schools, and certainly in all elementary geology texts.

The theory has the great advantage of being able to offer a plausible account of the origin of mountain ranges. The Himalayas, for example, could have been formed by the recent collision of the Indian subcontinent with the rest of Asia; the Urals might be the scar of some much more ancient collision between two plates. A mixture of oceanic crust and overlying sediments, together with sea water, might be carried by subduction from the Pacific Ocean region under the

Andean range, on the descending arm of a convection current into the hot region of the mantle. There, at depth, the former oceanic sediments, mixed with water, might fuse more easily than the usual material of the mantle, thus giving rise to magma. Such magmas, being more acidic in composition and less dense than the usual mantle material, would tend to rise up into the Andean chain, according to the principle of isostasy, forming the great volcanoes of that mountain range; likewise to the north in Central and North America.[28] If they failed to reach the surface, they might form batholiths, which could subsequently be revealed by erosion.

Also, recent work is trying to develop theories about distinct convection currents in the earth's core and mantle, relating the positions of 'hot spots' such as Hawaii and Iceland to places where convection currents are rising, whereas places such as Vietnam and Peru are related to localities where there are supposedly descending convection currents (Irvine, 1989). Such localities form a geometric pattern on the globe (such as might have pleased Elie de Beaumont). However, it is beyond the scope of this book to deal with the later development of plate-tectonic theory, although a few further points will be made in the following chapter.

Plate-tectonic theory is unquestionably a paradigm in the Kuhnian sense of the term. The vast majority of geologists today think about the earth through the lens of the plate tectonic theory. But there are a few who have stood out against it. As has been shown by Robert Muir Wood, in his interesting book *The Dark Side of the Earth* (1985), some Russian geologists, notably Vladimir Belousov (1962, 1968), were noticeably reluctant to accept the concepts of continental drift and plate tectonics.[29] Wood has tried to argue that plate tectonic theory was eventually accepted with such enthusiasm in the United States that it became a kind of flagship for American scholarship and was to some extent embroiled in cold-war politics. Thus Wood sees the mobilism of plate tectonics, in a mobile democratic society, pitted against the 'fixism' of Russian geology in the 1960s and 1970s, in an intensely conservative society.

This, I suggest, is drawing a pretty long bow. While I am sympathetic to historians who attend to the relationship between scientific theories and the social circumstances in which those theories are developed, I think it unlikely that the 'fixism' of Russian politics had its causal theoretical counterpart or consequence in Belousov's 'fixism'. A more parsimonious hypothesis is that Russian geologists in the cold-war period had less access to oceanographic data such as were compiled by the Scripps, Lamont and Woods Hole institutes, these data being gathered partly for military purposes or with military ends in the background. Moreover, the greater part of the former USSR is geologically stable, so that Soviet scientists were, in the main, not working in areas that might suggest mobilist theories. In contrast, Chinese geologists accepted mobilist ideas and plate tectonics relatively easily, apart from during the period of the Cultural

Revolution, when much scientific work stopped in China and contacts with Western scientists were severely restricted. However, some of the Chinese theories were significantly different from those developed in the Western world (Yang and Oldroyd, 1989). In the case of Belousov, he had his own theory of mountain building, developed from the notion of diapirism. And he had concrete objections to the mobilist theory, with its emphasis on lateral crustal movements. For example, Belousov objected that the plate tectonic theory did not account satisfactorily for the large vertical movements that seem to have taken place in and around continents, as for example in the area of the North Sea.

There were a few Western geologists who sought to resist the floodtide of plate tectonics. Son and father Arthur and Howard Meyerhoff (1972a, b) stood out for Dana's old idea of permanentism, trying to suggest that the magnetic striping of the sea floor might be attributed to varying chemical composition of strips of the sea floor. Perhaps predictably, Jeffreys didn't like plate tectonics any more than continental drift, and maintained his opposition into the last edition of his book, *The Earth* (1976).

However, the most interesting theory opposed to plate tectonics still current is that of earth expansion. This was being explored as quite an active research programme in the 1950s, when the old fixism was under question and when the arguments in favour of drift raised by Wegener and du Toit, such as the 'fit' of the continents and certain striking facts from biogeography, were not adequately answered by the geology of isthmian links, permanentism and a shrinking earth. Laszlo Egyed (1956, 1957, 1960), Professor at the Geophysical Institute of the Eötvos University, Budapest, contended, on the basis of evidence compiled by H. and G. Termier (1952), that the geological record suggested an overall diminution of sea level through time, which was attributable to earth expansion. To explain the expansion, he suggested that the inner core, the outer core and the mantle of the earth were three different forms (phases) of essentially the same ultrabasic material. Further, there was a slow, steady and irreversible transition from the material of the inner core to that of the outer core and then to the mantle. This was of lesser density, so that the whole process involved expansion, with an annual increase in the earth's radius of about 0.5 mm. By conservation of angular momentum, the increase in size produced a decrease of the earth's rate of rotation.[30] But the geomagnetic tests for the hypothesis that Egyed proposed for his theory proved inconclusive. The Hamburg physicist, Pascual Jordan (1966/1971), proposed an expanding earth as a natural 'accompaniment' of an expansion of the cosmos as a whole. Bruce Heezen (1959, 1960) at Lamont saw the rift valleys of Africa, deep-sea trenches and the valleys along midoceanic ridges as extensional features, but he represented the idea of an expanding earth as 'quite speculative'.

Perhaps the most obvious and in a sense the most intriguing arguments for expansion were produced by the (former) East German engineer Klaus Vogel.

Figure 11.13 Warren Carey and Klaus Vogel, holding a model to illustrate the expanding-earth theory (Carey, 1988: 268) (by permission of Herr Klaus Vogel)

The point was that the continents could not all be fitted snugly together to form a 'supercontinent' (Pangaea, Gondwana, Laurasia or whatever) on a globe of the dimensions of the present earth. But, if the earth were formerly much smaller, the fit was remarkably good. Suppose, for example, one draws the present continents, fitted together on a semi-inflated balloon, and then one blows the balloon to its full size, with the continents separating from one another and forming a set of distinct land masses.[31] This is something like what Vogel did. He can be seen in Figure 11.13 with one of his models: a small 'ancient' globe inside a larger transparent 'modern' globe. When seen thus, the theory certainly looks most plausible.

The man holding the globes with Vogel in Figure 11.13 is the Australian geologist, S. Warren Carey (b. 1912), Emeritus Professor of Geology at the University of Tasmania and, at the time of writing, one of the few remaining upholders of the expanding-earth theory. In 1992, I had the pleasure of attending a large meeting organized in his honour by the Geological Society of Australia at Ballarat, Victoria, on the occasion of Carey's eightieth birthday. At this meeting, speaker after speaker got up to pay their respects to the doyen of Australian geology but promptly went on to criticize his ideas. Only a handful of delegates, mostly his former students, as I understand, had much to say in favour of the earth-expansion theory. I conclude, then, that Carey's expansion theory is now in a state of mortal decline so far as the community of geologists is concerned, such that it should perhaps be called a 'degenerating research programme' (Lakatos, 1970). Yet it has recently been published, generously illustrated, by a prestigious American academic press (Carey, 1988). And I suppose, if we are to believe Lakatos's ideas about the dynamics of science, it may one day rise like a phoenix from the ashes and again capture attention. This is, I suspect, unlikely. Nevertheless, Carey's theory offers such an interesting way of thinking about the earth that I should like to give it some consideration. As a historian of science, I am not obliged to recount only the side of a controversy that is 'winning' at any given time.

Carey's doctoral research was conducted before the Second pre-World War in the physically difficult and geologically complex terrain of New Guinea. At that time, he became sympathetic to mobilist ideas, but for the sake of securing his degree he chose not to deploy them in his thesis. He was thinking of ways of mentally straightening out crumpled mountain ranges[32] and closing gaps now occupied by seas or oceans so as to yield Wegener's Pangaea. He did publish these ideas later (Carey, 1955) and presented them more fully in a well-known symposium organized at the University of Tasmania (Hobart) in 1956 (Carey, [1958] 1959). At this conference, however, Carey began to consider the possibility that a better fit for the continents might be achieved if the earth had formerly been smaller. He had already encountered the ideas of Egyed.

For present purposes, there is no need to take the reader through all the stages of the development of Carey's thinking, but reference may be made to his remarkable synthesis of his old age, *Theories of the Earth and Universe: a History of Dogma in the Earth Sciences* (Carey, 1988), which I think all geology students of today should read, just as earlier students should have read du Toit's *Wandering Continents* (1937) or Wegener's *Origin of Continents and Oceans* ([1929] 1966). Whether or not they believe Carey is not my concern; but it is interesting to see how phenomena can be interpreted from the perspective of a paradigm quite different from the one that currently prevails.

In line with some cosmological theorists,[33] Carey proposes a continual production of new matter, balancing that lost to us by the perpetual expansion of the universe. He further suggests that the optimum place for the creation of new matter may be at the centres of planets, stars or galaxies, which are loci of minimum potential energy. So the expansion of the earth may be due to the constant production of new matter at its interior. Iron may be the most likely first element, it is suggested, since it has the minimum energy per nucleon. But as the mass builds up there may be a kind of 'pressure-cooking' of the elements so that others, such as hydrogen, may be formed. The process gradually accelerates: the more matter there is, the more rapidly is new matter formed. Turning to the solar system as a whole, we can see different planets in different stages of evolution. Mercury is like the earth at an early stage, but giant Jupiter is on the verge of turning into a star.

Carey has nothing against the idea of magma arising along mid-oceanic ridges, moving sideways and acquiring a 'striped' palaeomagnetic pattern as it forms. He is also content with Wilson's transform faults. But, for Carey, subduction is simply a myth.[34] He conceives of the earth's surface as being made up of eight continental polygons, each several thousand kilometres across and meeting at the tectonically active zones. These are equivalent to the plates of orthodox theory, but Carey supposes that they extend right down through the mantle to the fluid core of the earth, so that they exist as huge polygonal prisms.[35] Superimposed on these, there is a second-order system of surface basins and swells, extending over the continental land masses and the ocean floors – for example, about ten for Africa. Superimposed on these, there are smaller polygons of crust marked off by substantial fault lines, sometimes sites of major seismic activity.[36] Within these, there are smaller pieces of the mosaic. Then we come to major joints, a few hundred metres apart, and finally there are the patterns of intersecting joints so often observed in the field, especially on wave-cut platforms. This hierarchy of structures has arisen, suggests Carey, as the earth adjusts itself to the pressures exerted from within as new matter is formed in the interior.

To account for mountain building, Carey, like Belousov, relies on the phenomenon of diapirism. With the earth's crust supposedly in tension from the pressure below, it gets thinned in some places. On the upper surface, it may form an oceanic trench in which sediments are deposited, as in geosyncline theory. But, with thinning of the crust, there is also the opportunity for upward diapiric movement of material from the mantle, which may pierce the crust and swell forth to form a mountain range. A Benioff–Wadati zone, on this theory, is a thrust plane up which an emerging diapir is sliding. Island arcs are formed where a diapir rises off a coast. The whole process has been speeding up significantly since the Mesozoic, for the more matter there is, supposedly the more there is produced. So the ocean basins are relatively new features. Therefore, the present is *not* the (simple) key to the past so far as earth history is concerned.

On top of all this we may then add what is perhaps Carey's most striking concept, namely huge global torsions. Two are to be considered: an equatorial torsion and a circum-Pacific torsion. It will be recalled that Suess proposed the idea of an ancient ocean (Tethys), which separated his two hypothetical primeval continents, Laurasia and Gondwana. Carey proposes, then, that there has, during the last two hundred million years, been more production of oceanic crust in the southern hemisphere than in the northern. Or, putting it another way, there is now more continental matter north of Tethys than to the south. This has supposedly led to a gigantic sinistral shear, roughly along the equatorial belt, as the northern hemisphere tends to lag behind the southern, under the stresses induced by the earth's rotation. If this displacement due to torsion is allowed for when trying to solve the jigsaw of the crustal fragments, then the pieces can be fitted together more satisfactorily, Carey claims.

But there is supposedly another great shear zone round the margins of the Pacific. The earth can be thought of as two hemispheres, one – the Pacific region – chiefly water, the other where most of the land is found. This asymmetry has, according to Carey's thinking, given rise to another great shear: a conjugate, dextral, great-circle shear, roughly perpendicular to the Tethyan shear. This must also be allowed for when playing the game of the crustal jigsaw. Carey's overall theory for the evolution of the earth is summarized in Figure 11.14, reproduced from his *magnum opus* of 1988.

The contrasts with the orthodoxies (dogmas, as Carey puts it; current paradigmatic way of thinking about the earth, as we say) of plate-tectonic theory are obvious. Plate tectonics has the Himalayas arising as a result of a 'collision' between India and the rest of Asia. Carey thinks otherwise. The celebrated 'Indus suture', where India is supposedly welded on to Tibet, is for Carey a mark of the Tethyan torsion. The Himalayas themselves are accounted for, not by compression, but by the diapiric model.

And whereas the plate tectonic theory envisages an 'endless' formation and subduction of the material of the ocean floors, so that it need not surprise if none of that floor is older than about two hundred million years, Carey objects that one might expect *some* more ancient floor to have escaped the subduction process if plate-tectonic theory were correct. For him, of course, the absence of earlier ocean-floor material is not a problem: he does not think that the ocean basins began to be formed until then. 'Subduction is a myth' is his motto.

Carey has also suggested how, in principle, his theory may be tested by direct measurement, using the laser reflectors that have been placed on the moon and on artificial satellites. At the time of writing, such tests have not been completed by the National Aeronautics and Space Administration (NASA), but we may look forward in the not too distant future to a direct empirical confirmation or refutation of the expanding-earth hypothesis, just as direct measurements have shown that the distance between Africa and America is increasing (which is, of course, compatible with both orthodox mobilism and the expanding-earth hypothesis).[37]

To conclude this chapter, I should like to say a few words about the role of biogeography in the formulation of plate-tectonic theory and also what the theory may now do for studies in this branch of science. Wegener himself placed

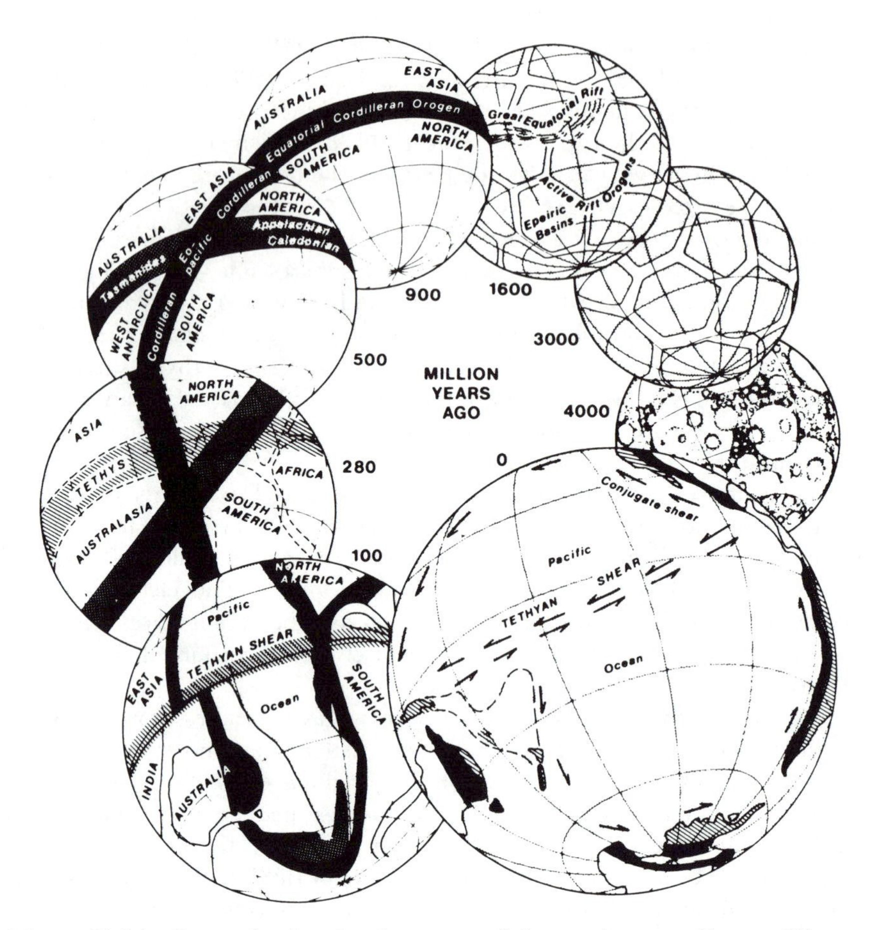

Figure 11.14 Stages in the development of the earth, according to Warren Carey (1988: 313) (by permission of Stanford University Press and S. Warren Carey)

considerable weight on arguments drawn from biogeography. For example, he referred (Wegener, 1966: 101) to the work of Arldt (1917), which mapped the world distribution of certain related garden snails, earthworms, pearl mussels, perch and mud-minnows, showing them to span oceans, such as the Atlantic, but not continents, such as America. It seemed unlikely that such creatures could have crossed land bridges, and the occurrence of similar freshwater fish in Europe and America was implausible according to the theory of isthmian links.

The occurrence of marsupials in Australia and South America with similar parasites was likewise difficult to explain. Any case where a bridge was required across climatic zones would seem to be particularly problematic as an explanatory device. Many such cross-zonal similarities are known, however: for example, the 'beech' and birch forests (*Nothofagus* and *Fagus*) of the cool/temperate zones of the southern and northern hemispheres respectively. Such bipolarities have been noted by many authorities (e.g. du Rietz, 1940). For a biologist or palaeontologist such facts may be more telling than all the high-powered calculations of geophysicists such as Jeffreys,[38] which seemingly proved that mobilism was impossible. Like Darwin faced with the calculations of Lord Kelvin concerning the age of the earth, and like Mr Micawber, the biogeographers were hopeful that something might turn up in the future.[39] As we have seen, mobilism did eventually prevail and bring sense to the chaotic data of biogeography – but only when the geophysicists, with their measuring instruments, said that it might do so.

But the new geology – or 'earth science', as they now like to call it, for it involves so much more than the old nineteenth-century petrology, palaeontology, field mapping and historical stratigraphy – has now begun to pay back its loans from the biological sciences, with interest. This is in the domain of what is called 'vicariance biogeography'.

The process of biological speciation may occur when parts of a species are geographically distant or separated from one another (allopatric speciation). This happened, for example, when the giant flightless moa birds of New Zealand evolved to great grazing animals, in the absence of mammals that occupy the 'grazing niche' in other parts of the world. The moas' ancestors, whatever they were, were certainly not giants and presumably they flew to New Zealand. Speciation may also occur when organisms are in the same general locality but occupy slightly different habitats, or are active at different seasons or different times of the day (sympatric speciation). Such speciation is generally initiated by behavioural changes within a population, as, for example, when the ancestors of otters began to adopt an aquatic lifestyle.

In some cases allopatric speciation is initiated by the formation of new geographical barriers – for example, the Panama Isthmus presumably divided a marine population that was formerly united, leading to the evolution of different marine species on either side of the isthmus.[40] This form of allopatric speciation is called 'vicariance'.

It can readily be appreciated that the phenomena of continental drift or plate tectonics could be favourable to vicariance biogeography. The separation of continents would be a ready cause of the formation of new barriers to interbreeding and would encourage speciation,[41] and the collision of continents could bring different populations into radical competition with one another, usually to the detriment of large numbers of species having to deal with new competitors. So some extinctions might be related to the mobility of continents.

It has been known at least since the nineteenth century that the earth's living organisms seem to be divided into zones. For example, A.R. Wallace, the co-discoverer with Darwin of the theory of the origin of species by natural selection, who spent a considerable amount of time in south-east Asia, noticed a remarkable difference between the oriental and Australian fauna. A remarkable biological boundary seemed to run east of the Philippines, south between Borneo and Celebes and then between Bali and Lombok in modern Indonesia. This remarkable division (monkeys to the north, kangaroos to the south, if you like) has come to be known as Wallace's line. Eventually, Wallace (1876) divided the earth into six zoogeographical zones.

Many other schemes of zonation have been suggested, but details are not required here.[42] The point is simply that plate tectonics begins to give new meaning to biological zonation, which is otherwise difficult to understand. For Darwin (1859: 352–8) (and Wallace), there were 'centres of origin'. Species appeared in limited areas and then spread as far as they were able to do so. Darwin gave much attention in his own research and in his writings to the ways in which the distribution of organisms might occur (on floating logs, on the muddy feet of birds, by floating seeds or whatever). Such phenomena are not excluded by vicariance biogeography, of course, but they are seen as processes subsidiary to the principal motor: the movement of continental masses.

'Vicarianism' was taken up with enthusiasm by some of the workers at the American Museum of Natural History (New York) in the 1960s and 1970s. They were interested in the ideas of a remarkable amateur writer, Leon Croizat, who had earlier worked out the principles independently in massive inchoate volumes published privately in Venezuela (Croizat, 1958, 1962).[43] Eventually collaborating with Croizat (Croizat et al., 1974), the Museum scientists conducted a programme of research in which they sought to synthesize biogeography and plate tectonics, dispensing in the process with the older idea of centres of origin.[44] The theory was also favoured by some biologists in New Zealand.

But there was another ingredient in the theoretical pudding: the so-called method of cladistic analysis of the German taxonomist Willi Hennig (1966). I should like to say a few words about this matter in order to show how scientists are now looking towards a higher-level synthesis involving plate tectonics, biogeography and systematics.

Hennig's ideas were by no means entirely new to science (Craw, 1992), but they were received with great interest when they became known in the English-speaking world through the translation of his major text in 1966, and Hennig is generally regarded as the founder of the system of biological classification known as cladistics or cladism. What Hennig offered was a classificatory scheme that was supposed to be in accordance with Darwinian principles, but would avoid certain difficulties arising from evolutionary convergence or problem cases such as the proper classification of birds or crocodiles. Evolutionary (phylogenetic) relationships were supposed to take precedence over apparent anatomical similarities (or differences).

Roughly, Hennig's method was as follows. One had first to distinguish between 'ancestral' and 'derived' characters. For example, has evolution proceeded from a hoof to a pentadactyl hand or foot, or vice versa? Considering the large number of five-digited creatures around, the limited number of hoofed animals and the specialized nature of hooves, it is reasonable to assume that the hoof is the derived character. It is, then, the cladist's job first to determine whether the various characters of a group of organisms being classified are ancestral or derived. This task accomplished, the classification scheme should then be constructed according to the number of derived characters the various species have in common: the more there are, the closer the evolutionary affinity. Hennig presented his results in the form of 'cladograms': branched figures (with only two branches at each point of branching), which sought to represent the pattern of shared derived characters as parsimoniously as possible (i.e. with the minimum number of *ad hoc* hypotheses to deal with aberrant data).

All this may seem somewhat remote from the concerns of this book, but the point to note is that vicariance biogeographers claim they can relate the existence of the biogeographic zones and the broad results of cladistics (taxonomy) to the separation and subsequent movement of continents according to the principles of plate tectonics (Nelson and Platnick, 1981; Nelson and Rosen, 1981; Briggs, 1987). Croizat himself (1962: 209) compared the development of vicariance patterns to the breaking of a sheet of glass by successive blows:

> [L]et us suppose that we have laid on the table . . . [a] piece of glass . . . and let us homologize this glass to a whole order of plants or birds. Let us hit this glass a blow in such a manner as but to crack it up. The sectors circumscribed by cracks following the first blow may here be understood to represent families. Continuing, we may crack the glass into genera, species, and subspecies to the point of finally having at the upper right hand corner a piece about 4 inches square representing a subspecies.

Thinking in this fashion, one may envisage the different groups of organisms as having come into being by the fracturing of continents, as a result of the processes of plate tectonics, with the breakup of Pangaea. Thus a grand synthesis is sought in which the results of studies in biogeography, taxonomy (cladistics) and plate tectonics all cohere. Such coherence is what is sought in science, and when it is found it can, I believe, be regarded as a mark of truth – or at least an indication that the researches are on the right track.

However, in the case sketched so briefly above, it cannot be said that the synthesis is either satisfactory or complete. Arguments about vicariance biogeography have been and are intense (Hull, 1988). There are too many complicating factors for all to fit together smoothly. After all, some organisms are certainly dispersed in the ways that Darwin suggested. But at least one may say that we have a vision of how plate tectonics can help to unify biology as well as geology. If that be so, we have the prospect, at least, of a way of thinking about the earth and its living inhabitants even grander than Wegener imagined.

Chapter 12

Some Grander Ways Of Thinking

In the previous chapter, I have given a sketch of the emergence of the theory of plate tectonics. By and large, it seems to do its job, explaining many phenomena that are otherwise puzzling. For example, consider palaeoclimates. As we saw in Chapter 7, while there is now a reasonably well-established astronomical theory to account for the fluctuating glaciations of the Pleistocene, it is known that in Australia the effects of this were only felt on the mainland in the highest mountain areas of the Kosciusko range. Yet there is clear evidence of glacial remains at low altitudes in Australia from Permian–Carboniferous and late Precambrian times. And there are indications of subtropical climates in Greenland at about the same time. Such phenomena cannot be explained by the astronomical theory of the ice ages alone, but they are intelligible in terms of mobilist theory. Australia, for example, must have been situated somewhere near the South Pole in Permian times, and has subsequently been moving northwards. Palaeoclimatological evidence was in fact adduced by Wegener in his original formulation of his theory, and the matter was re-examined in the 1960s with the renewed interest in mobilist theories. It was found to cohere well with the data on palaeomagnetism (Opdyke, 1962), providing further evidence assisting the emergence of the plate-tectonic paradigm.

So the plate-tectonic paradigm offers more synthesis, more coherence. But, as initially envisaged by Wegener, drift theory proposed the breakup of a single large supercontinent since the Mesozoic. Wegener was reticent as to why there was a supercontinent in the first place. Geological investigation and theorizing since the plate-tectonics revolution has been attending to this question. A recent survey article by Murphy and Nance (1992),[1] for example, gives some indication of the direction of contemporary thinking.

Suppose that, for some reason, we have a single supercontinent – a 'Pangaea' – the rest of the globe being covered with a 'superocean' (which has been called 'Panthalassa'). Pangaea might be unstable for two reasons: first, because it acts as a kind of insulating blanket on a portion of the heat-emitting earth; and, second, because of forces generated by the spinning of a lop-sided earth.[2] (In principle, both factors could be at work.) Such a supercontinent might then begin to divide, rather as Africa is presently splitting along her rift valleys, and as the Red Sea has apparently opened relatively recently in geological history.

Mantle material might be expected to upwell at the site of a rift, but this mafic material is denser than the previously present sial of the continental crust, so it subsides as it cools and solidifies, and provides a site where a new ocean may begin to form. We may call this an 'internal' ocean, in contrast with the 'external' Panthalassa. Such 'internal' oceans can extend by the continued upwelling and lateral movement of basic magma, as is occurring in the Atlantic at present from the mid-Atlantic ridge.

But, in time, the oldest 'internal' ocean floor (at the furthest remove from the mid-oceanic ridges) cools, to the extent that it becomes denser than the material of the asthenosphere, and then the process of subduction begins. It has been estimated that this process might take about two hundred million years to begin, which is, of course, conveniently close to the measured age of the oldest known rocks of the ocean floors. It is plausible to suggest, therefore, that subduction and mountain building may begin round the margins of the Atlantic fairly soon in the geological future.

One may reach a stage, then, when the original supercontinent is broken into fragments; but the heat that originally built up under the supercontinent is then largely dissipated, and further emitted heat can escape through the various ocean floors. But subduction is still proceeding, and, as Murphy and Nance put it, it can now 'reel in' the dispersed continents. So the intercontinental oceans disappear once again, and a new supercontinent may be formed. It is suggested that such a cycle may take about five hundred million years – two per billennium.

According to the Murphy/Nance model, there can be three kinds of process leading to the formation of mountains. First, there are those such as the Andes, where a subduction zone meets an advancing continent. Second, there can be mountain ranges such as those of the Himalayas, where two continents may have collided with each other. Third, there are island arcs such as are found in the Aleutians. In this third case, there could be a subduction zone in an 'exterior' ocean, adjacent to a retreating land mass. The rising melt therefore passes through oceanic crust to form a fringing archipelago, and becomes different in composition from that generated in a case such as the Andes. In some instances, the 'exterior' subduction is thought to pull off fragments of continental blocks, as perhaps in the example of Japan. But the ocean floor may convey the island arc chains or the microcontinents, such as Japan, back on to the main land mass, which thus grows in a kind of process of continental accretion (as envisaged long ago by geosyncline theorists such as Dana and Schuchert). Likewise, the collision of continental fragments can gradually lead to the formation of continental 'mosaics'. Even little Britain is a 'mosaic', according to the latest palaeogeographic atlas (Cope *et al.*, 1992). Various fragments of the country, such as the Hebrides, the Grampians, the southern uplands of Scotland, the Lake District, the midland platform, etc., may be fragments of different ages ('terranes') that have become 'parked' together by the vicissitudes of plate tectonics. Thus the 'United Kingdom' was formed. Great faults such as the Great Glen fault, which separates the northern highlands from the Grampians, document the coming together of these disparate terranes.

This grand theory offers a glimpse of a cyclic theory of geotectonics. It explains why there may have been great periods of orogeny at different times; and it may perhaps be related to great biological changes that occurred at, for example, the beginning of the Cambrian, when the first shelly fossils evolved at the time of a breakup of a supercontinent. This prompts the thought that the geotectonic cycles envisaged in plate-tectonic theory may be intimately connected with the history of life. Indeed, as will be indicated below, some theorists are now toying with the idea that life itself may be necessary for the cycles to occur. This leads us to consider what has come to be called the 'Gaia' hypothesis and the recent suggestions about the earth itself being (in a sense that is not wholly metaphorical) some kind of living entity.

To consider the history of the development of such ideas, albeit briefly, we must give some consideration to geochemistry and sedimentology – topics that may seem somewhat unglamorous compared with plate tectonics or evolutionary biology, and which have scarcely been treated by historians of science. Yet in the long run they may prove to be every bit as interesting and important as ways of thinking about the earth. In any case, they are now being integrated with plate tectonic theory.

As pointed out by Jacques Grinewald (1988), Hutton's cyclic theory may in a sense be regarded as the first manifestation of modern ideas of the earth as a huge living organism. Hutton imagined the earth as a 'macrocosm', with cyclic properties analogous to those of the 'microcosm' of the human body. Also, farmer Hutton recognized the importance of living organisms in the cyclic behaviour of the 'macrocosm'. But it was the Plutonist aspect of Hutton's theory that fared better than the cyclic one in the nineteenth century (Gregor, 1992). In 'classic magmatism', fresh magma was erupted from to time from the earth's interior, and the new rocks were gradually attacked by weathering and erosion so that sediments were deposited in the oceans. To us, this implies a cyclic process. But the cyclic aspect of Hutton's theory was lost sight of in the nineteenth century. Hutton's unconformities, so important to him as evidence for the cyclic character of terrestrial processes, were assimilated by catastrophist theory with its doctrine of linear progression. There were major cyclic theorists such as Lyell, but as we have seen, his non-progressionism didn't make much progress.

Other eighteenth-century cyclic mineral theories such as those of Buffon and Lamarck, non-mechanical and involving life as part of their 'theories of the earth', are only recently being reconsidered as interesting forerunners of modern ideas. Buffon (1783–8) supposed that five primitive 'glasses'[3] were formed as the primeval globe cooled. By various processes, including interaction with the gases of the atmosphere and the involvement of lime from living organisms, these primitive 'glasses' supposedly gave rise to the large range of minerals and gems found on the earth. Soil, from vegetable detritus, was thought to play an important part.

Lamarck, who was Buffon's protégé and from whom he derived many of his ideas, suggested that decayed matter from animals and plants served as starting-points for two main chains of mineral conversions, both ultimately leading to the formation of rock crystal or quartz (Lamarck, 1797/Year 5, table facing p. 349[4]). The mechanism whereby this might happen was not made clear, although it seemed to involve some kind of stripping away of extraneous matter from the pure elementary earth. So Lamarck envisaged a 'cyclic' conversion of one kind of mineral into another and sought to develop a unified approach to the study of the earth, the atmosphere and living organisms (coining the term biology for this last) (Lamarck, 1802/Year 10: 7–8). He insisted (ibid.: 151) that living organisms were constantly modifying the materials of the earth's crust. But Lamarck's chemistry was archaic. He rejected the new ideas on elements set forth by the powerful Lavoisier, preferring the ancient view that earth, air, fire and water could be interconverted. So, along with his evolutionary ideas, Lamarck's views on minerals, his biogeochemical ideas and his thinking on the 'co-evolution' of the earth and its living inhabitants were all ignored or rejected.

Lamarck's mineralogy, if so it may be called, was out of tune with the new chemistry. It was speculative rather than empirical. It was not a success. Yet I believe he recognized that life had to be considered as a fundamental feature of the earth's evolution, not just the evolution of animals and plants. In this regard, his thinking foreshadowed some of the most interesting twentieth-century ways of thinking about the earth.

To an extent in the tradition of Lamarck, but much more successful for the establishment of a unified, holistic theory of the earth, was the 'Romantic' scientist and explorer, Alexander von Humboldt (1769–1859). He performed heroic deeds in the jungles of Central and South America (Botting, 1973), and established the 'cognitive style' which has usefully been calleed 'Humboldtian science' (Cannon, 1978). It was really the forerunner of the sciences of ecology and biogeography. With his vast empirical knowledge of the earth, its living inhabitants and its physical circumstances, von Humboldt wished to achieve a synthetic or holistic view of the cosmos, showing particularly how different organisms were related to the conditions of temperature, light, salinity, soil type and so on. It was, needless to say, beyond the capacity of one man to produce such a synthesis, although von Humboldt did succeed in publishing a mammoth five-volume work, *Kosmos* (1845–62), which went some way towards achieving his goals, and he only closed his project when his life ended at an advanced age. No one could question Humboldt's empirical credentials.

As a result of his experiences during his voyage round the world in the years 1838–42 and his consequent familiarity with corals in the Pacific islands, Dana (another great traveller) began to think of the role of life in the formation of the earth, and supposed that the earth was proceeding progressively in the direction of 'cephalization'. However, this aspect of Dana's theory was modest compared with his total contribution to geology, and it has received little emphasis.

It will be recalled that the nineteenth-century neo-Neptunist Sterry Hunt endeavoured to give a chemical account of the history of the earth and the chemical processes, perhaps involving life, that might have led to the formation of, for example, the massive deposits of limestone or gypsum seen in many parts of the world. Hunt's model was unidirectional, not cyclic. Nevertheless, he was also drawing attention to important features of the earth's surface that tended to be overlooked or treated unfavourably by other nineteenth-century theorists. However, Hunt's English-language contemporaries, busily engaged in mapping, and with an essentially Huttonian magmatic theory, did not wish to see a resurgence of Neptunism in any shape or form, and his ideas made little progress, particularly after the *Eozoön* débâcle.

It was Eduard Suess who coined the term 'biosphere' near the end of his book *Die Entstehung der Alpen*, in the phrase *selbständige Biosphäre* or 'self-sustaining biosphere' (Suess, 1875: 159). The term took its place alongside the terms 'lithosphere' and 'hydrosphere', which Suess also coined in the same work. Life was at the interface between the atmosphere and the lithosphere. As we know from Chapter 8, Suess also attempted a monumental synthesis of geological knowledge in his *Das Antlitz der Erde* (Suess, 1883–1909), which can be seen as another grand synthesis in the Humboldtian style. The topic of the biosphere was again considered in the concluding chapter of Volume 4 of this great book (Suess, 1909: 739). Yet so far as English-language readers were concerned the concept was largely an irrelevancy. It was Suess's tectonic ideas that attracted most attention, both in his own day and in the eyes of subsequent historians.

Suess's ideas did, however, strike a resonant chord in France, being congenial to those who admired the work of Henri Bergson (1907/1911), with its vitalist tendencies. In particular, they appealed to the young Jesuit priest, Pierre Teilhard de Chardin (1881–1955), then Professor of Geology at the *Institut Catholique* in Paris, and later to become famous as a palaeoanthropologist[5] and, after the Second World War, as author of voguish mystical Christian writings (e.g. Teilhard de Chardin, 1969). Deploying Suess's notion of 'biosphere', Teilhard de Chardin went beyond the Austrian master to propound the notion of a 'noosphere' – a 'sphere' of consciousness or mentality – and, even more bizarrely, a 'Christosphere' – a sphere of spiritual morality. Ultimately, all were somehow to unify in a single transcendent 'Omega Point', which was Teilhard de Chardin's (unorthodox, perhaps even heretical) notion of God.

Such ideas may seem to lie beyond the scope of the present exposition. But they are not, in that they were discussed in Paris with the Bergsonian Catholic mathematician and philosopher, Edouard le Roy, and with the Russian geologist, Vladimir Ivanovich Vernadsky (1863–1945), then in Paris and now coming to be regarded as one of the major figures in twentieth-century ways of thinking about the earth. Vernadsky, it should be observed, had in a sense a much grander view of nature than did Teilhard de Chardin, in that the French priest was chiefly concerned with the physical, mental and moral evolution of mankind, whereas Vernadsky took all life for his canvas in his thoughts about the earth. It is poignant, however, that he recorded (Vernadsky, 1945: 5–6) that his development of a 'geological conception of the world', in which life and mind played critically important roles, came to him as a result of his experiences in the First World War when he was engaged in obtaining precise data on strategic raw materials for the Russian government. Mankind, he then realized, was becoming a mighty geological force.

Vernadsky studied chemistry in St Petersburg under the celebrated Dmitri Mendeleev and geography under the less renowned – but, so far as Vernadsky was concerned, more influential – Vasili Dokuchaev, who specialized in soil studies. Vernadsky's efforts were chiefly devoted to the study of Precambrian rocks and the occurrence of life forms in these rocks,[6] with the idea that living organisms might have acted as major geochemical agents.[7] From such studies, and developing ideas discussed with his French colleagues, Vernadsky propounded the idea of a global ecology – an ecology of the biosphere.[8] Life was a global phenomenon and had (as in von Humboldt) to be seen from a cosmic perspective. Like Lamarck before and like modern sedimentologists since, Vernadsky was interested in the grand cycles of the different elements – oxygen, nitrogen, carbon, sulphur and so forth – that occur in and on the earth (or more generally within the biosphere). Life-forms, he recognized, were responsible for the emergence of an atmosphere containing oxygen, and the upper shield of ozone was the protective outer layer of the biosphere. The various exchanges of the elements between sources and sinks should be studied quantitatively, not merely mystically, as did Teilhard de Chardin with his loose talk of 'noosphere' and 'Christosphere'. It was Vernadsky who coined the term 'biogeochemistry'; but he also deployed the idea of a 'noosphere'.[9]

According to his own account, Vernadsky (1944: 487) began his systematic biogeochemical enquiries in 1916. Their first major expositions, given in his books *Geokhimiya* (1924) (appearing also in French the same year as *La Géochimie*) and *The Biosphere* (1926 in Russian[10]), were to a considerable degree overwhelmed by the troubled state of Soviet biology and politics in the 1930s, and were further forgotten as a result of the events of the Second World War.

In his own country, his ideas were only revived in the 1960s, but they have since become highly regarded. They began to have some influence in the West before then as a result of the work at Yale of a Russian émigré, Alexander Petrunkevich, and Vernadsky's son, George (or Georgii Vladimirovich) who also worked at Yale. Besides, several of Vernadsky's earlier publications were available in French from an early stage. Thus his ideas have long been known in the West, compared with those of some of his countrymen.[11]

It was through Petrunkevich and Vernadsky's son that the Russian ideas became known to George Evelyn Hutchinson, Professor of Biology at Yale, and an important school of ecology and biogeochemistry came to be established at that university, with Vernadsky's ideas in view.[12] There was great interest in feedback mechanisms and self-regulating cycles. Also incorporated were the ideas of the Norwegian geochemist, V.[13]M. Goldschmidt (1888–1947), who had known Hutchinson's father. Goldschmidt too was deeply interested in the cycling of the different elements through the earth and biosphere.

Essentially, Vernadsky's work showed how life-forms penetrated into every corner of the globe, except the upper reaches of the atmosphere and the deep interior of the earth, and that life was involved in the formation of virtually all the sedimentary rocks – limestones, dolomites, coal, salt deposits, phosphate deposits, bauxite, ironstones, etc. – to the extent that it was impossible to think of the geological history of the earth in a meaningful way without considering the role of living organisms, be they ever so small and insignificant individually.

To be understood properly, then, the earth has to be thought of as a kind of system, of almost infinite complexity. Life is ubiquitous. We cannot understand the earth by thinking of it just as a machine. Rather, it has itself the self-regulating properties of a living organism. To this end, Vernadsky studied the cycles of the different elements in the biosphere. He wished to do this quantitatively, but, at the time and place that he was working, he hardly had the resources to undertake such a task.

In one of his last writings, Vernadsky (1945) speculated on an even wider front, suggesting that the granitic envelope of the earth is 'the area of former biospheres'. This notion seemed bizarre to his contemporaries and was not, at the time, given much attention. For, before the establishment of the theory of plate tectonics, the perpetual cycling of everything was not a grand concern of geologists, although they had, of course, long been familiar with and interested in such processes as the hydrological cycle (which was understood in a general way by Aristotle), and there had been major geological cyclists like Lyell.

But in fairly recent years, since the development of plate tectonics and with the enormous increase in sophistication of sedimentology, the rather speculative ideas of Vernadsky have come into their own. Thus we find, for example, the Norwegian petrographer and geochemist T.F.W. Barth writing (1962: 379) in the last line of his book: 'The diversification of igneous rocks is caused by sedimentary processes'. This seemingly outlandish statement, according to the tenets of classic magmatism, was put forward because by the 1960s it was evident – as Hutton had realized long before – that magma contains recycled matter, and the matter formed by igneous processes may differ according to the nature of the sediments that are recycled.

Only in fairly recent years has the vast programme that Vernadsky envisaged begun to bear fruit, in that details may now be filled in for the general idea that the earth should be regarded as a system, or rather a multiple system of systems, in which living organisms play a fundamental role – perhaps even being essential to the tectonic movement of plates. In this emerging synthesis, the work of the American geologist Robert Minard Garrels (1916–88) has been of fundamental significance, although it is only becoming known quite recently to the general public through the work of Peter Westbroek (1991). Better known is the work of James Lovelock, supported by Lynn Margulis, in their 'Gaia' hypothesis, which has captured the public imagination, and has been embraced by the environmentalist movement. I shall now attempt to indicate some of the leading ideas of these writers, in which we see emerging a remarkable synthesis of ideas from plate tectonics and sedimentology.

Garrels started his academic career at Northwestern University, Illinois, in 1942, took up the position of Chief of the Solid States Group of Geochemistry and Petrology Branch of the US Geological Survey in 1952, moved to a chair at Harvard in 1955, returned to Northwestern in 1965, moved to the Scripps Institute in 1969, then to a chair at Hawaii in 1972, back again to Northwestern in 1974 and finally to a chair at the University of South Florida in 1980 (MacKenzie, 1989). He was, like the elements he studied, constantly on the move.

One of the things that Sterry Hunt and other neo-Neptunists had tried to do back in the nineteenth century was establish a sequence of chemical reactions that might have occurred in the oceans in the earth's distant past and which might have produced the sequence of rocks that we now see in the stratigraphic column. But this goal was hopelessly premature. Knowledge of both geology and chemistry was insufficiently advanced to undertake such a task convincingly. But, in the twentieth century, there has been a better chance of attempting to investigate such speculative chemical geology both theoretically and empirically.

In his textbook *Evolution of Sedimentary Rocks* (1971: 294), Garrels and his co-author Fred MacKenzie suggested chemical reactions that might have occurred in the very early oceans. For example:[14]

$$\underset{}{14H_2O} + \underset{\text{(albite)}}{6NaAlSi_3O_8} + \underset{\text{(siderite)}}{3FeCO_3} + \underset{\text{(enstatite)}}{15MgSiO_3} + 3CaCl_2 =$$

$$6NaCl + \underset{\text{(chlorite)}}{3Mg_5Al_2Si_3O_{10}(OH)_8} + \underset{\text{(calcite)}}{3CaCO_3} + \underset{\text{(greenalite)}}{Fe_3Si_2O_5(OH)_4}$$

$$5MgSiO_3 + CaAl_2SiO_8 + CO_2 = Mg_5Al_2Si_3O_{10}(OH)_8 + CaCO_3 + 4SiO_2$$
(enstatite) (anorthite) (chlorite) (silica)

$$Fe_3Si_2O_5(OH)_4 + 3CO_2 = 3FeCO_3 + 2SiO_2 + 2H_2O$$
(greenalite) (siderite) (silica)

Such reactions, which might have been occurring in the early Precambrian (or Archaean), had an important role for ferrous iron, in the absence of atmospheric oxygen and when the concentration of carbon dioxide in the atmosphere was much greater than at present. The albite and anorthite (types of feldspar) might have originated from the leaching of volcanic rocks, with the precipitation of amorphous silica (chert). As can be seen, a complex sequence of reactions is envisaged, ultimately leading to the formation of chemical precipitates ('sediments') of silica, ferrous silicates, calcium carbonate and pyrites (not shown in the preceding equations). Such a chemical geology was just what had long before been envisaged by the neo-Neptunists, without, however, having the chemical knowledge to propose such reactions. It may be noted that what is proposed above in the hypothetical equations does not envisage any role for life; but the materials formed seem to agree with what is known for rocks older than about two billion years.

But somewhat younger rocks – say less than two billion years in age – display different characteristics. The sulphur appears in oxidized form, as sulphates in gypsum. The iron is now in its ferric condition, appearing as great deposits of haematite, instead of the ferrous compound, siderite. Something had happened: it was the results of the activities of living organisms, which, since their first appearance, have been inexorably influencing the whole process of the 'earth cycle'. It was living organisms that gave rise to an atmosphere containing free oxygen, which changed the nature of the chemical reactions occurring on earth forever[15] and which thus fundamentally and permanently altered the character of the cycles that we know of now through the work of the likes of Garrels.

The chemical processes that led to the formation of life are still not understood, although much research has been done on this fundamentally important question.[16] I shall, for the purposes of this book, assume that simple life forms began as early as 3.5 billion years ago, perhaps somewhat earlier,[17] on a planet now thought to be about 4.5 billion years old. As we shall see, there is reason to believe that without life the earth would have dried up back in the Archaean, and we would thus have become a dead planet like Mars.

But to return to Garrels. In his textbook with MacKenzie (1971), he described the evolution of the crust–ocean system as being analogous to a large factory: a factory with two major plants with a connecting conveyor system. The first plant was the site where magma is formed, the heat being generated by radioactivity. In the second plant, the sediments are cooked by burial, heat again being

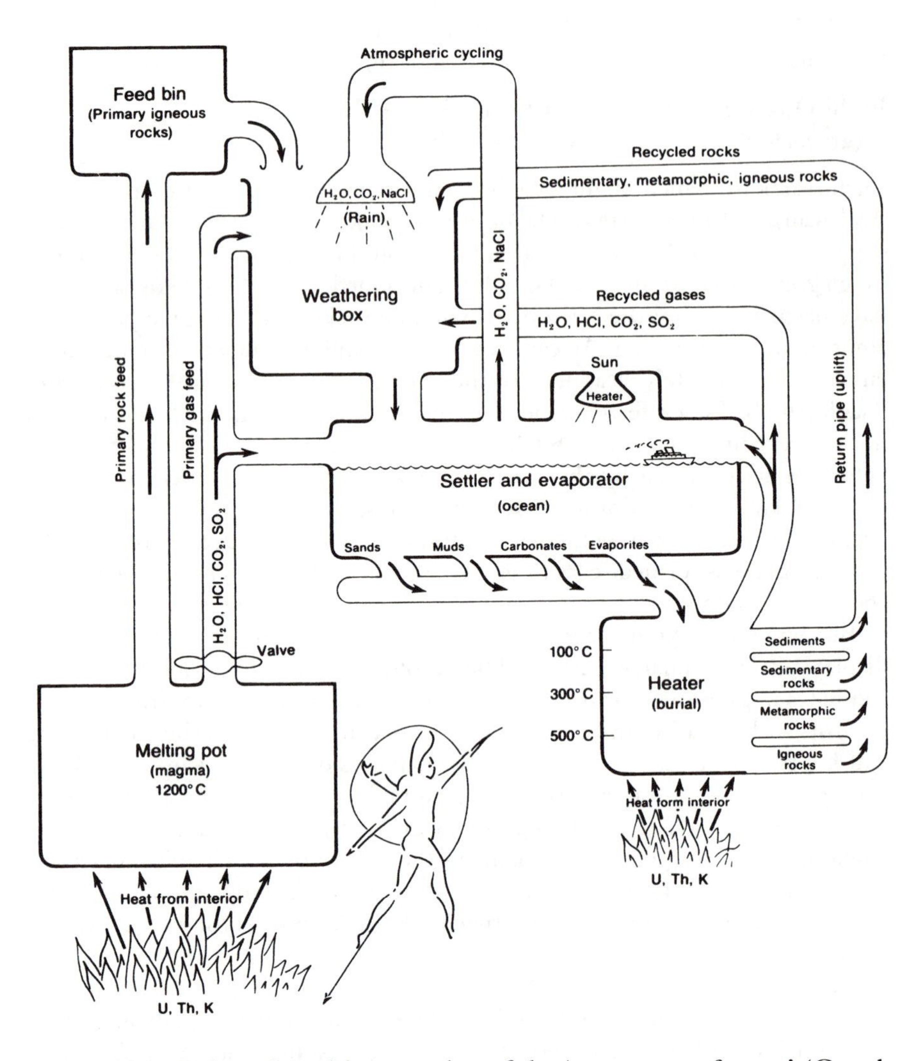

Figure 12.1 Robert Garrels' conception of the 'crust–ocean factory' (Garrels and MacKenzie, 1971: 330) (by permission of W.W. Norton and Co.)

provided by courtesy of radioactivity. The model (perhaps of a kind to be expected from authors who hailed from an industrialized country such as the United States) is shown in Figure 12.1.

Three years later, this time co-authoring with Edward Perry, Garrels produced a major paper, in which an attempt was made to quantify the things that were going on in some of the more important pipes, reservoirs and conveyor systems of the 'factory' (Garrels and Perry, 1974). The paper has been analysed by the Dutch geologist Peter Westbroek (1991), whose simplified version I have drawn on for the purposes of the present exposition.

Garrels and Perry (1974: 313) present a figure in which circles are used to depict various reservoirs of the major constituents of the biosphere/lithosphere/atmosphere/hydrosphere. (The noosphere is not included.) These are such materials as water, silica, calcium carbonate, carbon dioxide, oxygen, gypsum, iron pyrites, organic carbon (including that buried as coal or oil), iron/magnesium/calcium silicates, etc. The sizes of the circles are proportional to the estimated sizes of the various reservoirs, in molar proportions (not actual weights or volumes). It is, of course, a hazardous enterprise to make such global estimates, but by the 1970s geologists were in a position to do so with some confidence.

Garrels and Perry's diagram also showed what were taken to be the interconnecting pathways between the various reservoirs, with numbers proportional to the supposed rates of movement of the different elements from one reservoir to another. There might only be a small amount of carbon present in the carbon dioxide reservoir at any given time, but that carbon might be exchanging with other reservoirs very rapidly. In contrast, the movement of silicon in and out of the vast silica reservoir might be much slower. The role of living organisms was essential in the transfer of elements from one reservoir to another, as for example in respiration and photosynthesis.

The whole system appeared to be stable, otherwise life presumably would not have continued. But was it constant in shape or form over time? Was the size of the reservoirs (after the basic processes of life had been established, with an oxygen-bearing atmosphere formed and aerobic respiration the norm) constant, or did they somehow fluctuate?

A clue to this problem was provided for Garrels and Perry by the presumption that the ratios of the different types of isotopes of sulphur and carbon present in the reservoirs were related to the proportions stored in the different reservoirs. For example, sulphur can be present in both the gypsum ($CaSO_4$) and pyrites (FeS_2) reservoirs. When the proportion of the ^{34}S isotope increases in the gypsum deposits, this implied (to the geochemists) that more sulphur was locked up in the pyrites reservoir than in the gypsum, and vice versa for a decrease in ^{34}S. Measurement of the sulphur isotope ratio for gypsum deposits of different ages suggested that there were indeed swings of sulphur back and forth over time between the two reservoirs.

If sulphur is locked up as $CaSO_4$ rather than FeS_2, this implies that oxygen must have been withdrawn from somewhere in the global system. But the figures suggested that it could not have come from the atmosphere: there simply wasn't enough oxygen there to meet the requirement. The suggestion was, then, that the oxygen withdrawn from the atmosphere to make up the gypsum store must have been replenished from the limestone reservoir, with the accompanying formation of organic carbon (as in coal deposits).

What is being suggested here is a remarkable coupling of the oxygen, carbon and sulphur systems. Westbroek has used the ancient Stoic metaphor of breathing to describe the process. One can think of the great reservoirs as slowly yielding or taking in elements. But the process is not unidirectional. There is not a perpetual depletion of one reservoir to the benefit of the other. If the process goes too far in one direction, it eventually reverses and the balance is restored. It is like a giant pendulum swinging back and forth, but in a rather irregular fashion. To take a grand geological example, there was a massive accumulation of organic matter during the Carboniferous and Permian, which would, by the foregoing argument, have led to the accumulation of pyrites and depletion of the gypsum reservoir (as indeed one sees in the rocks). But eventually the pendulum swung back, and in reaction great deposits of gypsum were formed in the Permo-Triassic.

In subsequent work with Abraham Lerman, Garrels showed to his satisfaction that the carbon and sulphur cycles were indeed coupled with one another, which was necessary for the model to work (Garrels and Lerman, 1981, 1984). Recognition of this coupling was possible since (it was believed) the question of whether carbon was chiefly locked up in the limestone or the organic carbon reservoirs was related to the proportions of ^{13}C isotope present.[18] Thus we begin to see some meaning in the stratigraphic column – why there should be great shifts to the desert-like conditions of the Permo-Triassic from the huge coal-forming forests and peat swamps of the Carboniferous–Permian.

But perhaps there is an even deeper meaning. Once life was established as a geological agent of major importance, which had itself been responsible for the oxygen-containing atmosphere, the things that hold steady through all this 'deep breathing' of the great reservoirs are life, the water of the oceans and the atmosphere. All the changes that take place between the reservoirs are facilitated by the biosphere. Indeed, from a Vernadskyan perspective, they are all aspects of the functioning of the biosphere. As the great swings of the pendulum occur, they do so in such a way as to maintain a set of environmental circumstances on earth that make the continuation of life possible. Can it be, then, that life is the 'force' that is somehow determining and shaping the whole?

This takes us swiftly in the direction of the controversial 'Gaia' hypothesis of the English independent scientist James Lovelock (b. 1919), which I shall now briefly discuss, as it provides, I think, one of the most interesting and potentially fruitful way of thinking about the earth currently available.[19] Lovelock's two main popular expositions of his views are *Gaia: a New Look at Life on Earth* (1979) and *The Ages of Gaia: a Biography of Our Living Earth* (1988). His work has received the strong backing of the American microbiologist,

Lynn Margulis, who has done major work on theories of the origin of the structures of living cells.

What is Gaia? It is, we recall, the name used by the Greeks for their earth goddess. So in reintroducing Gaia at this point we come in a sense full circle in our thinking. We began with mythology, and perhaps we end there too. Anyway, by introducing the ancient term – the root of the 'geo' in the word geology itself – Lovelock was particularly felicitous in that, for whatever the reason, the term has 'caught on' and his books have become runaway best-sellers. They may often be found in the 'new-age' parts of bookshops, but they should, I believe, be accorded great respect. They are serious works, based on a profound understanding of the way the world works.

So Lovelock's name is almost a household word, perhaps in part by good fortune in coining the word Gaia, which, with its mythological connotations, seems to have appeal. Garrels, in contrast, has until recently been known only to the scientific community, although it is possible that now, through the exposition of his ideas by Westbroek, he will become known more widely.

But again, I ask, what is Gaia? To tell us that it is a name drawn from Greek mythography does not get us very far in the twentieth century. Fortunately, Lovelock (1987: 152) is explicit about the matter. He defines the Gaia hypothesis as follows:

> [It] postulates that the physical and chemical condition of the surface of the Earth, of the atmosphere, and of the oceans has been and is actively made fit and comfortable by the presence of life itself. This is in contrast to the conventional wisdom which held that life adapted to the planetary conditions as it and they evolved their separate ways.

I suppose most people in the post-Darwinian era probably do adhere to the 'conventional wisdom' on this matter, until they read Lovelock. From a 'mechanical' view of the earth, conditions may change from time to time, for reasons that are entirely beyond human control or the control of any organisms on the planet. Organisms just have to adapt or die out. For example, nothing can be done to alter the ellipticity of the earth's orbit or the inclination of its axis of rotation to the plane of the ecliptic. So, if the theory of glaciation being due to astronomical factors is correct, the cause is inexorable and so too are the resultant climatic changes. Life on earth just has to adapt as best it can.

But, on the Gaia hypothesis, the whole biosphere, the ecosystem, is one that deploys feedback processes which maintain approximately uniform conditions, such that life can be maximally sustained on the planet. So, just as all the actions of the cells in a body are coordinated in their activities in such a way as to make the body of the organism function as a viable whole, so too are all the activities of the organisms making up the biosphere coordinated so as to preserve life on earth. This process is quasi-purposive, or at least it appears to be so. The biosphere itself behaves like a living organism, capable of homoeostasis.

Lovelock tells what may be called a 'just-so story' as to how life may have originated on earth, and how it may have developed its atmosphere and its present stable condition. (Many other scientists have attempted the same task, although few have tried to tell the whole story, concentrating rather on just a corner of the narrative.) We need not follow all Lovelock's arguments, but a couple of examples as to how the system of Gaia seems to be self-regulating may be instructive.

Although numerous organisms live in sea water, a very high salt concentration is not conducive to life. But salt is removed from the sea in evaporite lagoons, and such lagoons are found in many places on continental margins. Also fossil deposits of evaporites (such as the Cheshire or Silesian salt deposits) are well known (and commercially important). For the existence of the evaporite lagoons, barriers of calcium carbonate are required. In the absence of life, this substance would be precipitated randomly on the floor of the ocean. But, in the presence of reef-forming organisms (e.g. limestone stromatolites, found as far back as the Precambrian), areas off shore would have been sealed off providing lagoons where evaporation and precipitation of salt could occur (Lovelock, 1988: 110–11).

A related example provided by modern coral reefs has been noted by Lovelock (ibid.: 124–5). It appears that coral organisms secrete a fatty substance which forms a molecular monolayer on the surface of the nearby ocean. This alters the surface tension of the water and modifies the behaviour of the waves in such a way that the wave damage to the reefs is reduced.

Such examples (or at least the second one) may also be explained by the normal processes of natural selection, but, taken all together and seeing the biosphere as an entity that has the properties of being self-stabilizing and capable of self-repair, it seems not unreasonable to coin a name for the totality of such processes: Gaia.[20]

There have been some hints in the literature that Gaia is coupled with the grand processes of plate tectonics in a remarkable way. For example, Anderson (1984) suggested that living organisms may have removed carbon dioxide from the primeval atmosphere, locking it up in limestone. The surface temperature of the earth would then have dropped (carbon dioxide being a 'greenhouse gas'), and perhaps the temperature of the upper mantle also. Then the boundary of the basalt/eclogite phase change in the mantle might have risen and plate-tectonic cycling of the mantle might have been initiated.[21] It is also possible that the deposition of limestone by micro-organisms has been so considerable that it has set in motion the processes of geosyncline formation, which have then triggered the grand convection motions of the mantle discussed in Chapter 11.

In a similar vein, it has been mentioned by Westbroek (1988) – although only as a tentative speculation – that there might be no plate tectonics on earth without the presence of life. There was formerly liquid water on Mars – arguably evidenced by the dried-up river channels observed on the surface of this 'dead' planet. If life had begun on Mars, then it too might have become a self-regulating physical/organic system, in which case the former waters might have been preserved. The origination of life is obviously a chancy affair. It happened on the earth, but apparently not on Mars. Suppose life had not got going on earth. Then there might be no oceans today, in which case there might perhaps be no plate-tectonic movement.

This suggestion has met with little favour from most geologists. After all, if the energy for the mantle convection is provided by radioactivity then one would hardly suppose that it would matter whether or not there was a biosphere. On the other hand, one might suggest that without water, retained on the earth by courtesy of living organisms, there would not be the surface inequalities between continents and oceans that would give rise to the exterior manifestations of the subterranean movements in the mantle.

I am not competent to adjudicate on questions of such complexity and magnitude. I am, however, persuaded that there is a coupling between the processes of plate tectonics and the biosphere. Which side of the couple is the chicken and which the egg is a nice question. Is the ponderous motion of the plates the driving force that brings about the swings between the reservoirs of gypsum, limestone, etc., about which Garrels and his co-workers have taught us? Garrels himself (Westbroek, 1991: 97) likened the earth to a huge grandfather clock, in which the rotation of one wheel determined the motions of all the others.[22] But which was the big wheel?

In the usual plate tectonics model, the radioactively generated heat producing the convection currents of the earth's interior is the 'big wheel', or rather the 'boiler' driving the wheel. The interchanges between the reservoirs, mediated by life, are merely the stabilized responses to the movement of the big wheel. We see the clock-face but not the internal wheel of the earth's mantle. Or, in so far as we do see it, the view is indistinct, obtained by courtesy of the 'wiggles' of seismograms and recondite mathematics. (In contrast, the great masses of lime- and silica-depositing organisms can actually be seen in the oceans by courtesy of satellite photographs.)

But, if Westbroek is correct in his speculation, there may have been no stabilizing wheel until the formation of the biosphere. Ultimately it was the origin of life that made the earth what it is. We are not merely inhabitants of a planet, adapting ourselves to it. Life has verily constructed the planet. It would not behave as it does, even in its deep interior, were it not for life. However, the origin of life on earth could, of course, still be regarded as an 'accident'.

We are thus faced with a grand division in ways of thinking about the earth. Some theologians, notably Hugh Montefiore (1985), Bishop of Birmingham, have sought to find theological significance in Lovelock's work, which is also highly compatible with the ideas of the distinguished Australian biologist and writer on theological topics, Charles Birch, in, for example, his recent book *On Purpose* (1990). For Birch, the feedback process, providing homoeostasis, is itself a manifestation of deity. Or rather it is itself divine, and hence worthy of worship. This theological position is called 'panentheism' (to be distinguished from pantheism, which sees every piece of matter in the universe as divine).

Lovelock has not accepted the suggestion that his picture of nature has theological implications. He writes (Lovelock, 1988: 205): 'I am happy with the thought that the Universe has properties that make the emergence of life and Gaia inevitable. But I react to the assertion that it was created with a purpose.' 'Purpose', of course, has been the watchword (or catchword?) of Christian theology from its beginning, and goes back further into the religion's Judaic and Greek roots. Lovelock cannot see purpose in it all; and no more can I. But we share the view that there can be, and has been, beauty in the scheme of things. And, likewise, I share his encouraging belief that the earth has been made 'comfortable' for life by the self-sustaining activity of organisms, be they ever so small and humble. There may be, therefore, a pleasing positive side to the Darwinian coin, even if the other side is the more famous 'nature red in tooth and claw'.

I am less happy, however, with the idea that the emergence of Gaia was inevitable. It could only be inevitable in so far as the origin of life was inevitable, and this seems to me an unwarranted assumption. Further, as Westbroek (1991: 213) has pointed out, there is a significant difference between the science of Garrels and that of Lovelock. Garrels' theories are embedded in empirical science and tested by the isotope data, which demonstrate the coupling of the carbon and sulphur cycles.[23] Lovelock's point of departure is an idea – the analogy between ordinary living organisms and the earth as a living entity. It is not obvious how one can test the Gaia hypothesis, and if it is untestable its scientificity is in doubt.

Westbroek (ibid.: 216) has described Garrels' model as one that cannot get off the ground. It is perhaps seeing only a part of the whole. In contrast, Lovelock's beautifully appealing holistic concepts 'hover above the surface and cannot land', in that they are difficult to test empirically. It would be pleasant to see the two approaches synthesized. That, however, would be the subject for another book. This text is concerned with the history of the way people (geologists and natural philosophers) have thought about the earth, not with the ultimate meaning of existence, the origin of life or science's future. However, it is geology that has thrown up grand metaphysical questions once again, as it has done in the past and will doubtless do in the future.

We cannot conclude this sketch of the history of geologists' ways of thinking about the earth without considering one other very grand question (which may also seem to have theological significance), namely the question of the earth's origin. This problem is chiefly within the domain of astronomy and cosmology/cosmogony, and so may not seem to have a proper place in the present enquiry. Indeed, it will be recalled that Lyell, in his historical introduction to the *Principles of Geology* (Lyell, 1830–3), argued that it was only when cosmogony was cleared from the stables of geology that the study of the earth could take on its appropriate character and could join the domain of science, rather than unbridled speculation or metaphysics. But questions of cosmogony have been of great interest to some geologists, even in the nineteenth century. And, in recent years, empirical geology and geochemistry have begun to throw light on questions about the origin of the earth and cosmogony more generally. We cannot, therefore, conclude without saying something about such grand questions.

In the eighteenth century, two broad theories were developed to account for the origin of the earth. The first developed by Buffon (1749) envisaged the earth and the other planets as having originated with a collision between a comet and the sun. This catastrophic collision led to the emission of a plume of matter which condensed into the several planets and their satellites.

The second theory, usually called the nebular hypothesis, was proposed independently by the Prussian philosopher Immanuel Kant (1724–1804) in 1755 and by the French mathematician, astronomer and philosopher Pierre-Simon Laplace (1749–1827) in 1796. Kant's theory, which was based largely on the Newtonian idea of universal gravitation and Thomas Wright's idea of the Milky Way as a mass of stars, exerted rather little influence in his day as most of the stocks of the book which propounded his theory were destroyed in a fire. It envisaged an initial inchoate mass of matter, which was not distributed entirely uniformly in space at the 'beginning of time'. Hence, the matter began to coalesce under gravity, becoming more ordered as, by the spread of the area of accretion, suns, planets and satellites were formed.[24] But, as the modern translator of his book, S.L. Jaki (Kant, 1981), points out, no adequate causal explanation was given as to why there should be circular motion in the system. Putting it another way, where did the angular momentum come from?

Laplace's version of the nebular hypothesis was inspired in part by the then recent telescopic observations of nebulae made by Sir William Herschel (1738–1822) and by meditating on the nature of the rings of Saturn. Laplace supposed that there had initially been a sun at the centre of a large rotating cloud of gaseous matter, the origin of which was not explained. As it cooled due to the loss of radiant heat, the globe would supposedly contract, but as it did so it would necessarily spin more rapidly in accordance with the principle of conservation of angular momentum. Eventually, the outer 'rim' would be spinning so fast that it would become detached from the remainder of the spinning cloud (like a spray of mud flying off a rapidly rotating wheel). There would then be a rotating sun-centred cloud, with a spinning ring, like the outermost of Saturn's rings.

The process might be repeated, and there would then be a second inner ring formed; and so on. The material of each ring was then supposed – by parting at its weakest point – to coalesce into a single globe of gaseous matter, which on further condensing and cooling would form a planet. Satellites might be formed by second-order repetitions of the process. According to this view, the earth would have originated as a spheroid of hot gas, which would have cooled to a state of liquidity and then turned into its present state.

The nebular theory still has its admirers and was certainly popular in the nineteenth century, but it suffers from severe difficulties, as has long been known. In the whole process of the formation of a solar system, one would expect angular momentum to be conserved. That is, the angular momentum of the original system should equal the present angular momentum of the sun, together with the sum of the angular momenta of all the planets and satellites.

But this appears not to be the case. The sun has a high mass, of course, but only has angular momentum by virtue of its axial rotation. The planets are comparatively light, but they have large angular momenta by virtue of their large orbital motions. Now suppose Laplace's mechanism for the formation of a planet were operating. The ring spun off (the 'mud off the wheel') should only have the angular momentum that was originally associated with the outer – planet-to-be – portion of the rotating 'nebula'. And the same would apply as the subsequent planets were formed. Yet, as said, the actual solar system has nearly all the angular momentum (actually about 98 per cent) 'in' the planets. The sun seems to be turning too slowly for its mass. This seems to be a fatal objection to the Laplacian model.

Since Laplace's day, numerous other theories for the formation of the solar system have been proposed,[25] but these come chiefly within the domain of astronomy or cosmogony, and I shall not try to analyse them here. We shall, however, look at some theories that have been propounded by geologists, and which have given rise to 'geological' ways of thinking about the earth. Pride of place must be given to the 'planetesimal' theory of the distinguished American geologist T.C. Chamberlin (see p. 182), developed in association with the Chicago astronomer Forest R. Moulton (1900, 1905). In his day, Chamberlin was regarded as the most powerful 'force' in American geology.

Chamberlin sought to establish a fundamental basis for thinking about the earth, by developing his planetesimal theory and then trying to make sense of the distributions of continents, oceans and mountain ranges according to this theory and general physical principles. In a paper of 1900 (Chamberlin, 1900), in his text with R.D. Salisbury (Chamberlin and Salisbury, 1906) and in his book *The Origin of the Earth* (Chamberlin, 1916), Chamberlin subjected Laplace's theory to damaging criticism, arguing as much on the basis of the kinetic theory of gases (about which Laplace knew nothing, of course) as on that of problems concerned with angular momenta. Chamberlin also utilized recent photographs of nebulae and 'solar prominences' from the Mount Wilson, Lick and Yerkes observatories.

From time to time the sun can be seen to expel great masses of intensely hot matter ('prominences'). Chamberlin and Moulton supposed that these might have been on a vaster scale in the past, and one of them might have been dragged out by the gravitational attraction of some passing star, which did not, however, actually collide with the sun or form a double-star arrangement. The dragged-out matter ('knots', in Chamberlin's terminology) might then have accreted on cooling to give small solid bodies, for which the term 'planetesimals' was coined. They were so called because they might have acquired an orbital motion around the sun as a result of the attraction of the star supposedly moving by. With regular motion round the sun, the planetesimals (miniscule planets) were different from randomly moving meteors. And, because there were a great number of different initial elliptical orbits, there would be many collisions. So accretion of the planetesimals would supposedly occur.

Chamberlin (1916: 117, 125, 128, 131) produced photographs of nebulae which suggested that such processes leading to the formation of solar systems were occurring elsewhere in the universe. In particular, 'knots' seemed to be visible in the arms of spiral nebulae. So the theory seemed to have some support on the basis of 'the present is the key to the past'.

Chamberlin's ventures into cosmology/cosmogony were interesting to astronomers, but were largely ignored by geologists, who did not see that their subject really required to go back quite so far in its principles. Undaunted, having produced what was to him a satisfactory basic theory for the origin of the earth, Chamberlin then set out to show how the fundamental structure of the earth, the distribution of continents and oceans and the formation of mountains might be understood according to this theory. This takes us back to themes we have considered earlier in this book.

The planetesimals supposedly 'plunge[d] into the earth-knot of the nebula' (Chamberlin, 1916: 159). An atmosphere formed, and waters gathered in the hollows of the surface. As the materials of the 'knot' and the planetesimals interacted, some – such as metals and their oxides and basic silicates – cohered and became denser matter, moving toward the centre of the embryonic earth. The more elastic collisions led to the production of less dense matter, which tended to move towards the surface. The atmosphere would have been formed in part from elements such as nitrogen and oxygen[26] present in the original 'knot', some of which might be released by volcanic action if trapped in the earth's interior. And some of the planetesimals might have provided further gaseous matter. The oceans might have been formed from water molecules by analogous processes. Henceforth, thought Chamberlin, there were continuous interactions and 'competitive struggles' between the lithosphere, the hydrosphere and the atmosphere. This 'prolonged struggle' was the 'very soul of geologic history' (ibid.: 169).[27]

Two major agencies shaped the progress of the 'struggle': gravitation and rotation. And since the regions of planetesimals traversed by the 'earth knot' (or the growing-by-accretion earth nucleus) were supposedly not uniform, the rotation of the earth was not necessarily steady at first. Changes in rates of rotation would, of course, induce deformations in the earth's structure. This might, thought Chamberlin (ibid.: 175), have something to do with the 'pulsations' in earth history that were being discussed by several theorists at about that time, as we have seen in Chapter 8.

Now suppose, with Chamberlin, that the earth, in its early history, from time to time increased or reduced its rate of rotation. With an increase, the equatorial regions would tend to 'swell' and be under tension, while the polar regions would fall and suffer compression. The opposite effects would occur with a decrease in rate of rotation. At some intermediate latitudes, which Chamberlin estimated might be at about 30°N and 30°S, there would be minimum effects. These were what he called 'fulcrum zones'.

As is well known, the stresses produced in cooling basalt produce hexagonal fractures, such as are seen in the Giant's Causeway in Northern Ireland. Such fractures relieve the strain with the minimum fracturing of the rock. Likewise, suggested Chamberlin, the stresses generated at the poles by the alterations in the rate of rotation would generate huge fractures, which might be at 120° to one another, like the angles at which three columns of basalt meet at a point. So three such 'fissure tracts' might be expected to extend from the poles to the 'fulcrum zones'.

Supposing, then, that we have three fissure tracts extending meridionally from the South Pole to the fulcrum latitude of 30°S. This carves the southern hemisphere, up to that latitude, into three great triangles, with their apexes at the pole and their bases along the line of latitude of 30°S. Chamberlin then imagined that there were three further triangles extending from these bases to apexes along the line of latitude 30°N; and these apex points reciprocally defined three further triangles with sides running meridionally to the North Pole. It followed, therefore, that for the equatorial belt (from 30°S to 30°N) there would be a system of six triangles, three with their apexes pointing south and three with their apexes pointing north, arranged as it were in sawtooth fashion. This, for Chamberlin was the supposed basic controlling structural framework for the earth. Note that an adjacent pair of equatorial triangles would define an obliquely aligned quadrilateral (running from 30°S to 30°N).

These fissure tracts supposedly penetrated deep into the earth's interior, and hence governed the whole subsequent development of the earth. They provided the 'embryonic framework on which the shaping agencies built their systematic and their adventitious growths' (ibid.: 200) – including the accretion of additional planetesimal matter. It was further supposed that, by chance, the original earth had more dense matter in the southern hemisphere, so that the general level of the solid earth was lower there. In consequence, the southern hemisphere was predominantly oceanic, and its structure came to determine the total global structure. In this manner, Chamberlin endeavoured to develop a grand theory of the earth. At least it explained, albeit crudely, why the land zones in the equatorial regions tend to run obliquely, for his model had obliquely running equatorial fissure zones, which might, paired up, define obliquely aligned continental blocks. The whole theory was speculative and the accompanying diagrams were not very convincing.[28] To get anywhere with the theory, Chamberlin had to make any number of *ad hoc* hypotheses (such as that the Alaskan region is a fairly recent development, which happened not to be determined by the primeval structure of the globe). It did not, in fact, follow the maxim Chamberlin adopted as the epigraph for the chapter where he laid out his structural theory: 'As the twig is bent [so] the tree is inclined.'

Although it is now largely forgotten, I have described Chamberlin's theory in a little detail because in the idea of planetesimals we have the forerunner of a doctrine that is again attracting attention in thoughts about the earth. While most Western geologists were attending to other matters in the mid-twentieth century, some Russian theorists were developing a version of the planetesimal hypothesis, which, however, was also somewhat Kantian or Laplacian in character, and sought to deal with some fundamental problems that faced Chamberlin's astronomy.[29] From 1940, Otto Yul'evich Shmidt (1891–1956) was the leading theorist,[30] and he was followed up by his student U.S. Safranov and others. The essence of his theory was first published in 1944.

In considering the problem of the origin of the solar system, Shmidt (1958: 20[31]) pointed out that any theory should account for the fact that the planets' orbits are almost circular, lie approximately in one plane and revolve round the sun in the same direction; that their distances from the sun seem to follow a numerical pattern (e.g. Bode's law[32]); that they seem to divide into two groups (Mercury, Venus, Earth and Mars; and Jupiter, Saturn, Uranus and Neptune); and that most of the angular momentum of the whole system is associated with the planets rather than the sun. To account for these facts, Shmidt hypothesized that the solar system had originated when the sun passed through a nebular cloud of dust, ice, gases, such as methane or carbon dioxide, and other matter. The various particles supposedly collided with one another, averaging out their velocities. That is, their relative velocities decreased. But, with a loss in mechanical energy due to the inelastic collisions, heat was generated; and, with the angular momentum of the rotating nebula being retained, the dust cloud began to flatten. Accretion of the particles under the action of gravity occurred, so that first asteroid-sized bodies and eventually planets were formed. The planets nearer to the sun were heated to the extent that they did not have many volatile materials, whereas the cold, more distant planets could retain gaseous atmospheres. The earth's atmosphere was not, in Shmidt's view, primeval, but had been formed subsequently by degassing from the interior. Shmidt (1958: 118) agreed with Vernadsky that the formation and composition of the atmosphere was, to a considerable extent, due to the activities of living organisms.

As Claude Allègre (1992: 110), Professor of Earth Sciences at the *Université de Paris*, has described Shmidt's theory in lively fashion: '[T]he formation of planets is like a gigantic snowball fight. The balls bounce off, break apart, or stick together, but in the end they are rolled up into one enormous ball, a planet-ball that has gathered up all the snowflakes in the surrounding area.' It should be noted that for Shmidt's 'lively' theory the question of angular momentum, which had dogged the Laplacian model,[33] was not a problem. The parent nebula envisaged by Shmidt had its own initial angular momentum, as did the sun. There was no reason, therefore, why, after the formation of a solar system, the momentum of the planets should not be much greater than that of the sun. Also, Shmidt was able to produce calculations for the distances between the planets and the sun which agreed quite well with observations – indeed, better than Bode's empirical generalization. Shmidt's theory was, then, somewhat similar to that of Chamberlin and Moulton, but also different. His nebular particles were not initially moving round the sun: they only became 'planetesimals' over time.

Under wartime circumstances and subsequently in the period of the cold war, Western scientists did not at first take notice of Shmidt's theory.[34] But, after the first lunar explorations and the consequent development of the science of selenology, it was taken up and developed with the help of computer simulations (which govern a good deal of geological thinking today). The moon's surface was found to be pitted with craters large and small, showing evidence of constant meteoriitc impacts, and the density of the craters was seemingly a rough measure of the age of any given area of the lunar surface, assuming a constant rate of impact. But ages of the lunar rocks could also be determined radiometrically, and when crater density was plotted against actual age it appeared that the rate of bombardment had decreased exponentially since the moon first formed.

This finding was connected to the Soviet theory of planetary accretion by George Wetherill (1976), of the Department of Terrestrial Magnetism at the Carnegie Institution, and this led to numerous attempts to generate computer simulations of the process of accretion. These seemed to work quite well and also to satisfy the requirement of a hot interior for the earth, for each impact generates a great amount of heat. But the original 'protosolar nebula' – a descendant of the kind of entity envisaged by Kant and Laplace, not the vastly greater nebulae made up of innumerable stars that are seen with giant telescopes – was, according to the theory, itself hot. So, as sources of heat, we have the primeval heat of the 'protosolar nebula', the heat generated by impacts, the heat generated by radioactivity and the heat that may be generated by the differentiation of the iron/nickel core of the earth from the ultrabasic material of the mantle. (With the sinking of the heavy iron and nickel to the core, there would have been a decrease in the earth's rotational energy, compensated by the production of heat energy.) This assumes, then, that the internal layers of the earth's interior have come into being by processes of differentiation subsequent to its formation, rather than by the successive accession of different kinds of material.

Such thinking has been strongly influenced by recent studies of meteorites, particularly the so-called chondrites – which fall on the earth as a sort of 'cosmic sediment', thought to be derived from nebular material. The chondrites contain small spheres called chondrules, typically made up of peridotite and/or iron and nickel, and some fine-grained matrix, chiefly olivine and pyroxene. Such meteoritic materials are plausible candidates for having been the sources of the principal materials of which the body of the earth is believed to be made. Interestingly, individual chondrules have been found to be differentiated into internal cores of iron and exterior peridotite – like terrestrial microcosms.

Shmidt's theory of the origin and subsequent differentiation of the body of the earth is, in its updated form, thought by some to be consonant with the doctrines of plate tectonics. There is enough heat available within the earth, according to the theoretical modelling, to allow the slow convective movements of the earth's interior required by plate-tectonic theory. The oceans and atmosphere supposedly formed at an early stage from the degassing of the mantle. Life (perhaps) stabilized the oceans and (almost certainly) converted the initial atmosphere into its present oxygen-bearing form.

As for extraterrestrial impacts, these are entirely compatible with the theory. As is well known, many geologists today believe that during the earth's geological history there may have been occasional massive collisions with large objects falling on to the earth from outer space (Albritton, 1989; Huggett, 1989; Ager, 1993; Glen, 1994). And these may have been responsible for some of the great extinctions that seem to be evidenced by the stratigraphic record, most spectacularly the demise of the dinosaurs.[35] But that is yet another story that requires another book.[36] What we have here is probably sufficient to be going on with.

Chapter 13

Some Concluding Thoughts

Thinking About the Earth began in the realm of mythopoeic cultures, where thoughts about the earth were taken on trust, being socially reinforced in myths and legends. Indeed 'terrestrial thoughts' were necessarily refracted through the lens of myth.[1] They were satisfactory, no doubt, if one was not concerned to have those thoughts 'correspond' with the way things are in the world or actually have been in the earth's past. They might serve a social purpose by giving a culture a sense of its identity (in its own special myths and legends) and could help to make the world an intelligible place, even if not one that was particularly comforting or comfortable – as in the Babylonian myths about mighty Marduk. Does the modern account of the earth that geology offers have a mythical aspect, as Feyerabend (1970) might have us believe, or has that been squeezed out of the system?

One can easily get scientistic about such questions, and hold that the latest results of scientific research are the best and that the best is right. But, as is patently clear from the narrative of this book, ideas about the earth have been changing constantly ever since the beginnings of geological science in the seventeenth and eighteenth centuries. Indeed, they changed much more rapidly after the establishment of science than in preceding centuries. So, if the history of science teaches us anything, it should be that the latest ideas may not be correct in terms of correspondence with the way things are in the world. But this thought leads us rapidly in the direction of the quicksands of relativism in our philosophy of science. Feyerabend would have been pleased.[2]

Such thoughts are reinforced when we have regard to the fact that most people are, with the best will in the world, not able to make an informed critical assessment of what they read in books or are taught in schools and undergraduate classes at universities. And, by the time the tiny minority who struggle through to engage in scientific research have enough knowledge to be in a position to make evaluative judgements of the prevailing paradigms, they may be so strongly acculturated or so bound by threads of social obligation, that it is difficult indeed for them to think independently on major theoretical issues. That this is so is revealed by the study of the history of any science, including geology. So perhaps the knowledge of our best PhD graduates does have certain mythical characteristics? There is such a lot that every scientist must take on trust. Perhaps like the (mythical?) Tannese boy who asked awkward questions about the myths of his village, s/he may look foolish or be derided if fundamental beliefs are queried. For Feyerabend, there were indeed mythical features to science (or more specifically empiricism), not so different from those of mythopoeic cultures.

Typically, myths operate through personifications and with the help of analogies, metaphors and similes, and their language is characteristically allegorical. The metaphorical/allegorical language necessarily frames the resultant thinking. Is it so with modern ideas about the earth? That modern theories of the earth use metaphorical language is undoubtedly the case. The concept of 'plate' is a simple example which we may consider for a moment. Is it just an instance of a 'dead metaphor' – like 'skyscraper' for example? (We know perfectly well what we mean by this word, and we are not going to think that the large buildings of modern cities literally 'scrape' the sky just because we use the word so commonly.)

I think that, for geologists at any rate, the word 'plate' is in somewhat the same position as 'skyscraper'. It is a dead (or almost dead) metaphor. Geologists know (or think they know) what plates are, almost as well as the commuter knows what a skyscraper is. Plates are defined or 'observed' by the results of seismic and gravimetric surveys, by bathymetry and magnetometry, by aerial and ground mapping and so on. They are real objects so far as nearly all present-day geologists are concerned. They are not just metaphorical entities; they are as real as skyscrapers. However, their reality cannot be established by the test favoured by Hacking (1983): they cannot be moved around at will, as electrons can be 'sprayed' in a television tube. They can't be built or blown up like skyscrapers. They are not so real as that.

There is also the problem that there is still an alternative theoretical model in the market place, even though its stocks do not stand high at present – the expanding-earth theory. And for this theory's leading exponent, Warren Carey, 'subduction' – one of the principal terrestrial processes according to plate-tectonic theory – 'is a myth'. It has been suggested, I believe, by some of the more gung-ho members of the nuclear industry, that nuclear waste might be dropped into an oceanic trench, where subduction is supposedly occurring, and the dangerous materials would all be conveniently absorbed into the bowels of the earth. But (supposing the rate of subduction were theoretically appropriate to the task in hand) would one really wish to take such risks, pouring dangerous materials into a 'theoretical entity'? Would you want to put all your money on a mythical process?

Of course, when Carey calls subduction a myth, he is deploying rhetoric in the context of a scientific controversy, for he is still to some extent – and certainly has been – involved in a long-time debate, where rhetoric has been freely used. Just because Carey calls subduction a mythical process, it does not mean that subduction does not occur. It may, rather, simply be the case that it is so taken for granted now that evidence that may point against it is ignored or swept under the carpet (see p. 344). If that is so, it is a serious matter, but it is only one that can be rectified by the challenge of upholders of the expanding earth theory or any other theory that may be propounded that will deal satisfactorily with the phenomena that plate-tectonic theory handles, together with any cases where that theory fails – perhaps the problem of the distance between Easter Island and the South American coast. It is a scientific question, to be fought out within the agonistic field of the scientific community. So the fact that subduction has been accused of mythic status does not mean that plate tectonic theory is a myth or fabric of illusion. The issue is not one of semantics or rhetoric but of the measurable distances between land masses.

To put my cards on the table, so far as I am concerned the truthfulness of scientific theories is to be judged by their consistency and their coherence. For me, these are marks or indicators of truth, and I like to think that science approaches truth by a kind of self-correcting process. When the pieces of the theoretical jigsaw fit together, then one can have confidence that the various pieces are telling the truth. But to form a judgement on that question, so far as plate-tectonic theory is concerned, I should prefer the opinion of a geologist to that of a historian. The trouble then, of course, is that the geologist will be more strongly acculturated in favour of the plate tectonic paradigm than will the historian of geology. So what to think? We have to take the opinion of the experts on trust; and so we go round the epistemological merry-go-round once again. There, perhaps, I'd better leave readers to enjoy the ride and to make their own judgements as to what they like to think.

The question of myth has been raised in another way in *Thinking About the Earth*, when we considered Lovelock's Gaia hypothesis. Much of the story of thoughts about the earth has to do with analysis, a precondition for synthesis. We have more and more minute studies of microregions of the earth's crust – say a single canton or a single valley of Switzerland or the geology of a particular sheet of the geological map of a country. We have more and more refined chemical analyses of particular minerals or rocks. We have the use of the most delicate scientific instruments (without their users necessarily knowing how they work within their black boxes). Now Lovelock transcends all this and offers us a grand synthetic, holistic view. What are we to make of it? Is it a myth? Why did he accept a name from Greek mythology as a sobriquet for his theory?

There can be no question that there is a crying need for theories about the way the earth and living things behave together as a whole. For human activities now span the whole earth, and we need to know how the totality works if it is to remain 'comfortable' for us. (It is, of course, the ecological message of Gaia theory that has had the greatest impact in these times of ecological crisis.) It might be an attractive historical hypothesis to suppose that in the writings of one such as Lovelock humans have rediscovered a world that they should never have lost. Does not Lovelock propose an 'organic' view of the world that is somehow analogous to that of the ancients and up to the time of the seventeenth-century scientific revolution? Analogous perhaps, but the similarity is weak. For the ancients and for a Renaissance man such as Paracelsus, the order of the world was understood by seeing it as similar to that of living organisms. But to say that did not carry one very far, either pragmatically or conceptually. It was really just a hopeful crutch to the understanding.

In seventeenth-century physico-theology, the organic theory of the earth was gradually discarded and the terrestrial order was explained as the activities of a master craftsman, whose activities were such that the mechanical details of things – earths, rocks, fossils, etc. – could be worked out by the natural philosopher. This was the period of analysis, and in a sense it has continued thus to the present. If one strips away the idea of the divine artificer, which (or who) really becomes superfluous so far as the study of the mechanical details of the workings of the earth as a machine is concerned, then one has a mechanical cosmos, for which the presence or absence of a god is not a scientific issue (as it was in the days of physico-theology) but one of faith and personal religion.

With Gaia, we have once again a 'living' earth. But it is not regarded thus because an animal or plant is the nearest thing with which we may hopefully compare it to assist understanding. It is in a certain sense 'live' because it has some of the characteristics of ordinary living organisms, notably self-repair. By the same token, however, if it is 'live' then it may be 'killed', and Lovelock has warned of the dangers of this.

Of course, Gaia, even if 'true' – even if there is a self-preserving earth and biosphere – does not have to be worshipped. Indeed, I suggest that it is naïve to do so. However, from a pragmatic point of view, it may be advantageous to worship Gaia. We are less likely to 'kill' 'Her' if we think 'She' is divine. Here I am edging towards the philosophy of the deep green movement (although I dare say deep green environmentalists would deplore such 'pragmatic' greenness).

Anyway, so far as I'm concerned, the exciting prospect extended by the new geology is that of establishing a synthesis so grand that it will link theories about the structure and inner workings of the earth (as in plate tectonic theory) with ideas in sedimentology, in such a way that we may understand how the living organisms of the biosphere actually interact with the earth so as to affect its inner workings. Such understanding would be achieved, not by moving the parts around at will (as Hacking 'sprays his electrons' to assure himself of their reality), but by our being an integral part of the total system – a system whose workings have been studied minutely in an analytic manner, and then comprehended synthetically as a whole. Perhaps we could judge the truth of our understanding not by moving the bits and pieces around at will, *à la* Hacking, but by our successful preservation of the totality as a 'living entity'. If this synthesis of understanding can be achieved, and the whole sustained, it will be a very grand thing indeed. We may then have a coherent understanding that will seem as intellectually satisfying as one of the ancient myths – even those grandest of myths: those of the monotheistic religions.

But possibly this is mere musing. We must see how the new geology develops. Meanwhile, I like to think that some understanding of the totality can be achieved by examining the history of the way ideas on the matter have developed – mostly all the little analyses, with the occasional big synthesis. And this is what I have sought to offer. The new ways of thinking can, I hope, be understood more clearly and accurately when seen in their historical perspective. Will there be an ultimate grand synthesis? Possibly (probably?) not, if the lessons of history can teach us anything.

And a final thought: if we do kill Gaia, it will be a slow process. So we need historical records to tell us whether She is gaining health or in a state of mortal decline. History does have uses.

Suggestions for Further Reading

Anyone contemplating a serious study of the history of geology requires a good bibliography, and this is supplied in a most comprehensive fashion by the massive ongoing compilation, *Geologists and the History of Geology*, by William Sarjeant (1980, 1987, and further volumes are in the press at the time of writing). Susan Thompson's *Chronology of Geological Thinking from Antiquity to 1899* (1988) is also extremely useful. The journal *Earth Sciences History* is exclusively devoted to the history of geology and includes a list of recent 'interesting publications' (i.e. writings on the history of geology) in each issue, as well as useful book reviews. In the following paragraphs, I shall only refer to books, and chiefly ones in English.

First, one should mention a judiciously selected anthology of numerous key geological writings – *A Source Book in Geology*, edited by Mather and Mason (1939) – which provides English translations from German, French and Italian authors, as well as British and American texts. I have often been pleased to find things there that I had been looking for elsewhere for some time. For an anthology of geological texts of philosophical interest, see Albritton's *Philosophy of Geohistory* (1975); see also his edited collection of papers dealing with important historical and philosophical aspects of geology, entitled *Fabric of Geology* (Albritton, 1963). Another good anthology is Preston Cloud's *Adventures in Earth History* (1970). Probably the most influential volume of collected papers devoted to the history of geology has been Schneer's *Toward a History of Geology* (1969), to which several references have been made in the present book.

For early ideas about the earth, within the tradition of Western thought, there is a well-known study by the American geologist, F.D. Adams: *The Birth and Development of the Geological Sciences* (1938/1954). This book is regarded by historians as somewhat dated, for the way in which it tended to represent early ideas as 'quaint', but it contains a wealth of information that is otherwise difficult to locate. More recent studies, such as those of Vitaliano's *Legends of the Earth* (1976) and Greene's *Natural Knowledge in Preclassical Antiquity* (1992), concentrate particularly on relating myths to actual geological events. Ancient geology is analysed in great detail in Volume 1 of Ellenberger's *Histoire* (1988). He has carried through his study in similar detail to the early nineteenth century in the second volume (Ellenberger, 1994).

There are two important nineteenth-century histories of geology: Geikie's *Founders of Geology* ([1897, 1905], 1962) and von Zittel's *History of Geology and Palaeontology to the End of the Nineteenth Century* (1899/1901/1962). Geikie's book is exceptionally readable, but is somewhat nationalistic in tone and 'takes sides' over the Vulcanist–Neptunist dispute. Von Zittel's text, though rather dry, has the virtue of providing good coverage of nineteenth-century Continental writers.

For more recent general histories of geology, I have liked Albritton's *Abyss of Time* (1980), Faul and Faul's *It Began with a Stone* (1983) and Gohau's *History of Geology* (1990). Rachel Laudan's volume, *From Mineralogy to Geology* (1987), gives a good coverage, with emphasis on chemistry and mineralogy, for the period 1650–1830, giving appropriate attention to the importance of Werner and the Wernerian school, but also analysing Lyell's geology in some detail. For Italian readers, approximately the same period is covered by Nicoletta Morello in her *Macchina della terra* (1979), though in less detail.

The professionalization of the history of science largely occurred after the Second World War, and one of the first topics tackled was the relationship between geology and theology. A classic text from the 1950s was Charles Gillispie's *Genesis and Geology* (1951/1959). It is notorious for its 'Whiggery' and for the way it treated physico-theologists with disdain. But it opened up the intellectual history of ideas about the earth in a fascinating and entertaining fashion, showing how geology could be a science of great interest to historians of ideas. Haber's *Age of the World: Moses to Darwin* (1959) dealt to some extent with the same topic but in a less flamboyant style, as did Toulmin and Goodfield's *Discovery of Time* (1965/1967) and Rossi's *The Dark Abyss of Time* (1979/1984), which particularly related ideas about the earth to those about history and to the philosophy of Vico. For nineteenth-century debates about the earth's age, see Burchfield's study *Lord Kelvin and the Age of the Earth* (1975).

More recently, the question of ideas about time (linear or circular?) is treated in a most interesting fashion by Gould (*Time's Arrow Time's Cycle*, 1987) with respect to Burnet, Hutton and Lyell. The same issue had been tackled more broadly by Jaki in his *Science and Creation* (1974), but with reference chiefly to cosmology rather than geology. The philosophical issues pertaining to uniformitarianism and catastrophism have been teased out in Hooykaas's influential study *Natural Law and Divine Miracle: the Principle of Uniformity in Geology, Biology, and Theology* (1963); see also Guntau's *Der Aktualismus in den geologischen Wissenschaften* (1967).

As implied above, many of the studies of the history of geology to date have attended particularly to the earlier stages of 'thinking about the earth'. But, as its title implies, Mott Greene's *Geology in the Nineteenth Century* (1982) focuses on the nineteenth century, with special reference to the history of mountain building. At the moment, there is nothing with a similar broad coverage for the twentieth century.

The development of the subdivisions of the stratigraphic column is treated in Berry's *Growth of a Prehistoric Time Scale* (1968). Influential books dealing with particular aspects of the establishment of the stratigraphic column are Rudwick's *Great Devonian Controversy* (1985) and Secord's *Controversy in Victorian Geology* (1986), which deal with the debates associated with the Devonian and with the Cambrian and Silurian systems respectively. These books have provided a valuable 'research site' for those interested in the social dimensions of geology and in scientific controversies. Rudwick's book, in particular, has interested specialists in the sociology of scientific knowledge, and is remarkable for its representation of the stages of a controversy by means of diagrams. He has also explored his interest in visual imagery in *Scenes from Deep Time: Early Pictorial Representations of the Prehistoric World* (Rudwick, 1992).

My book, *The Highlands Controversy* (Oldroyd, 1990a), has also focused on the fine details of a geological controversy, drawing on ideas from Bruno Latour's *Science in Action* (1987) and the concept of the scientific community as an 'agonistic field'. Indeed, geological controversies seem to attract considerable attention. See, for example, Read's *Granite Controversy* (1957); Hallam's *Great Geological Controversies* (1983), Müller *et al.*'s *Controversies in Modern Geology* (1991) and Glen's *Mass-extinction Debates* (1994). The early social history of British geology is examined in a youthful work by the distinguished historian of science and medicine – and much else besides – Roy Porter: *The Making of Geology: Earth Science in Britain, 1660–1815* (1977).

However, of all the topics in the history of geology, it is the plate-tectonics revolution that has attracted by far the greatest share of attention. Monographic studies of particular interest and value are: Hallam's *Revolution in the Earth Sciences: From Continental Drift to Plate Tectonics* (1973) – which sees the episode as an exemplification of the ideas of Thomas Kuhn; Marvin's *Continental Drift* (1973); Glen's *Road to Jaramillo* (1982); Menard's *Ocean of Truth* (1986); and Le Grand's *Drifting Continents and Shifting Theories* (1988) – which sees the controversy from the perspective of the philosophy of Larry Laudan, as expounded in his *Progress and Its Problems* (Laudan, 1977). Wood's *Dark Side of the Earth* (1985) has endeavoured to represent the intellectual positions adopted by the participants in the controversy as being influenced by their social situations.

Other books that deal with the recent history of geology in ways that I have found particularly instructive are Westbroek's *Life as a Geological Force* (1991), Hsü's *Challenger at Sea* (1992), and Allègre's *From Stone to Star* (1992). A collection of papers, entitled *Eustacy* and edited by Dott (1992), which deals with the history of that concept, also brings us toward the present, but goes back as far as de Maillet in the eighteenth century. The influence of these books may be seen in my own 'thinking about the earth'.

The various branches of geology have been studied by historians in a very uneven fashion. Palaeontology is relatively well served in such volumes as Rudwick's *Meaning of Fossils* (1972), Desmond's *Hot-blooded Dinosaurs* (1975), Bowler's *Fossils and Progress* (1976), Halstead's *Hunting the Past* (1982) (a popular book which deals with general geology as well as the search for the origin of life), Buffetaut's *Short History of Vertebrate Palaeontology* (1987), Gould's *Wonderful Life* (1989) and Yvette Gayrard-Valy's *Story of Fossils* (1994) – an introductory but admirably illustrated little volume. Geomorphology is also well represented in, for example, Chorley *et al.*'s *History of the Study of Landforms* (1964), Herries Davies' *Earth in Decay* (1969), Tinkler's *Short History of Geomorphology* (1985) and a collection of papers edited by him entitled *History of Geomorphology* (Tinkler, 1989). Imbrie and Imbrie's *Ice Ages* (1979) gives a valuable account of the history of ideas about glaciation, with particular reference to the astronomical theory.

Stratigraphy as a whole is not dealt with very adequately, although the 'controversial' texts mentioned above, as well as Berry's little volume, should be mentioned. However, much useful information on the principles of stratigraphy, with mention of historical matters, is to be found in Ager's popular *Nature of the Stratigraphical Record* (1973/1981). Two exemplary histories of geological mapping in Ireland, written by Herries Davies (1983, 1995) are not, to my knowledge, matched by similar studies of other countries.

Writings on the histories of the various branches of geophysics are scattered, although the American Geophysical Union is publishing a series of volumes on this field, under the general editorship of C. Stewart Gillmor. The situation for the histories of mineralogy and petrology, geochemistry, etc., is not good. For petrology there is an old text by Loewinson-Lessing, entitled *Historical Survey of Petrology* (1954), and an even older book by Crook – *History of the Theory of Ore Deposits* (1933) – contains useful information bearing on this branch of geology. Tomkeieff's *Dictionary of Petrology* (1983) provides a comprehensive list of mineralogical and petrological terms, with references to the publications where these terms were first used. Laudan's *From Mineralogy to Geology* (1987) has some material on the early history of mineralogy and crystallography. On this topic, see also Burke's *Origins of the Science of Crystals* (1966) and Emerton's *Scientific Reinterpretation of Form* (1984). A detailed study of geochemistry for the nineteenth century, as it developed from the Vulcanist–Neptunist dispute, is available in German: Fritscher's *Vulkanismusstreit und Geochemie* (1991). But I know of nothing similar in English.

Numerous biographies of individual geologists have been published. Just a few may be mentioned – for example, Wilson's *Charles Lyell, the Years to 1841*

(1972), Rupke's *Great Chain of History: William Buckland and the English School of Geology* (1983), Schwarzbach's *Alfred Wegener* (1986), Lapo's *Traces of Bygone Biospheres* (1982/1987) (on Vernadsky), Roger's *Buffon* (1989), Dean's *James Hutton* (1992) or Vaccari's *Giovanni Arduino* (1993). There is ample scope for further biographical, or perhaps autobiographical, studies. But at the time of writing the great need seems to be to synthesize the extensive journal literature on twentieth-century geology in book form; to extend studies of the history of tectonics and the earth's interior into the twentieth century; and to compose synthetic accounts of the histories of volcanism, petrology, mineralogy, geophysics and geochemistry. Economic geology also needs more attention. I should also like to see detailed studies of particular regions – say the Alps, the Andes or Japan (in English, please) – and perhaps of the various geological systems. I can imagine, for example, that a book on the history of studies of the Jurassic system could be most interesting. But such tasks remain to be realized in the future.

Notes

Introduction

1 Readers and reviewers will, no doubt, be quick to notice that there are rather many of my own publications cited in the bibliography. This is not intended as self-glorification, but reflects the fact that the present book is partly a synthesis of material that I have written elsewhere in scattered publications.
2 Geological maps have the task of trying to depict 'four-dimensional events' in two dimensions. They are tricky things.

Chapter 1

1 The boundary lines between myth, legend, folk-tales and folklore are indistinct. Folklore, which was in the past transmitted orally and represented the unwritten traditions of a people, although of course it may appear in books today, can be taken to include the other three categories. Folk-tales are commonly not regarded as true by the people that tell them, whereas myths and legends would have been regarded as true by their tellers. Myths generally have a sacred character, whereas legends may be regarded as sacred or secular. Legends generally have humans in the major roles, whereas in myths gods or animals are the major characters (Vitaliano, 1973: 3). I believe that there can be modern as well as ancient creation myths, such as that Australia became a nation on the battlefield of Gallipoli. Philosophers have their myths too. One of them is that the first beginnings of critical enquiry and rational thought occurred with the Greeks (Popper, 1958–9). The case against this has recently been cogently argued by Greene (1992).
2 There are in a sense certain religious practices associated with science too. For good reason, the great nineteenth-century museums have been called cathedrals of science.
3 It is a nice question whether the scientist who argues against a strongly entrenched viewpoint (or paradigm in the terminology of Thomas Kuhn) is in a position similar to or different from the Tanna boy whom we imagined as querying his tribal myths. Indeed, a scientist who takes a view different from all the rest may well be thought mad, be driven from the tribe and go hungry. On the other hand, there is an endless debate going on at numerous levels within the scientific community. Each scientist has to acquire tacit knowledge as to the acceptable balance between healthy criticism and conceptual

'madness'. Incidentally, modern theology has in some quarters become a 'seething mass of argument', with much critical discussion. In the process, where the debate is most intense the mythic character of the religion is, I suggest, declining.

4 Which god of the Hindu pantheon was meant, I do not know.

5 'When above the heaven had not [yet] been named' (Sproul, 1979: 91).

6 The texts of other creation myths from the Middle East are given by Sproul. See also Arrhenius (1909).

7 These are thought to be two main constituent sources for Genesis: the older (eighth or ninth century BC) 'Yahwist '('Jehovist') document (*J*), and the younger (fourth century BC) 'Priestly' document (*P*).

8 There are, however, thought to have been 'two Hesiods' – first the author of *Works and Days* and then the author of the *Theogony*. Both came after Homer.

9 For example, Cronos, the young and dangerous god of Time, spake thus:

> 'Mother, I will undertake to do this deed, for I reverence not our father of evil name.'
> So he said: and vast Earth rejoiced greatly in spirit, and set and hid him in an ambush, and put in his hands a jagged sickle, and revealed to him the whole plot.
> And Heaven came, bringing on night and longing for love, and he lay about Earth spreading himself full upon her. Then the son from his ambush stretched forth his left hand and in his right took the great long sickle with jagged teeth, and swiftly lopped off his own father's members and cast them away to fall behind him (Hesiod, 1914: 91, 93.)

10 A recent dating of the seven 'lean years' in Egypt, mentioned in the Bible, has been made by examining the growth rings in the ancient bristle-cone pines in California, and it has been suggested by Colin Humphries that they can, at 1628 BC, be related to the famous eruption of Santorini, which would have clouded the whole earth.

11 Quite apart from the possible geological events that may have lain behind Hesiod's account, the names of the goddess-nymphs supposedly born to the earth mother (Gaia) are interesting. Hesiod gives them as Coeus, Crius, Hyperion, Iapetus, Theia, Rhea, Themis, Mnemosyne, Phoebe and Tethys. Of these, Iapetus and Tethys have been utilized in modern geology for hypothetical tracts of water present on the surface of the earth in earlier times. And the goddess Gaia provides the origin of the term geology, not to mention the recent 'theory of Gaia' (see Chapter 12).

12 For a survey of theories of myth, see particularly Kirk (1970).

13 The major surviving texts of the presocratic philosophers (or descriptions of their ideas given by classical writers) are conveniently collected in Kirk and Raven (1957). Whether the presocratics were in fact the first to engage in critical discussion in the manner so admired by Popper is not really known. As mentioned, Popper's story of the presocratics perhaps represents Western philosophy's 'creation myth'.

14 Popper has himself been a strong advocate of the virtues of critical discussion for both science and philosophy. It is no surprise, therefore, that he should have emphasized particularly this aspect of the work of the Presocratics.

15 Traditionally, Thales has been regarded as the first Western philosopher of whom we have knowledge. And he has been represented as a kind of absent-minded professor (Guthrie, 1962: 39). But in the view of Greene (1992, ch. 5) Thales was in all probability a man of many parts, with a strong practical bent.

16 Here, I suggest, Heraclitus was beginning to fall into the trap that Anaximander, with his *apeiron*, had sought to avoid.

17 One might, it is true, also claim some empirical support for the theory, as did the later Roman poet Lucretius, in his poem, *De rerum natura*. He pointed, for example, to the apparent dance of motes of dust in a sunbeam.
18 This is not to say that Dalton's views were purely the result of empirical considerations. He was, in fact, heavily influenced by the traditions of the Newtonian matter theory, which contained a strong metaphysical dimension.
19 A great deal of information on the folkloric tradition of floods was collected together by Sir James Frazer (1919, vol. 1, ch. 4).
20 The Japanese term, used by geologists, for destructive 'tidal waves' produced by earthquakes. (Such waves do not, of course, have anything to do with tides in the usual sense.)
21 The most recent information I had available to me at the time of writing this chapter was a newspaper report (in *The Observer*) stating that a scientific team had discovered a large boat-shaped object at an altitude of 7,000 feet, about twenty miles from Mount Ararat. It was to be excavated in the summer of 1994. The American leader of the team of investigators, Mr David Fasold had suggested a provisional explanatory hypothesis similar to that of Suess. The 'boat', supposedly identified by radar imagery, was said to be immediately below the mountain of Al Judi, regarded in the Koran as the resting place of Noah's Ark (Wroe, 1994). Fasold did make an expedition to the site in 1994, but in company with the Australian geologist Ian Plimer. The journey proved to be a fiasco for Fasold, and he retracted his views. His 'ark' was a geological structure of curious appearance: a syncline of hard rock standing out at the bottom of a valley. It certainly bears some resemblance to a boat, and is claimed to have dimensions similar to those detailed in the Bible. Also, it appears that there are place-names in the neighbourhood which suggest association with the Noachian legend. But one can turn the whole argument round. The place-names and dimensions need not necessarily evidence the Noachian legend. On the contrary, the legend may have grown up to account for the existence of the curious boat-shaped structure. An interesting TV programme broadcast in Australia in 1994 showed a disappointed Fasold admitting defeat at the hands of Plimer. However, other 'arkeologists' are not wanting.
22 This idea is highly reminiscent of Stoic ideas and may be a Stoic accretion to Pythagorean doctrine. See p. 18 for a discussion of the ideas of the Stoic Seneca.
23 There are certain silicified logs occurring in the Egyptian desert which are aligned in such a way that they have some resemblance to the remains of ships, and these could have been the source of the idea that is mentioned by Ovid.
24 Consultation of a topographical map for this part of Greece suggests to me, however, that it is a case of 'antecedent drainage'. That is, the pre-existing river probably cut its own gorge, as the hills slowly rose between the plain and the sea.
25 'Powder of the cave' (from the Arabic): arsenic sulphide.
26 Hard, red, ochreous iron ore.
27 The possibility of the continued meaningful use of the metaphorical term 'body' in this context should be noted. It indicates the survival of the old 'organic' metaphor. Likewise, mineral 'veins'.
28 This may be taken to be prior to AD 79, the year of the great earthquake of Vesuvius, as this could hardly have escaped mention in *Aetna*.
29 The anonymous *Aetna* gives the closest approximation we have to such a treatise. But it was in the form of a didactic poem rather than a scientific treatise. The idea of earthquakes being due to subterranean winds was retained for many centuries. In *1 Henry IV*, III. i, Shakespeare has Hotspur say to the boastful Glendower:

'O, then the earth shook to see the heavens on fire,
And not in fear of your nativity.
Diseased nature oftentimes breaks forth
In strange eruptions; oft the teeming earth
Is with a kind of colic pinch'd and vex'd
By the imprisonment of unruly wind
Within her womb; which, for enlargement striving,
Shakes the old beldame-earth and topples down
Steeples and moss-grown towers.'

30 At what is now Aswan. Syênê gives its name to the important rock type, syenite.

31 Anaximander's notion of a disc-shaped earth did not last long. The Pythagoreans had the earth as spherical, which followed from the fact that the earth casts a circular shadow on the moon during an eclipse.

32 Neoplatonism was essentially the creation of Plotinus (AD *c.* 205–70). It utilized the Platonic doctrine of archetypal 'ideas', but assumed that they originated as ideas in the mind of God. The ideas were also, so to speak, in a 'chain' or hierarchy, with the idea of the 'good' at the summit. Further, the 'higher' ideas were supposed to give rise to the lower ones, by a process that was variously likened to the emanation of heat or light or by multiple reflections in mirrors.

33 There are, for example, two interesting 'T-shaped' manuscript maps of the world held in the *Stiftsbibliothek* at St Gallen, Switzerland (*Handschrift* No. 237, p. 1 and *Cod. Sang.* 236, p. 89), dated AD *c.*800 and 850 respectively. The first divides the world (depicted as a circle) crudely into four parts (Africa, Asia, Europe and Uninhabited Earth); the second omits the Uninhabited Earth. There are two principal rivers, the Nile and the Don, and two seas, the 'Paludes' and the Mediterranean. The representation is structured geometrically rather than realistically, and appears to be trying to gesture towards the Christian symbol χρ. The offspring of the three sons of Noah (Ham, Shem and Japheth) are associated with the three continents Africa, Asia and Europe respectively. The knowledge indicated in these 'maps' is symbolic and theological as much as geographical. One must acknowledge that it would be difficult to form a satisfactory geographical knowledge of the earth and its inhabitants from land-locked Switzerland. But such 'T-maps' of the world were produced quite frequently at that time. (Whether the two that I have seen at St Gallen originated in Switzerland, I do not know.)

34 Some medieval Latin translations of Aristotle's *Meteorologica* contained a section entitled '*De mineralibus*', additional to the authentic fourth book. Following the researches of Holmyard and Mandeville (Avicenna, 1927), it appears that this was written by Avicenna, being an alternative version of parts of his *Book of Remedy* (*Kitâb al-shifâ*) (written in 1021–3).

35 The discovery of the precession of the equinoxes, which is due to a secular change in the earth's axis of rotation, like that of a spinning top, is attributed to Hipparchus (*fl.* 146–127 BC).

36 The modern figure for the period of the precession of the equinoxes is about 24,000 years, not 36,000.

37 The sun and moon are depicted as small spheres, and a few stars can be seen indistinctly also.

38 Right at the centre of the earth was Satan, frozen in a block of ice, but with a trifaced head (vermilion, whitish yellow and black) perpetually chewing at the world's supposed three worst sinners: Judas, Brutus and Cassius. See Dante Alighieri (1950: 182–3).

39 Could this have been a poke at authors such as Ristoro d'Arezzo?

40 Examples are provided by such authors as Isidore of Seville, Albertus Magnus or Vincent of Beauvais.

41 First published in French as *Les mots et les choses* (Foucault, 1966).

42 For example, certain plants might be known to grow well together, but the mongoose was antipathetic to snakes.

43 The mandrake or the ginseng plant, having some similarities to human anatomy, might be thought relevant to (say) human fertility. The correspondences could thus be read by the adept, who could recognize the appropriate signs or 'signatures'. (When in Tanna in 1991, a villager proudly showed me two huge yams, one of which was shaped somewhat like the female body while the other was male in appearance. It appeared that they were to be used in some village festival the following week.)

44 But this information, he said, was derived from a fellow-naturalist correspondent, Johannes Kentmann.

45 Also styled Aureolus Philippus Theophrastus Bombast of Hohenheim.

46 These were property-conferring 'principles' rather than the common-or-garden substances that we know as salt, sulphur and mercury, which were merely the closest approximations among everyday substances to the three philosophical principles. 'Salt' was supposedly the principle that conferred the properties of salinity, solubility and astringency; 'Sulphur' was responsible for inflammability and odour, while 'Mercury' caused metalleity, mobility, fusibility and shininess. (I capitalize the names of the principles of the *tria prima*, to distinguish them from the salt, sulphur and mercury that we may find bottled up in a chemical store-room.

47 The *archeus* could be thought of as God's 'workman' or artificer, who put into operation the divine plan. There was a similar gnostic concept of *archon*, and the idea perhaps had its origin in Plato's 'demiurge', who supposedly put together the cosmos according to the divine plan. Paracelsus also used the notion of an *archeus* in his medical theory. He supposed, for example, that there was a little archeus sitting in the stomach and responsible for digestion.

48 The fact that the Paracelsian corpus was translated into English by A.E. Waite in the nineteenth century is also significant. The work was done at a time when there was a considerable revival of interest in mysticism and the occult.

49 Or 'formula', 'blueprint' or 'architectonic plan'.

50 For further details, see Oldroyd (1974c).

51 The origins of this theory were to be found in Paracelsus (Jevons, 1964).

52 To a modern geologist this has the flavour of hydrothermal mineralization.

53 Likewise, an artisan applies form to matter with his tools.

54 This concept had been used by Aristotle in his writing on physiological processes such as digestion.

55 There were considerable differences in the meaning of the word 'earth' in different contexts. It could mean the planet Earth; soil; or the Aristotelian element, of which various earthy substances found in the ground were but approximations.

56 Blood, yellow bile, black bile and phlegm.

57 Before Paracelsus added Salt to make the *tria prima*, Arab chemical theory had invoked the notion of 'philosophical' Sulphur and Mercury as the two essential chemical ingredients.

58 The process may derive from Stoic accounts of the evolution of the cosmos, as each step was envisaged as involving a decrease in 'tension'.

59 No exact translation is possible. An approximation might be production, extension or development.

60 Paracelsus had likened the sea to the canopy of a forest, with trunks and roots below being analogous to supposed subterranean rivers.

61 For further details of Becher's mineral theory, see Oldroyd (1974b) and Laudan (1987).

62 In the eighteenth century, the best known of these 'property-conferring principles' was named 'phlogiston', the idea being developed by Becher's follower, Stahl, who was a pupil of the alchemist G.W. Wedel of Jena. Inflammable substances supposedly contained this principle, which conferred the property of inflammability to substances in which it was present. Likewise, there might be a magnetic 'principle' in a lodestone, responsible for the stone's ability to attract iron. But, just as one would not expect to extract a handful of magnetism from a magnet, so one would never expect to extract a handful of phlogiston from (say) charcoal. Even so, as Stahl developed the theory, it came to be thought that phlogiston could be transferred from one substance to another, and the theory was of some utility in explaining metallurgical processes and combustion in general.

63 Some modern commentators have not been inclined to agree with Lyell on this. See, for example, Taylor (1992).

Chapter 2

1 Or as an analogue of the program for a computer, as modern cognitive science would have it.

2 But, in Italy in particular, some of the Renaissance scholars, such as Leonardo, Alberti or Brunelleschi, were already highly competent in practical matters.

3 The clock analogy was by no means fresh in the seventeenth century. As mentioned in Chapter 1, it had been used as far back as the time of Cicero in his work *On the Nature of God*, although at that early date the water clock was considered, not the remarkable mechanical clocks which were already quite common by the seventeenth century.

4 In fact, this was by no means entirely the case. Leibniz complained against Newton that his perfectly hard corpuscles, which never wore out despite all their mechanical interactions, would suffer infinite accelerations at every impact if they were completely undeformable.

5 A well-known example is offered by Robert Boyle, who supposed that the corpuscles of gases were like little coiled springs or fleeces of wool. So the macro-behaviour of a gas under pressure was thought to be explicable in terms of the micro-properties of the hypothetical constituent particles.

6 This objection was held against Newton's gravitational theory. But the success of his theory in its mathematical aspects, for both terrestrial and celestial mechanics, was such that the objection was not pressed very strongly. In his *Opticks*, Newton gave a clear pronouncement on his matter theory, which was that matter was made of 'solid, massy, hard, impenetrable, movable particles'.

7 How, for example, might the springiness of Boyle's 'woolly' corpuscles (see footnote 5) be explained?

8 If this process were repeated at two levels, so to speak, then one might get a moon circulating round a planet, circulating round a sun.

9 Not vacua, of course, but regions occupied by air.
10 For further details, see Oldroyd (1974d).
11 Indeed, religion might be said to have been a central feature of the scientific revolution, in that it was essential that the new philosophy should be compatible with Christian doctrine.
12 Rabbinical chronology also yielded a very short time-scale for the period since the Creation. And so too did a Roman Catholic calculation based upon the Clementine edition of the Vulgage version of the Bible.
13 This was a work of massive scholarship (776 pages plus tables and plates). The book also contained interesting maps indicating the geographical knowledge of the time. Raleigh did not think that mountains were produced by the Flood. Mount Ararat had to be there for the ark to land on.
14 For details see Haber (1959), Dean (1981) and Gould (1993).
15 Indeed, some Oxford University Press editions of the Bible retained Usher's date for the origin if the earth until 1910. So an overall age of the earth of about 6,000 years was 'officially sanctioned' for many years and presumably believed by numerous churchgoers in Britain, although by 1910 it was surely an acute anachronism.
16 For example, limestone forms quicklime on heating.
17 A kind of creature related to the modern squid, which left a fossil somewhat resembling a bullet or unbarbed spearhead.
18 Some 'mechanists', such as Woodward, also produced grand collections and classifications in the manner of the natural historians.
19 The list consisted of 1,766 specimens, including corals and trilobites, as well as more 'ordinary' shells.
20 In central Australia, when there is a large flood after many dry years, large fish appear in the rivers as if by magic. I should be pleased to know where they may have been hiding in all the years of drought.
21 In the process, Newton offered some interesting arguments about the dating of the events of ancient Greece. Knowing the rate of precession of the equinoxes, he knew which constellations were visible in the Mediterranean skies in any given century. Hence he was able to provide a rough estimate of the date when the constellations were named, which was done, according to Greek tradition, at the time of the Argonaut expedition. This led Newton to propose a change in the date of this expedition.
22 Kepler's laws of planetary motion: (1) Planets move in ellipses round the sun, with the sun at one of the foci of each ellipse. (2) the area swept out by a planet in its elliptical orbit is proportional to the 'time of sweeping'. (3) For each planet T^2 is proportional to R^3 where T is the period of the planet's orbit and R is the semidiameter of its elliptical orbit.
23 For elliptical orbits, this is not an easy task, according to the method Newton used, although it is simple enough if one makes the approximation that the planets move in circular orbits. In fact, the calculation is rarely done these days by Newton's method. For my reconstruction of his calculation, see Oldroyd (1990b).
24 There would be no difference in result if the earth were perfectly spherical.
25 They did not emerge without much controversy, recently analysed by Illiffe (1993). See also Berthon and Robinson (1991, ch. 6).
26 The quotation is continued in Chapter 6, p. 132.

Chapter 3

1 Such questions are, however, likely to be the concern of the modern mineralogist.
2 For a discussion of the history of the term, see Dean (1979). He points out that the word 'geologia' was used by several authors in the seventeenth century, and a sixteenth-century Italian usage by Aldrovandi has been reported. According to Dean, the word 'geology' was used by Benjamin Martin in 1735 in reference to a 'philosophical view' of 'the Nature, Make, Parts, and Productions of the *Globe of Earth* on which we live'. It covered geography, hydrography, phytography and zoography. But the term 'geology', as we would recognize it in use today, only began to emerge in the 1780s, following the usage of Continental authors such as de Saussure and de Luc in the 1770s.
3 The irregularities corresponded to mountains and ocean basins, but the overall shape was spherical.
4 For a more detailed account of Hooke's theory, see Oldroyd (1972b). It is just possible that Hooke's theory was inspired by his reading of Buridan (see Chapter 1).
5 Euhemeris of Messene (*c.* 300 BC) was a writer who supposed that the actions of gods, as related in myth, were obscure accounts of the doings of real human beings.
6 His original name was Nils Steensen, but he is almost always referred to by the Latinized version.
7 The work to which the *Prodromus* was intended as an introduction never appeared. There has, incidentally, been discussion in the literature as to whether Steno may have derived some of his ideas from Hooke, for he visited London shortly before the *Prodromus* was published and may have heard of the Englishman's ideas. Hooke himself accused Steno of plagiarism (Oldroyd, 1989: 217).
8 The candidate site is the valley of the Era River (a tributary of the Arno), near Volterra. According to the views of modern geology, the structure is in fact one involving block faulting, or what geologists term a 'horst and graben' structure. The river valley corresponds to a down-faulted structure. But there is much slumping of strata in the region, and this may have suggested Steno's model section.
9 In fact, according to widely accepted opinion, the union of scholar and craftsman occurred earlier in Renaissance Italy, in the work of artists such as Leonardo or Alberti or architects such as Brunelleschi.
10 'Booklet on ores' and 'Booklet on assaying'. The original author may have been Ulrich Rülein von Kalbe, mayor of the mining town of Freiberg at the end of the fifteenth century.
11 The anonymous author wrote of the union of quicksilver and sulphur in ore formation, being the female and male principles respectively. Different metals were also supposedly produced under the influence of different planets. So the theory of ore bodies relied on astrological considerations.
12 When Louis XV ran out of money for the project it was continued as a private venture by the Cassinis, and emerged in 182 sheets as '*La carte de Cassini*'. The scale was 1 : 86, 400.
13 One may also remark that Guettard's interpretation of layering suggests that Steno's linking of layering with temporal order was not necessarily something that was manifestly obvious to someone thinking about the earth.
14 These may, not unreasonably, in my view, be regarded as early representations of what came to be called 'cyclothems' in the twentieth century. See Chapter 8. On Lavoisier's geological work, see further Rappaport (1968, 1973).
15 The most recent flows, of the 'Third and last epoch', are depicted with the pebble-shaped

marks. Earlier lava flows, no longer connected to volcanic vents, are exemplified by the *Montagne de la Serre*.

16 There were only seventeen subdivisions in the German edition (Lehmann, 1756). The thirty-first unit represented the 'primitive mountains'.

17 Lehmann called it the *wahre rothe todte* rock.

18 Füchsel's figures were keyed as follows: 25 = *Ambitus fundamentum* (fundamental or basement rock); *A* = 24 = *rothe tode Lager* ('red dead layer'); *B* = 23 = *Steinkohlicht Geburge* (coal stone); *c* = 22 = *Alaunschiefer* (alum shale); *D* = 20 = *schwarz-blau Schalgeburge* (black-blue limestone); *F* = 19 = *roth Schalgeburge* (red limestone); *E* = 18 = *weisslich Schalgeburge* (whitish limestone); *e* = 17 = *Sandfloetz* (layered sandstone); *G* = 16 = '*das Flötze*'; *h* = 15 = *grau Gyps oder Alabasterlager* (grey gypsum or alabaster layer); *H* = 14 = *Brennkalchgeburge* (sulphur rock); *i* = 13 = *Glühsand* ('glowsand'); *J* = 12 = *Sandgeburge* (massive sandstone); *K* = 11 = *rothe Gypslager* (red gypsum); *K* = 10 = *Muschelkalch* (fossiliferous limestone).

19 Of course, the practical problem is the reverse of this; one has to induce the internal layered structure of the earth's crust from the mappable outcrops of the strata at the surface.

20 I have collated and summarized some of the more important of the classificatory systems of the period in tabular form in Oldroyd (1974a).

21 This is the first section of such complexity known to me. A very stylized section of a Somerset coalmining area (which, however, showed an interesting unconformity) was published by John Strachey (1719). In Sweden, the naturalist Carolus Linnaeus produced a geological cross-section of a mountain (kinne-Kulle) in southern Sweden, which he visited in 1746 (Merriam, 1992).

22 In this figure, according to Arduino's key, *A* represents schist, *a* veins of iron-like mineral, *B* arenaceous, quartzose micaceous sandstone, *H* fossiliferous marble, *L* a white or red rock (dolomite?), *M* a calcareous stone with flints, *N* and *P* a black ferruginous stone, *q* an arenaceous calcareous stone with marine fossils. The section was first published by Stegagno (1929, Plate VIII).

23 The term Secondary was used well into the nineteenth century. Later, in the 1840s and at the suggestion of John Phillips, the 'palaeontological' terms (Palaeozoic and Mesozoic) superseded Primary and Secondary. But the name Tretiary has has been retained to this day. The change in terminology reflected the shift from lithology to palaeontology as the basis for stratigraphy.

24 Of course, that understanding is in direct historical line of descent from the work of Arduino.

25 This useful formulation of the principle is due to Archibald Geikie, not Lyell himself.

26 It will be noted that Arduino's lowest rocks were called Primary, not primitive. There is a difference in that the Primitive rocks were thought to date back to the very origin of the earth. Arduino's northward journey in Italy did not bring him into granite country, to judge by his section, reproduced in Figure 3.7.

27 See also Hedberg (1969a,b).

Chapter 4

1 Already Chapter 3, we have seen fire and water theorists – Moro, for example, having preference for fire, Lehmann for water.

2 A German edition was published the same year; also a French edition in 1859.
3 It should be noted, however, that he had definite interests in mining engineering and, more generally, in ideas about the nature of the earth. So his views concerning the earth were not developed simply as an outcome of his employment as court historian.
4 The source of this water was not explained satisfactorily. However, modem theories of the origin of the earth, atmosphere and oceans also contemplate a 'degassing' process.
5 Becher's idea was not implausible, given that many fossils are mineralized (e.g. with pyrites), so that their substance is quite different from that of any living forms. (Moreover, many fossils are quite different from the forms of living organisms.)
6 This view would have appealed to a historian who was seeking to trace back the whole history of the globe. The Bible was inevitably the source of much of his earlier historical information.
7 For further discussion of Leibniz's theory, his sources and the manner in which his theory fitted into the thought of his time, see Ariew (1991) and Ellenberger (1994: 137–48).
8 De Maillet also envisaged an alternative sequence of developments in which more water might be gathered on the earth, such that the oceans would become so deep as to become uninhabitable once again. For a helpful synopsis of de Maillet's cosmology and his theory as a whole, see Carozzi (1992b).
9 Only thirty-six were completed in Buffon's lifetime. The series was continued by other authors after his death, the last volume appearing in 1804.
10 For Buffon's life and work, see particularly Roger (1989).
11 Though published in 1749, it bears a date 3 October 1744. It was updated somewhat between then and the year of publication.
12 Heaps of moist pyrite can become strongly heated as a result of the action of sulphur-loving bacteria.
13 Whether there was a causal connection between Pallas and Buffon I do not know.
14 As will be seen in Chapter 11, the notion of land bridges to explain the peculiarities of biogeography continued well into the twentieth century, with the theory of 'isthmian links'. But such twentieth-century theorists had no direct historical connection with Buffon.
15 This states that the rate of cooling of a hot body is proportional to the temperature difference between the body and its surroundings.
16 Buffon's ideas were criticized by the clerics in the Sorbonne when first adumbrated in 1749. But, being a man of independent means and powerful social position, Buffon was not impeded in his thinking and his work, and he proceeded on his merry way to develop yet more radical ideas. Even so, he was circumspect in what the actually published, and did not give public utterance to his notebook speculations about the possible age of the earth.
17 Hutton did not marry, but is known to have fathered an illegitimate son, which may led to his retirement from city life for a number of years.
18 As a scientist/farmer, Hutton was not content to let the process proceed unassisted, but sought to make improvements in the soils on his farms by the suitable addition of marl, manures and fertilizer. For Hutton as a farmer and for the influence of his farming interests on the development of his thinking about the earth, see Jones (1985).
19 This will be known to anyone using a bicycle-pump.
20 In modern terms, Hutton was thinking of the formation of what are called 'batholiths'.
21 A somewhat similar view was taken by the Italian Giovanni Targioni-Tozzetti (1712–83). He wrote:

'[W]hen one thinks one has reached the ultimate limit of diversity, at the point where one may say with confidence that [at least] some part of the globe is, so to speak, virgin, and in its initial state, as created by the most omnipotent and wise author of nature, one must acknowledge that our capacity to comprehend is inadequate, and [even] that part is not as it was in the beginning but is made up of other, superimposed broken parts, each of which, if properly observed, is also not in a primaeval state' (Targione-Tozzetti, 1777, vol.2: 189).

I am grateful to Ezio Vaccari for this reference. It has a very 'Huttonian' ring and suggests that many ideas that have been traditionally ascribed specifically to Hutton were in fact developed elsewhere than Scotland in the second half of the eighteenth century. But more research on this matter is needed.

22 While '*p*' would be the nest of arguments in Hutton's theory of the earth, '*q*' would be the existence of strata lying unconformable to one another. It will be recalled that Robert Hooke had predicted slow changes in the direction of the meridian on the basis of his theory, but he failed to find what he predicted. (He never succeeded in establishing '*q*'.) In the case of theories such as that of Descartes, the logical connection between grand theory (the hypothesis of vortices) and facts observed (say inclined strata in mountains) was excessively weak. It was a 'just-so story', which did not offer an empirical prediction but a 'postdiction'. In the case of Steno, it is not known whether he made predictions about the arrangement of the Tuscan strata in advance of seeing those strata.

23 Usually called the 'country rock'.

24 This substantial memoir was actually read before the Society in 1785.

25 Published only two years before Hutton's death, the book was imperfectly edited. Much of the second volume consists of extended and somewhat undigested (i.e. untranslated) extracts from French authors such as de Saussure. The whole text has long been regarded as verbose and repetitive, and the lack of direct impact of Hutton's work was formerly attributed to his poor literary style. It should be remarked, however, that Hutton's text does in fact contain numerous firmly stated and impressive passages, and in Scotland at least his ideas made a considerable impact.

26 In the version edited by Craig *et al.* (1978), the picture is entitled 'Detailed E – W Section, Northern Granite, Isle of Arran, Strathclyde'. Hutton gave a detailed description of his visit to Arran in 1787 in Chapter 9 of the unpublished third volume of his *Theory of the Earth.* This text is of great interest as it shows that Hutton regarded the island as a geological exemplar: 'such as to lead to knowledge of all that is necessary in the production of the land or the surface of the earth, as a habitable world' (Hutton, 1899: 199). Further, after describing the nature and distribution of the island's rocks (the 'mineral natural history'), he also gave a sketch of the geological history of Arran and even its possible future (ibid.: 262).

27 Hutton's theory was, in contrast, often called the 'Vulcanist' theory.

28 But data for the nineteenth century do not all fit Foucault's analysis equally satisfactorily. See p. 199.

29 Such a procedure was perhaps best exemplified by Linnaeus's procedures for botanical classification. See further the comments on Linnaeus's crystallography in Chapter 9 (p.193).

30 The *Kurze Klassifikation* did not actually contain the theory *per se*. It set forth the order of the rocks as produced according to the theory. This theory was taught to the students at Freiberg and may best be known by examination of Werner's manuscripts and from surviving copies of his students' lecture notes. The theory has been examined in detail

by Alexander M. Ospovat, and may be found stated in his translation of the *Kurze Klassifikation* (Werner, 1971).

31 Various versions of the list can be found in Werner's manuscripts and in the publications of his students.

32 Today, the phenomenon is attributed to rises in the level of the land after the removal of the great load of ice in the Baltic region, subsequent to the last ice age.

33 As will be discussed in Chapter 9, however, there were significant attempts in the nineteenth century to re-establish a Wernerian-style geognosy, in, for example, the work of Sterry Hunt.

34 For example, a given red sandstone does not, by virtue of its lithology, proclaim itself as belonging to the New Red Sandstone (Permian or Triassic) or the Old Red Sandstone (Devonian). Such a determination has to be made by consideration of fossils as well as rock type.

35 The earlier works of Lehmann and Füchsel, although undoubtedly conceptual antecedents of Werner's ideas, did not achieve paradigmatic status. There was no 'Lehmannian' or 'Füchselian' radiation (although some of Werner's ideas were derived from Lehmann).

36 This is commonly called the Vulcanist–Neptunist dispute.

37 But for Hutton the granitic veins in Glen Tilt did clinch the matter.

38 It is nice question whether I, in criticizing Geikie's historiography, am myself guilty of historiographical anachronism and Whiggery.

39 Perhaps an earlier example is provided by Robert Hooke, who wondered whether the craters on the moon were due to external impacts or internal volcanoes. He tried heating a pasty mix of plaster until it bubbled like hot porridge, and also threw objects at it to see whether the result looked like the lunar 'craters'.

40 On this, see Lawrence (1977).

Chapter 5

1 If the historian is desirous of carrying matters to extremes, it may be noted that fragments of a map, drawn on papyrus from ancient Egypt, have survived, which give some rudimentary 'geological' information pertaining to gold deposits and building stones. See McMahon (1992).

2 Certain geological sections drawn in the eighteenth century, such as those of Lehmann or Arduino (see Chapter 3), did, however, convey a sense of the past history of the earth.

3 This was dated 1774, in a book published in 1775. The map was entitled '*Geographischer-Plan der Gefürsteten Graffschafft Henneberg Chur Saechsischen Antheils enthaelt die Aemter Schleusingen Suhl and Kuhn-dorff mit Bennshausen. Gezeichnet und herausgegeben 1774 von Fried. Gottlob Glaeser*'.

4 I am indebted to Andrew Grout for this reference.

5 The rock is constituted of myriads of small spheres of limestone, so that the whole has an appearance somewhat similar to that of fish roe.

6 For the published version, see Phillips (1844: 30).

7 Something similar to this procedure was recommended by Werner and Jameson before Smith first issued his map in 1815. They intended the deepening of colour to indicate the relative positions of the strata in their vertical order. See Jameson (1809–10: 156–7).

8 This document, which is in poor condition, being so heavily vanished as to be almost illegible, survives in the archives of the Geological Society, London.

9 Also held by the Geological Society.
10 He certainly used fossils, and parts of his own collection still survive in the bowels of the British Museum of Natural History.
11 Smith MSS, Geology Department, Oxford University. Extract from an addendum to notes prepared by Smith for a lecture at Scarborough in August 1824. See Edmonds (1975).
12 In Edinburgh, the Professor of Natural History, Robert Jameson, a former student of Werner, expounded his teacher's geognostic principles from the professorial chair, and established a Wernerian Society to propagate the German's theories. Many and vigorous were the debates between the Huttonians and the Wernerians.
13 Smith did not accept the Huttonian idea of valleys being carved by weathering and erosion. Rather, he supposed that they were the products of the original processes of deposition. An 'outlier' was not, therefore, regarded as an erosional remnant, formerly connected to the main body of strata of the type concerned. It was a 'splat' produced by the original deposition, like an isolated patch of foam on a beach.
14 The map turned up in an American archive. It is published in Torrens (1994). (Farey did not use Smith's technique of graded coloration.)
15 I refer to the work of Hooke's later years. In his earlier days, it will be recalled (see p. 61), he thought that fossils might be sufficient for this purpose.
16 Such correlations were not entirely new. Phillipe Buache (1700–73), who assisted with Guettard's map of France (see Chapter 3), showed a '*bande marneuse*' extending across the Channel (Guettard, 1751, Plate 31; Taylor, 1985: 20; also see Figure 3.2).
17 I have argued elsewhere (Oldroyd, 1979) that this historicization of geology occurred at about the same time as the emergence of a historical view of the law and other human institutions. Thus it is tempting to see the historicization of geology as a manifestation of a much larger-scale way of thinking that began to manifest itself in the late eighteenth/early nineteenth centuries.
18 One would not, of course, expect fossils of a particular age to be identical on opposite sides of the globe, or even much closer than this (although some are). Even so, some correlation may be possible, because the general characteristics of fossils of a particular age may be similar: and, by step-by-step lateral correlations, the whole may in principle be analysed. At least, that is the goal of historical geology.
19 De la Beche was, however, drummed out for insubordination after two years.
20 He also geologized on the Continent before going out to Jamaica.
21 In fact, as described in detail by Herries Davies (1983, 1995), things got going a little earlier in Ireland, but this lies somewhat beyond the point I wish too develop here.
22 After dealing with Devon, De la Beche surveyed Cornwall, and then jumped across the Bristol Channel to work in the economically important region of South Wales. Thereafter, work proceeded quite systematically, with areas of economic importance being given particular attention. The Scottish survey began in 1854 and a Scottish Branch of the Survey was established in 1867.
23 Thus in modern parlance it ran from the Upper Palaeozoic to the top of the Mesozoic.
24 This statement requires some qualification. There is in Germany a sandy unit which is partly white and partly red. The latter the German miners called the *rothe tode Liegende* or 'red dead layer'. Werner referred to this unit as the *aelter rother Sandstein* or Old Red Sandstone. According to today's thinking, this German unit belongs to the Triassic System. But Jameson (1808; 159), in applying the German geognosy to British rocks, made a mistaken

correlation and called a common British red sandstone unit the Old Red Sandstone. This was satisfactory as a determinate unit and the name has been retained. However, the mistaken correlation illustrates the problems that can arise when lithological rather than palaeontological criteria are used. The correlations were cleared up in Buckland and Conybeare (1824).

25 On the other hand, as will be discussed in Chapter 8, there are grounds for thinking that the subdivisions we use are 'natural', representing very large-scale changes in the stratigraphic record.

26 On Murchison, see particularly Thackray (1978), Rudwick (1985), Secord (1986), Stafford (1989) and Oldroyd (1990a).

27 See particularly Rudwick (1985) and Secord (1986).

28 After the establishment of the Survey, the surveyors regarded themselves as the 'professionals', while all the other geologists, from university professors to local fossil collectors, were termed 'amateurs'. The amateurs did not reject the label.

29 De la Beche relied more on structures and lithologies and, although his battle proved unsuccessful, later commentators agree that he was by no means wholly mistaken. In fact, as has been shown by Rudwick (1985), both sides altered their opinion so as to adopt an intermediate position as the controversy proceeded. Incidentally, not all the rocks of the Old Red Sandstone are red and not all are sandstones.

30 For some reason, Murchison never seemed to get so excited about the extension of the other two systems which he was involved in naming: the Devonian and the Permian. Murchison also suggested the idea of the 'fundamental gneiss' (of the north-west Highlands of Scotland) as the base of the stratigraphical column in Britain. But he never seemed to be particularly keen to see it spreading here, there and everywhere. The reason, I believe, was that he wanted to see a relatively simple, intelligible and straight forward stratigraphical column established for Britain. For this, only one base was desired. Thus Murchison resisted the efforts of 'amateur' geologists to discover 'Archaean' or Precambrian rocks in the southern part of Britain.

31 The Ordovician rocks were well characterized by the fossils that they contained: graptolites.

32 Murchison became Director General in 1855 and retained this position until his death in 1871. His influence on the organization lasted long after that, however.

33 Based on Topley (1885), Koehne (1915), Sarton (1919), Merrill (1920), Greenly and Williams (1930), Johns (1976) and Socolow (1988) and on responses to letters of enquiry. I am grateful to Peter Schimkat for information about German surveys.

34 This being so, it is hardly surprising that geologists have long had recourse to colour to aid them in this complex task.

Chapter 6

1 Cf. Shakespeare, *The Tempest*, I. II. Probing his daughter's memory, Prospero asks her: 'What seest thou in the dark backward and abysm of time?'

2 Or, to be more specific, fossils of organisms with hard parts.

3 Some critics regard this state of affairs as a strong argument against Darwinian theory. See, for example, Milton (1992: 105–7). Geologists, however, may see no mystery here, recognizing lithologically distinct beds within the Gault (Wells, 1938: 184), so that sedimentation may not have been continuous.

4 The terms 'catastrophists' and 'uniformitarians' were coined later by William Whewell (1832: 126), in a famous review of the second volume of Lyell's *Principles of Geology* (1832). Whewell coined the word with geologists such as Cuvier specifically in mind.
5 Cuvier's evidence for catastrophic events within human history was based on other records from antiquity besides the Bible.
6 For example, in his 'Introduction' to the 1813 translation (*Essay on the Theory of the Earth*) of Cuvier's *Discours préliminaire*, the Wernerian Robert Jameson was pleased to note that the 'Mosaic account . . . coincide[d] with the various phenomena observable in the mineral kingdom' (Cuvier, 1813: v).
7 Cuvier did not himself discuss theological questions explicitly in his geological work, although he had much to say about the earliest records from human history and the light that they threw on the earth's past.
8 One may think, for example, of the Reverends John Playfair, John Henslow, Adam Sedgwick, William Buckland, William Conybeare, Charles Kingsley, Thomas Bonney, John Blake, Charles Callaway, William Symonds, Henry Crosskey and numerous others.
9 Hutton's ideas were chiefly promoted by his friend and biographer John Playfair (1748–1819) (Playfair, [1802] 1964).
10 By the early nineteenth century the term 'natural theology' was preferred.
11 Because of the way the Oxford curriculum was structured at that time, the science lectures were optional extras, and science was not part of the regular degree. Buckland, however, succeeded in drawing large audiences, of both staff and students; and he took his classes on field excursions around Oxford, which proved most popular. Buckland's special interest was palaeontology, and particularly cave deposits. He also examined the clays and gravels of southern England, regarding them as evidence of the occurrence of the Noachian Deluge – *Reliquiae diluvianae*, as he called them (Buckland, 1823). The cave remains, with bones evidently gnawed by hyenas, etc., suggested that animals had inhabited these places and were not necessarily entombed therein by the Deluge. Buckland's treatise (1823) was chiefly about cave deposits, so the book's title was not really consonant with its theoretical contents. The cave remains were not really 'relics of the deluge'. However, Buckland certainly thought there was ample empirical evidence for a Deluge on the basis of clays and gravels over much of southern Britain. See Rupke (1983: 39 and *passim*).
12 Such an idea was implicit in the work of Desmarest also. See Figure 3.4.
13 It might have been argued, of course, that the sea level had risen and fallen by a corresponding amount. But there were few indications of a general sea rise and sea fall in the Mediterranean since Roman times; and such changes, had they occurred, would presumably have been recorded by historians.
14 There were, however, certain assumptions here that were not accepted by all geologists of that period. Leopold von Buch, in particular, was not persuaded that volcanic cones such as that of Etna were built up in the way Lyell supposed. See p. 168.
15 An excellent film, 'Lyell on Etna', has been made (1974) by M.J.S. Rudwick for the Open University (UK), which sets out the foregoing argument. It shows the places that Lyell visited and how he reasoned about what he saw.
16 Although Lyell presented his 'method' as somewhat novel, it had been deployed by, for example, Hutton and Scrope without their making so much philosophical noise about the matter.
17 Newton did not always practice what he preached in this regard; but that is not the point at issue here.

18 For this distinction, see Gould (1965).
19 This was the way the principle of uniformitarianism came to be conveniently stated by the historian of geology Archibald Geikie ([1905] 1962: 299).
20 Lyell made an exception for man, which he thought was a fairly recent and very special exception to the general rule about the uniformity of the earth and its inhabitants.
21 Lyell was encouraged by the discovery of mammalian remains in the Jurassic; but that wasn't really good enough. He excused the failure to find Palaeozoic mammals by suggesting (Lyell, 1830: 148) that geological enquiries to date had chiefly been in Europe and North America, where there had been large Palaeozoic oceans – unlikely places for mammals too be preserved. But Palaeozoic fish were known.
22 The recent extermination of the dodo might exemplify extinctions, even though, in this case, humans were the cause.
23 Lyell found particularly good evidence for this when he visited Niagara Falls in 1842. Some of the cut-back by the Falls had occurred in living memory, allowing the present to be used as a key to the past, as his philosophy of geology required. See Lyell (1845, vol. 1, ch. 2).
24 As said, Cuvier did regard the earth as very old. But catastrophism was not solely a product of Cuvier's theorizing.
25 In Lyell's day 'transformism' was regarded as politically radical, for the idea of betterment or change might get into the wrong hands – the underprivileged working class. And Lyell was not one to wish to encourage such sentiments.
26 In my own experience, I have seen this most clearly on the island of Tanna, mentioned in the 'Introduction'. But Darwin did not visit Vanuatu.
27 The Great Barrier Reef of Queensland is one such example, thought to mark a great line of fault along the coast, with the earth's crust sinking on the seaward side of the fault. A sinking submarine volcano could give rise to a circular coral atoll.
28 Notably the 'principle of divergence' (see Ospovat, 1981).
29 It is to be noted, however, that quite complex organisms such as trilobites appear near the bottom of the Cambrian.
30 In modern palaeontology, this has led some theorists to entertain the idea that most evolutionary change is concentrated in short bursts ('punctuations'), while the most common situation is that of 'stasis'. Controversy continues on this issue.
31 This phrase was Scrope's. It has been suggested by Rudwick (1974) that the idea of 'drafts of time' may have appealed to Scrope by virtue of his occupation as a banker.
32 The case is interesting from the standpoint of elementary philosophy of science in that Kelvin was making an auxiliary hypothesis which was unwarranted, namely that there was no ongoing source of heat in the earth.

Chapter 7

1 Elongated rounded hills of gravel, etc.
2 Long serpentine ridges of gravel, sand and water-worn stones.
3 Also, in 1824 Jens Esmark saw evidence of former extended glaciers in Norway.
4 Venetz's memoir was originally read as early as 1821. He worked with Perraudin in the Val de Bagnes in 1815–18.
5 The idea that the different races of mankind were created separately; opposed to monogenism, which saw all races as descended from a single stock.

6 William Buckland: entry made in the Visitor's Book at the Goat Hotel, Beddgelert, North Wales, on 16 October 1841.

7 Earlier, the astronomer Sir John Herschel (1830) had suggested that changes in the eccentricity of the earth's orbit might have had climatic influences in the earth's past.

8 Humboldt was citing Poisson (1836), who was drawing on a theorem of Lambert. Essentially the same point had been made earlier by Herschel (1830).

9 See Irons (1896).

10 Croll was not a great deal of use to Geikie as an outdoor surveyor, and it was to the Director's great credit that he was willing to support him in the organization – as theoretician and philosopher. However, Croll performed useful indoor administrative service. It is said that he was held in awe by his colleagues, who could not, however, understand the details of his arguments. Croll was in many ways a poignant figure. He suffered brain damage at the age of 59 and had to leave the Survey without a pension, never having passed the requisite Civil Service examinations.

11 An old Scottish name for 'boulder-clay', the stiff unstratified clay containing boulders left by retreating glaciers.

12 He was father-in-law to Alfred Wegener, celebrated for this theory of continental drift (see Chapter 11).

13 See Imbrie and Imbrie (1979) for a detailed account.

14 'Eustacy'. For papers on the history of the concept of eustacy, see Dott (1992).

15 Minute marine shelly organisms.

16 When rocks are formed they acquire a magnetism according to the prevailing magnetic field at the time of formation. This was discovered by a geophysicist, Bernard Brunhes, in 1906, examining the magnetism of bricks manufactured in a French brickyard (Imbrie and Imbrie, 1979: 147).

17 An accumulation of wind-blown dust characteristic of deposits formed adjacent to glaciers or cold deserts.

18 In this figure *AA* represented granite: *aa* 'dikes and eruptive beds'; *BB* 'non-conformable rocks'; and *CC* 'rocks of Carboniferous age'.

19 The form of the land profile for a graded river system, such as is shown in Figure 7.3, was one such law (Gilbert, 1877: 116).

20 Gilbert's *Report on the Geology of the Henry Mountains* is also of interest and importance because of his study of laccoliths (large, mushroom-shaped, igneous intrusions).

21 This was only one sheet of a three-part panoramic view of the scenery.

22 As a point of personal interest, there is just such a waterfall in my garden in Sydney's North Shore.

23 However, some regard his influence as too strong, even baneful. The uncritical acceptance of Davis's ideas in Australia was not always helpful to the progress of geomorphology. For modifications of Davisian views of peneplanation and uplift in eastern Australia, see Scott (1977, ch.9).

Chapter 8

1 The idea is thought to have been inspired, in part at least, by his friend von Humboldt's accounts of the eruption of the new volcano of Jorullo in Mexico in 1759. In his *Cosmos* (1880–3), von Humboldt described a contemporary account as follows:

> [O]n the day of the eruption itself the flat soil was seen to rise perpendicularly, and the whole became more or less inflated, so that blisters appeared, of which the largest is now the volcano. These inflated blisters, of various sizes, and partly of a tolerably regular, conical form, subsequently burst, and threw boiling hot earthy mud from their orifices, which are still found, at an immense distance, covered with black stony masses (von Humboldt, 1883, vol. 5: 313).

2 Such structures are part of the theoretical armoury of modern geology, being called laccoliths (Gk *lakkos* = cistern). The classic case is provided by the Henry Mountains in Utah, described by Gilbert (1875: 19) and mentioned in the previous chapter.

3 Noticing the parallelism between the eastern coastlines of South America and the western coast of Africa, and the similarities of the geologies of the distant continents, von Humboldt (1801) also suggested that the Atlantic might be regarded as an immense valley.

4 The Wernerian theory did not collapse completely. For discussion of 'Neo-Wernerianism', see Chapter 9.

5 But Lyell was travelling in a part of Switzerland where the order of the strata is now known to have been disturbed by great earth movements and 'overthrusts'.

6 The rocks in this region are very remarkable. At the top of one of the ranges, above the village of Elm, at a place called the Segnis Pass, one finds a hard, partly metamorphosed rock overlying a limestone, which overlies slaty strata. The limestone has weathered away at one spot, literally giving a hole through the mountain, through which the sun shines on only two days of the year lighting up the tower of the village church in the valley below. The contact between the metamorphic rock and the limestone is today recognized as a thrust plane, and can easily be seen from a considerable distance. It dips down so that the contact eventually comes out at the level of the valley floor a few miles to the north. This conveniently sited contact was discovered by Escher von der Linth in 1840, and is marked by a plaque put up by the *Naturforschender Gesellschaft des Kanton Glarus*, recording the spot as the place where Alpine geology was founded.

7 The contraction hypothesis had earlier been propounded by Adam Sedgwick, Charles Babbage, W.W. Mather, Henry De la Beche and Constant Prévost, not to mention Elie de Beaumont. For Prévost's major exposition, see Prévost (1839–40).

8 Hall thought he was dealing with sediments which had a total thickness of this order of magnitude in the Appalachians.

9 Hall's concept of geosyncline was established in this paper, though he did not actually coin the term. Hall discussed his idea in a paper presented to the American Association for the Advancement of Science in Montreal in 1857, but the text was not published until 1883. Ironically, the term 'geosyncline', customarily and appropriately linked with the name of Hall, was suggested by his opponent Dana. In a paper 'On the origin of mountains' he used the term 'geosynclinal' (Dana, 1873b: 430). Ten years later, in his *New Text-book of Geology* (1883), Dana used the currently favoured term 'geosyncline'. It should be noted that a geosyncline might contain many subsidiary component synclinals or anticlinals. Hall stated in his original argument (1859: 70) that the sinking and production of a 'great synclinal axis' would be accompanied by numerous synclinal and anticlinal axes. (On the history of the terminological question, see Mark (1992).)

10 In fact, investigations in the twentieth century, based on the microscopic examination of the daily and annual growth rings in corals, suggest that the number of days in the year is increasing rather than decreasing. So the earth is rotating more slowly now than in the distant past. The retardation is ascribed to tidal friction.

11 For further information on the early history of the concept of geosynclines, see Hsü (1973), Dott (174, 1978, 1979) and Greene (1982).
12 A term used by Stille.
13 After careful examination of the historical record, Suess (1904–24, vol. 2: 392) dismissed the example of the Temple of Serapis at Pozzuoli (see p. 136) as being due to a history of marine incursions into the area of the temple and a local effect rather than the result of successive elevations and subsidences of the land.
14 An example might be the worldwide occurrence of graptolites as fossils of major importance in the Ordovician.
15 For Suess's work on this concept, see Hallam (1992).
16 Indeed, the idea may be traced back to de Maillet (Carozzi, 1992b), though not with respect to glacial theory.
17 The terms Alpine, Hercynian and Caledonian are used by geologists to refer to three great periods of mountain building in Europe. 'Caledonian' refers to a large-scale series of earth movements that occurred during Lower Palaeozoic times and led to the formation of a mountain system extending from Ireland through Scotland to Scandinavia, as well as mountains along the Atlantic coast of North America. The name 'Hercynian' was formerly used to refer to the wooded mountain region of middle Germany (the Harz region). Von Buch used the term *hercynisch* to refer to a system of north-west-trending faults and folds through central Europe. For Bertrand, these were related to mountain building movements that occurred in late Carboniferous and early Permian times. The term Alpine refers to the mountain building that occurred during Tertiary times in the Alpine regions of Switzerland, etc.
18 This reputation was considerable. By working on the observations and ideas of Heim and Escher von der Linth and using analogies from the work of Jules Gosselet (1880) in Belgium, Bertrand suggested – without actually having been to Switzerland – that the structure of the Glarus Canton might be understood in terms of a single thrust of material from the south, rather than a double, pincer-like movements from north and south as envisaged by Heim. This interpretation permanently superseded that of Heim, although Heim's interpretation of the way that folds could pass over into thrust faults has also been fully accepted.
19 A somewhat similar map had been drawn earlier by the German geologist, Melchior Neumayr (see Figure 11.7). However, Neumayr's continental areas were smaller and the areas of ocean correspondingly larger, so that the general impressions created by the maps of Neumayr and Haug are different. In Volume 2 of his *Face of the Earth*, Suess envisaged two large primeval continental masses which he termed 'Atlantis' and 'Gondwána land' (Suess, 1906: 254). These were thought to have been separated initially by a sea aligned east – west: the 'Tethys'. Greenland represented a remnant of Atlantis, while parts of Australia, Africa and India were remnants of Gondwanaland.
20 For an attractive account of the exploits of this Survey, see Berthon and Robinson (1991, ch. 8, 'Measuring India').
21 The latitudes differed by 5° 23′ 37.058″ according to determination by astronomical means, whereas the difference according to geodetic survey (by triangulation) was 5° 23′ 42.294″.
22 In the words of Dutton (1892: 53):

> If the earth were composed of homogeneous matter its normal figure of equilibrium without strain would be a true spheroid of revolution; but if heterogeneous, if some

> parts were denser or lighter than others, its normal figure would no longer be spheroidal. Where the lighter matter was accumulated there would be a tendency to flatten or depress the surface. For this condition of equilibrium of figure, to which gravitation tends to reduce a planetary body, irrespective of whether it be homogeneous or not, I propose the name isostasy.

Dutton stated (ibid.: 58) that the essential features of the theory might be found in the writings of Charles Babbage and Sir John Herschel in the 1830s. See Babbage, 1838: 23.

23 Chamberlin is celebrated for his paper 'On the method of multiple working hypotheses' (1897). He urged that geologists, or scientists in general, should allow their minds to entertain a large range of speculative ideas, so that they should not be unduly attached to any one of them.

24 Diastrophism = crustal movements, faulting, folding, elevation, depression, mountain formation, etc.

25 A term that denotes the sum of the lithological and palaeontological characters exhibited by a deposit, which indicate the conditions under which that deposit was formed (e.g. alluvial facies, glacial facies, etc.).

26 Consider a sloping shoreline. Different kinds of deposits will be accumulated on the sloping surface at any given point in time. If, then, a marine incursion or regression is occurring, sediments of the same kind of material will be deposited at different times. And sediments of different kinds will be deposited at the same time at different places. Clearly, the stratigrapher must bear this principle in mind in making correlations. 'Walther's law' has been formulated thus: 'Contiguous facies in vertical sequence were laterally contiguous in the area of deposition. I.e. the vertical sequence at any given locality is made up of lithofacies which have migrated laterally.'

27 Pallasite = a type of stony-iron meteorite, containing some olivine, named after P.S. Pallas (see p. 83), who discovered such a meteorite in Siberia in 1772.

28 Rocks constituted chiefly of silicon (and oxygen), with a large proportion of magnesium.

29 Rocks constituted chiefly of silicon (and oxygen), with a large proportion of aluminium.

30 The terms 'sial' and 'sima' had been deployed by Suess in his *Face of the Earth* (1904–24, vol.3: xxiv, 626).

31 It has been shown in Johnson and McKerrow (1991) and (1992) that modern data for marine incursions and regressions for the Silurian correlate reasonably well with those proposed by Grabau. Grabau's ideas on recurrent facies changes began to receive acceptance from the late 1940s, particularly through the efforts of the Chicago geologist F.J. Pettijohn.

32 One can think, for example, of Oswald Spengler's influential *The Decline of the West* (1926). This work was mostly written before the First World War, being published in German in 1918.

33 For details see, for example, Heckel (1984), Langenheim and Nelson (1992) and Buchanan and Maples (1992).

34 Chapter 10 is devoted to an overview of the early history of seismology, with reference to the development of ideas about the earth's deep interior. By tracing the waves generated by earthquakes through the earth, seismologists are able to make 'well-educated guesses' about the inner constitution of the earth. But, by means of explosions detonated on the earth's surface and looking for reflections of these from the surfaces of the different layers of the underlying strata, stratigraphers now have a tool by means of which they can chart

the internal structure of the upper layers of the earth's crust. For marine work, explosions can be generated by guns using compressed air to propel missiles, towed behind ships.

Chapter 9

1 An invaluable chronology and bibliography of the development of petrology has, however, recently been published by Yoder (1993) and could provide an excellent starting-point for a thorough study of the history of petrology. See also Loewinson-Lessing (1954).
2 The *Oxford English Dictionary* gives the dates of first usage as 1690 and 1811 respectively. But a date of 1651 is given for petrography.
3 There are problem cases, such as corals, lichens and slime moulds.
4 In modern biology, the nature of species is a highly controversial matter and there are hotly contested criteria for the determination of species.
5 Obviously this criterion fails for fossils.
6 This assumes (incorrectly), as did Linnaeus, that a quartz crystal consists of a hexagonal prism, with hexagonal-based pyramids at each end of the prism.
7 Geikie was quoting from the work of the Edinburgh Wernerian Robert Jameson, *Elements of Geognosy* (1808). A similar style was evinced in Jameson's *Treatise on the External Characters of Minerals* (1805).
8 I am not aware that Romé de l'Isle took anything from Steno, but it is certainly possible that he did.
9 In so doing he was following a line of development that is in keeping with the history of natural history as described by Foucault (1970). That is, towards the end of the eighteenth century, naturalists began to concern themselves with more than external form in their classifications. They considered internal constitution also. Haüy was doing precisely this.
10 Below this level were the chemical constituents (*'les premiers principes ou les élémens des corps'* (Haüy, 1784: 49), the nature of which could not be revealed by cleavage experiments but might be discovered by chemical analysis. The *molécule intégrante* was the supposed smallest part of the substance that was of the same nature as a whole, and as such it was a precursor of the modern 'molecule'. However, Haüy's method of analysis by fracturing crystals only produced cleavage fragments, which did not necessarily reveal the crystals' essential basic forms. It should be noted further that, although Haüy is remembered chiefly for his crystal theory, his mineralogical classification (Haüy, 1801) was based on chemical lines. For example, his first class was 'acidiferous substances composed of an acid united with an earth or an alkali, and sometimes to both'. In this class, the first order was 'earthy acidiferous substances composed of an acid untied with an earth'. Of this, the first genus was lime, within which group were the species 'carbonated lime', 'phosphated lime', 'fluated lime', 'sulphated lime', 'nitrated lime' and 'arsenated lime'. Thus the new chemistry allowed the taxonomic practices of eighteenth-century natural history to be applied to the mineral kingdom. But such knowledge was not very useful to the geologist seeking to identify a rock in the field.
11 That is, certain arbitrary rules were proposed for the mental construction of crystals from the several shapes of micro 'building blocks'.
12 Actually, Haüy's theory was a little more complicated than that which I have described. The 'primitive form' of a crystal might be revealed by its mechanical division. There

were supposedly six such 'primitive forms' (the parallelipipedon, the octahedron, the tetrahedron, the regular hexagonal prism, the dodecahedron with rhombic faces and the dodecahedron with triangular faces. It was these shapes that were important. They were not necessarily the smallest crystallographic units. These were the *molécules intégrantes*, of which there were only three kinds: the tetrahedron, the triangular prism and the parallelipipedon. By appropriate combinations, the *molécules intégrantes* could generate the primitive forms, and these could generate the great mass of secondary forms. However, it should be remarked that in fact the cleavage fragments obtained by Haüy were not of the same form as the unbroken crystals of calcite.

13 The lithological category of sedimentary rocks was formally introduced by Brongniart (1827: 34), but the general idea of sediments becoming lithified goes back much further than that.

14 Wallerius referred to *lapides mixti*, generally constituted of quartz, feldspar and mica; but only one of these, the *saxum mixtum*, specifically had these constituents. In the French edition (Wallerius, 1753), the actual term 'granite' appeared, but in reference to cases where only two of the constituent minerals appeared.

15 For further information on the origin of rock names, with bibliographical details, see Tomkeieff (1983). See also Rudwick (1969).

16 Such rocks have traditionally been called 'acidic', silica (SiO_2) being the acid anhydride of the weak acid, silicic acid. However, it should not be thought that silica turns litmus red. It is almost insoluble in water, though soluble in hot alkali. The proportion of silica in a rock was determinable by the end of the eighteenth century by methods such as those mentioned in Chapter 3. These days, the term felsic is often preferred to acidic.

17 The word 'dyke' is an old Scottish term for 'wall'.

18 One will not, however, so far as I am aware, find precisely this table in any textbook of the period. It is a generalized picture that satisfies the ideas expressed in those text books.

19 Berzelius (1814: 27) introduced the term 'siliciate', which we now use in the shortened form 'silicate'.

20 This was a ten-point scale, from the softest minerals, such as talc, to the hardest, such as diamond. Minerals at each point on the scale can scratch ones with lower Mohs numbers or be scratched by ones with higher numbers. Hardness is still a useful factor to consider when identifying minerals in the field.

21 This was anonymous, but was very probably written by Jameson himself.

22 The word 'trap' derived from a Swedish word meaning 'stair'. The term was used very loosely at that time, often meaning basalt or trachyte. Layers of laval flows of these kinds of rocks often formed 'stairs', one on top of another: hence the name.

23 In fact, the *London Encyclopaedia* article already showed signs of severe theoretical 'stress'. In the list given, items 6 to 12 were given as 'subordinate', while the others were designated as 'principal'. But the temporal and geometrical relationship of the former to the latter was obscure. Moreover, rocks such as granite and 'sienite' were also reported in the 'transition series'; and the notion of a general sequence for the *Floetz* rocks had virtually collapsed, in that different specific sequences of strata had to be listed for different parts of Europe.

24 Harker (1932: 203) defined foliation as 'a more or less pronounced aggregation of particular constituent minerals of . . . [a] metamorphosed rock into lenticles or streaks or inconsistent bands, often very rich in some one mineral and contrasting with contiguous lenticles or streaks rich in other minerals'. But, he pointed out, foliation was not one and the same

as schistosity. As the intensity of metamorphism increased, schistosity might decline, whereas foliation might increase. Foliation involved segregation of material and might result form local solution, diffusion and recrystallization.

25 Early on, the first Director, Henry De la Beche (1834: 300), had taken a rather Wernerian view, suggesting that igneous rocks might have been formed in association with materials thrown down from a heated ocean, and that schists and gneisses might somehow be stratified deposits, the laminations being associated with some kind of depositional process (ibid.: 293).

26 *Drüse* = gland. The term 'druse' derives from an old German mining term for an irregular rock cavity encrusted by minerals on the cavity walls. The crystals in the cavities are often well formed and prized by collectors.

27 For an account of Hunt's style of geology, see Brock (1979). Hunt's more important papers up to 1875 are conveniently collected in his *Chemical and Geological Essays* (Hunt, 1875). Among his later papers, perhaps his most important was Hunt (1885).

28 Earlier, Bischof (1854–5, vol. 1: 252, 344) had suggested that the large deposits of carbon and sulphur in ancient rocks were possibly organic in origin. Hunt acknowledged his indebtedness to this suggestion.

29 A crystal was sliced diagonally and then the two halves were struck together again with Canada balsam. A ray of light entering a calcite crystal divides into two polarized rays. Nicol cut the crystal in such a way that one of the rays was diverted away by total internal reflection. The other passed straight through the (composite) crystal. Thus a beam of polarized light could be created, using a 'Nicol prism'. Such light could be directed through a thin section of some material, held on the stage of a microscope, and the crystalline materials would rotate the plane of polarization of the incident light. The amount of rotation could be determined (for a given standard thickness of crystal) by using a second Nicol prism located below the thin section as 'analyser'. The rotation is characteristic for a particular kind of mineral, and so too are the colours produced by the crystals observed in thin section using polarized light. Thus we have the polarizing microscope – an instrument of fundamental importance for petrography.

30 Later the term 'dynamic metamorphism' was preferred.

31 The term was introduced by Archibald Geikie, not Lapworth.

32 The term was introduced by Suess (1888, 1909, vol. 4: 551) as 'batholites'. A stock is a smaller-scale structure than a batholith, but may be connected to an underlying batholith. The case cited at Arran is strictly a stock rather than a batholith.

33 But, according to some views (Hamilton and Myers, 1967), batholiths are generally quite thin, though initially formed at depth. They may (it is suggested) have crystallized under covers consisting chiefly of their own volcanic ejecta.

34 The idea was that the material of the earth was not homogeneous initially. The 'lava lakes' therefore represented the residues of material that had originally been more fusible. Also, there might be vapours generated by the heat in the fluid lakes which could produce elevations and foldings of strata or volcanic action. Such ideas were later challenged by American geologists such as Dutton (1880: 116, 125).

35 Hutton, it should be observed, was familiar with what we now call metamorphic rocks. In this *Theory of the Earth* (1795, vol. 1: 318), he wrote: '[T]he waved structure of the stratified alpine stone, which, though it has not been made to flow, has been brought to a great degree of softness, so as to have the original straight lines of its stratification changed to those undulated or waving lines which are in some cases much incurvated.'

36 The term 'granitification' had earlier been used by T. Virlet d'Aoust (1846–7: 500). Virlet d'Aoust thought that there was no evidence of the existence of truly 'primitive' or original rocks on the surface of the globe. His 'granitification' – a metamorphic process – did not, he thought, have to take place at a high temperature.
37 Sederholm specifically pointed out that the rocks designated as gneisses in Finland were quite unlike the granitic gneisses of Germany.
38 On the coast of the Gulf of Finland, not far from Helsinki.
39 The photographs in Sederholm's own publications are not of good enough quality for reproduction here.
40 From the Greek word for serum, lymph or pus.
41 The sequence of changes on the right of Figure 9.3 would also involve some loss of iron and magnesium.
42 According to Tomkeieff (1983: 358), this term was introduced by K.F. Naumann as far back as 1826 (*Lehrbuch der Mineralogie*: 209) in reference to the chemical transformation of one kind of mineral into another. In various publications in the 1850s, C.J. Sainte-Claire Deville had proposed the idea of changes brought about by mineralizing agents, that is, gases that brought in new mineral matter and carried off the old.
43 Two key publications are Read (1957) and Gilluly (1948). The latter consists of a series of papers presented by the opposing sides of the controversy and a lengthy reportage of the discussion that occurred at the meeting of the Geological Society of America where the debate occurred.
44 A kind of trachytic rock. The Greek *propolos* means a servant who goes before. Later investigations showed that von Richthofen's propylites were an alteration product of andesite.
45 See p. 197.
46 Foyaite is a kind of rock of the syenite group, named after a place in Portugal (Foya) where it occurs.
47 See also Brøgger (1894).
48 So named after the work of Charles Soret (1879–80). He investigated the case of a copper sulphate solution, and found 17.3 per cent of the salt in the cooler part of the mixture at 20°C and 14.0 per cent in the hotter part at 80°C. Although the phenomenon is named the Soret principle, it was actually first discovered by C. Ludwig (1856).
49 The idea had also previously been adumbrated by the British surveyor J.G. Goodchild (1892). The idea was also proposed independently a little later by authors such as J. Barrell and N.V. Ussing.
50 However, Darwin (1844: 123) discussed the possibility of the separation of crystals under gravity leading to the differentiation of trachyte and basalt. He found no clear evidence for such a process occurring in granitic plutons, although he did note that basalt dykes were often found intersecting granite formations. There were other geologists such as F. Loewinson-Lessing and M. Schweig thinking along similar lines to Daly at about the same time.
51 Salic = silica and alumina: rocks with plentiful quartz and feldspars. Femic = ferro-magnesian: rocks with plentiful olivines and pyroxenes.
52 A small quantity of the molten material was let fall into a dish of mercury below a vertically orientated furnace, so that the abrupt cooling fixed the kinds and concentrations of the materials in equilibrium with one another at the high temperature. The method had been devised not long before by Shepherd et al. (1909).

53 Feldspars are particularly apt to behave in this fashion. They can display ranges of composition, according to the proportions of sodium, potassium and calcium ions present, so that in some cases the distinctions between feldspars in the series is essentially arbitrary.

54 'Magmatic stoping' is a process of igneous intrusion in which (according to theory) a magma gradually eats its way upwards, breaking off blocks of country rock, which sink down into the magma.

55 Such structures often produce doming of strata, and oil may be trapped under the domes. Hence oil prospectors are often on the lookout for salt domes. In fact, the concept of diapirism was first introduced in relation to saltdomes. See Mrazec (1907) and Rogers (1918).

Chapter 10

1 A small vibratory motion of the earth's axis of rotation, of period about nineteen years, additional to that involved in the larger movement that causes the precession of the equinoxes. Whereas the phenomenon of precession was known in antiquity, nutation was only noticed in the eighteenth century.

2 A general instrument used to record seismic waves is called a seismograph and the record made is the seismogram (cf. telegraph/telegram). A seismometer is an instrument that receives seismic impulses and records their intensity and duration.

3 The instrument itself, designed by Chang Heng, has not survived, but it is thought to have consisted of a heavy pendulum within a large jar. There were eight mobile arms inside the jar which, according to the direction of the seismic impulse, would release mechanically one of eight balls held in the outside of the jar. The falling ball was supposed to be caught in the mouth of one of the eight model frogs, appropriately arranged around the jar.

4 On Mallet, see Cox (1982), Melville and Wood (1987) and Dean (1991).

5 The bottom of the spring made contact with a dish of mercury and thus completed an electrical circuit when disturbed by an earth tremor.

6 It was unfortunate that the first global network of seismic observatories, established by Milne in the 1890s, had instruments that soon became obsolete (Agnew, 1989: 1198).

7 Oldham's interest in earthquakes was first stimulated by Mallet, with whom he came into contact in his younger days.

8 There are two other kinds of waves that arrive considerably after the *P* and the *S* waves. These are the so-called 'Rayleigh waves', first described by Lord Rayleigh in 1887; and the Love waves, described by A.E.H. Love in 1911. The Rayleigh waves travel like water waves whereas the Love waves have lateral vibrations. Both these kinds of waves are transmitted through the surface layer of the earth, rather than its interior. Having further to travel than the *P* and *S* waves, they arrive significantly later.

9 I.e. the angle between the position of the earthquake, the earth's centre and the position of the recording station.

10 In modern diagrams, the trajectories of earthquake waves are represented by curved lines, since the waves are refracted as the temperature and pressure change with depth in the earth's interior. Oldham simply ignored this consideration in his diagrams and used straight lines. But Oldham's basic idea was fine.

11 Gutenberg worked first at Frankfurt-on-Main,and then moved to Pasadena, California.

12 Part of the delay in the publication of Lehmann's work was due to the fact that she had to spend some time in the years 1930–3 trying to determine which of the European observatories were furnishing data that might be relied on for her calculations (Lehmann, 1987).

13 This paper must surely hold the Guinness record for the brevity of its title: simply '*P'*'. The symbol *P'* was used by seismologists as shorthand to represent so-called *PKP* waves: that is, those *P* waves which pass through the outer part of the earth's interior (now called the mantle), then through the outer core and then through the mantle again. The German expressions *Steinmantel* (or *Mantel*) and *Kern* have been used since Wiechert (1896). The English terms 'crust' and 'core' came into general use with Daly (1940). The French terms are *écorce, manteau* and *noyeau.*

14 This important paper was written in Serbo-Croat and has not, therefore, been read by the present author. However, a careful exposition of its contents, with partial translation, is to be found in Rothe (1924).

15 Mohorovicic had only recently established Wiechert seismographs in Croatia before he examined the signals from the Kulpa Valley earthquake. The new instrument thus paid rapid dividends.

16 In the late 1950s and the 1960s there was a fruitless attempt (the so-called Moho project or project Mohole) to drill through the earth's crust, right down to the Mohorovicic discontinuity and into the asthenosphere, to have a direct look at what might be found there (see Bascom, 1961). The scheme, which was impracticable and unjustifiable from the point of view of the ratio of costs to potential scientific benefits, was eventually abandoned, having cost $US30 million. The whole scheme was apparently dreamed up in part in order attract popular interest and funding for geologists. Though not justified in economic terms, the project, if successful, would have yielded results of enormous interest to geologists. The material of the oceanic floors is basalt. The underlying material of the mantle was thought to be peridotite or eclogite. Eclogite has the same overall chemical composition as basalt or gabbro, but being formed under very high pressure is made of compact minerals of low volume/mass ratio. Another kind of basic rock, peridotite, has a different chemical composition from basalt or gabbro. It has less silica, less aluminium, less calcium oxide and more magnesium oxide than basalt. It is also denser than basalt. Because of their similar chemical compositions, basalt might be expected to convert too eclogite under very high pressure, not peridotite. Now, if one were a theorist who believed in 'fixism', with up-and-down motions of continents or ocean floors, rather than continental drift or mobilism (see Chapter 11), then if basalt turned into eclogite the rock would sink with its increase in density and subsidence would allow the accumulation of geosynclinal sediments. If eclogite turned into basalt, there would be a volume increase and ocean basins might be converted to mountain ranges. Thus a mantle of eclogite was congenial to the 'fixist' theorists. In fact, in places where ancient sea floor appeared to have been converted to mountains, peridotite, not eclogite, was found beneath the basalt and gabbro. This finding was therefore compatible with 'mobilist' rather than 'fixist' doctrine. If one could drill directly into the mantle and determine whether it was made of peridotite or eclogite, one would have information that would be highly relevant to the debate between fixists and mobilists (see Hsü, 1992; 14–15).

17 In this paper, the decrease in velocity for the *P* and *S* waves below the Mohorovicic discontinuity was suggested to be only between 0.5 and 3%.

18 There was thus a significant 'intellectual' spin-off from the dreadful period of frequent atomic testing after the Second World War, so far as understanding the nature of the earth's interior was concerned. But in my judgement, for what it is worth, the scientific advantages for geophysics were far outweighed by the time, effort and money devoted to the weapons programme and provide no warrant for that programme. But see Bolt (1976).
19 General theoretical models could also be generated according to considerations of the earth's density, shape and the compressibility, viscosity and temperature of its supposed constituents at different depths and pressures. An early example of such modelling is provided by Wiechert (1897). Wave velocities and travel times could be calculated for different portions of the earth's interior, according to the model adopted, and the theory checked against the data obtained from seismometers (Weichert and Zoeppritz, 1907).
20 But Benioff (1954) did not refer to Wadati's work.
21 Of course, we may have some confidence that they know their business from the work they accomplish at a pragmatic level in relation to earthquakes.
22 But, in this view, the entities of astronomy are not known to be real. We cannot move galaxies, black holes or whatever around at our discretion. Hacking has taken this on the chin and has acknowledged that so far as his philosophy of science is concerned we cannot know that galaxies are real.
23 The example was first used by Heidegger.
24 That is why there are so many of them in this book. There could have been very many more if space had allowed.
25 See p. 150.

Chapter 11

1 I am not unsympathetic to this interpretation myself. For Kuhnian interpretations, see particularly Hallam (1973), Laudan (1980) and Stewart (1990). For other writings on the philosophy of science based on the history of continental drift and plate tectonics, see, for example, Frankel (1979), 1982) and Le Grand (1988).
2 Interestingly, Dana's argument was suggested in part on observations of the surface of the moon, parts of which appeared to be studded with volcanoes while other parts were free of them. This suggested an irregular cooling for the moon, and the earth might be thought to have cooled in a similar fashion. The continental areas supposedly solidified first, so that subsequent contraction affected other areas most. Hence the oceanic depressions were formed. Marine deposits might be found on continental areas because in earlier times the ocean basins were not deep enough to hold all the water. See Dana (1846).
3 For example, the marsupials of Australia and South America. Darwin found that the data of biogeography did not sit well with Lyellian theory, and this was one of the reasons why he was moved to develop his evolutionary theory. To explain similarities in distant organisms, however, Darwin had to consider possibilities of long-distance transport of organisms by flotation, carriage by birds, etc.
4 If one is interested in precursors of ideas, one might see this as a forerunner of the concept of subduction of modern plate-tectonic theory. But to do so is anachronistic. There was no direct historical connection between Ampferer's suggestion and the later development of plate tectonics.

5 For further details of early drift theories, see, for example, Marvin (1973). The claim sometimes made that the first person to notice the 'fit' of the continents of South America and Africa was Sir Francis Bacon has been rejected by Carozzi (1970).
6 It is known, however, that his attention had been drawn to such matters and the general problems then facing geology, back in 1901 when he was a student. See Flügel (1984).
7 We may be reminded of Hooke's somewhat similar suggestion for the cause of earthquakes proposed in the seventeenth century. There is no historical connection between the two theorists, however.
8 Ritter Lecture, Scripps Institution of Oceanography, March 1994. See also Oreskes (1990).
9 Available in back pocket of Argand (1977). The map prepared for the Brussels conference was a development of an earlier one completed in 1913, without regard, of course, to ideas of drift.
10 However, students of that era – myself included – were advised too read, but not believe, these heretical notions in Holmes's otherwise perennially popular text.
11 So called because it contained 'Laurentian' rocks from the Canadian Shield and rocks from Asia.
12 See also du Toit (1927).
13 Indeed, in the second edition of his *Physics of the Earth's Crust*, Fisher (1891, Appendix: 17) had the idea of convection currents rising under oceans and pressing the crust towards the bordering continents, thus producing mountain ranges. But this aspect of his thinking was no more successful in gaining support than his views on the layers of the earth's interior.
14 Wegener himself set much store by trying to determine the actual movement of Greenland away from America by direct survey methods. He believed that he had found such evidence, but his data were discredited after his death (poignantly during the failure of an expedition in Greenland).
15 This theory received strong support at that time from the work of the geodesist, William Bowie (1927).
16 For his crossing of the Atlantic, Willis did make his 'bridge' run along the mid-Atlantic ridge for same distance.
17 The history of oceanography in relation to the development of ways of thinking about the earth has been well treated by Menard (1986) and Hsü (1992).
18 But, as with other theorists of the time, Vening Meinesz had the hot tops of his convection currents located under the continents.
19 In Chapter 10, I gave some account of the early development of instruments devised for investigation of seismic phenomena. I did so because I specifically wished to emphasize the role of scientific instruments in the development of ways of thinking about the earth. But, for the present purposes, I shall 'take as read' the development of instruments for measuring gravity anomalies, magnetic fields, ocean depths, etc. We shall, as do many scientists themselves, treat such instruments as useful and available 'black boxes'.
20 The existence of these had been known in a general way form the nineteenth century, when the first telegraphic cables were laid across the Atlantic and following the investigations of the British research vessel *Challenger* in 1873. This work simply involved dropping a weighted line to take soundings every hundred miles. In the years 1925–7, the German vessel *Meteor* first used echo-sounding equipment to obtain an approximate profile of the Atlantic floor.

21 It is interesting that the 'rifts' in the mid-Atlantic ridge were first suggested by Marie Tharp, draftsman (or draftsperson) to one of Ewing's expeditions in 1952, who was collating the information being collected. Her suggestion was at first discounted by Heezen as mere 'girl talk' (Wertenbaker, 1974: 144). See Tharp and Frankel (1986).

22 'Guyots' were so named in honour of the Swiss oceanographer, Arnold Guyot. In the Afar-depression of Ethiopia, we find a part of the Red Sea floor that is now exposed and dried out. There, in a quasi-lunar landscape, one may see guyots in the open air.

23 According to White *et al.* (1970), the term 'subduction' was first used by the Swiss geologist André Amstutz, in a sense not unlike that in which it is now used in plate tectonics. The idea could be linked to the fault planes discovered earlier by Wadati and Benioff, mentioned in Chapter 10.

24 Matthews was Vine's supervisor.

25 Bullard stated (1975: 21) that his co-workers did most of the computational work necessary to establish the goodness of fit. Wegener, it should be remarked, had also insisted on the importance of matching continental shelves rather than coastlines.

26 The omission of Iceland was justified for Bullard by the fact that it was a geologically new terrain.

27 But these authors supposed at that time that there were regular periodic reversals in geomagnetic polarity.

28 The extrusive rock typically making up the Andean range is appropriately called andesite, being so named by von Buch (1836b: 209).

29 I mention Belousov chiefly because significant quantities of his work have been translated into English, and he is better known to English-language historians than other Russian geologists. I confess my inability to read Russian.

30 Examination of daily and annual growth rings of corals suggested that the number of days in the year was greater in Devonian times than at present. This slowing done of the earth's rotation was usually attributed to tidal friction, not increase in size of the earth.

31 The arguments for goodness of fit on a globe of reduced size (about 80 per cent of that of the present) have been presented by other writers, for example H.G. Owen (1976, 1981), of the British Museum (Natural History).

32 For these, Carey coined the name of 'orocline'. His work has been characterized by the invention of a number of neologisms.

33 E.g. Hoyle (1950). Such ideas are not currently favoured by cosmologists.

34 In contrast, Owen (1981) has suggested that both expansion and ocean-floor spreading/subduction processes may be occurring.

35 This idea has some parallel with the much earlier theory suggested by Chamberlin at the beginning of the century (see Chapter 12).

36 I have, just at the time of writing this chapter, had the opportunity of visiting what would, I presume, serve as an example of one such fault zone. It is the 'Dent fault', which separates the Yorkshire Dales from the English Lake District. This famous geological structure was first identified by Adam Sedgwick as a young man. He hailed from the tiny but beautiful village of Dent in the Dales.

37 According to Carey (pers. comm., 5 March 1994), the National Aeronautics and Space Administration (NASA) has been making measurements between Easter Island and Peru. According to plate-tectonic theory, there is only the Andean subduction zone between, and the intervening distance should be decreasing. By earth-expansion theory, it should be increasing. The results were not made available to Carey by NASA (in 1992) as they

were thought to be 'anomalous'. Search was under way for a hitherto undetected spreading zone in the area. The situation was intriguing, to say the least.

38 It should be observed, however, that in the fifth edition of his classic work *The Earth* (Jeffreys, 1970), published when the pendulum had swung decisively in the direction of mobilist theory, Jeffreys contended that the main arguments in favour of mobilism were palaeontological. He could not renounce his earlier physical arguments.

39 The sagacious Darwin was not without an answer directly. During the ice ages, he suggested, the climate might have been quite cool, even at the tropics. So cool/temperate organisms could have passed from one hemisphere to another. But polarities in similar latitudes across great tracts of ocean required all the 'adhoccery' of isthmian links.

40 The similarities on opposite sides of the isthmus are greater at the generic level than at the species level. This is consistent with the fact that speciation has proceeded since the oceans were separated by the formation of the isthmus.

41 According to suggestions recently published by Yves Coppens (1994), the rifting in Africa probably led to the production of a drier, savannah type of climate in eastern Africa, fairly sharply separated from the densely forested areas of equatorial west Africa. This could have led to the divergence of apes from hominids.

42 For a brief survey, see Briggs (1987), or, for more historical treatment, see Browne (1983).

43 Croizat envisaged 'tracks': regions occupied by a species. A set of 'tracks' for a set of species might together from a generalized track, or a biota. (A track was not a migration route, according to Croizat's thinking.) His hypothesis was that the history, or 'phylogeny', of land masses coincided approximately with the history, or phylogeny, of the world's organisms. This notion was very acceptable to some biologists after the establishment of plate tectonics. But others were less taken by Croizat's leanings towards orthogenesis – the idea that once a species had started evolving in some way there was a natural tendency for it to continue in that fashion.

44 Darwin's confidant, J.D. Hooker (1853), had had the idea of a 'pan-Austral' biota, prior to the introduction of Darwin's theory in 1859. Wallace (1876) followed Darwin and denied Hooker's suggestion.

Chapter 12

1 See also Murphy and Nance (1991).

2 This idea is essentially the same as that of Joly. Murphy and Nance use the homely analogy of a book lying on an electric blanket. The idea of earth and sea being 'unbalanced' reminds me of the speculations of Buridan, discussed in Chapter 1, not to mention those of Carey, considered in Chapter 11.

3 Quartz, jasper, feldspar, schorl and mica.

4 The table was entitled: '*Tableau des corps bruts ou exposition des principales substances minérales, disposée dans un ordre relatif au progrès des altérations qu'ont subi les dépouilles des corps vivans, et successivement leurs différens produits*'.

5 Teilhard de Chardin carried out important work in China on Peking man, and was also mixed up to some degree with the notorious Piltdown affair.

6 Such studies seem to have made Vernadsky a 'hyper uniformitarian'. In his book *The Biosphere*, he laid it down, as it were axiomatically, that:

'(1)No traces of the creation of an organism out of matter has ever existed, through all geological periods. (2) No period in which the Earth has been deprived of life has ever been detected. (3) Consequently (a) living matter today is genetically connected with all the living matter of preceding ages, and (b) conditions have never failed to be favourable to life on Earth, i.e. they have never differed vastly from those of the present day. (4) Throughout all geological ages the chemical effects of living matter on its surroundings have not changed, so that the composition of the crust and of living matter are the same today as they have always been. (5) The total mass of living matter has never differed considerably form its present value. (6) Whatever may be the phenomena of life, the energy evolved by organisms is mainly (perhaps entirely) the radiant energy of the Sun' (Vernadsky, 1986: 15).

7 The total scope of Vernadsky's work was massive, however, as is shown in Vinogradov (1963). He travelled widely, studying in Germany, France and Britain and also visiting America and many parts of the Soviet Union. His worked ranged from basic mineralogy to the study of radioactive elements.

8 The characteristics of the biosphere, as understood by Vernadsky, were set out in English in almost axiomatic form in Vernadsky (1944, 1945).

9 This term was originally coined by Le Roy (1927: 196) in reference to the terrestrial envelope that was being created by the activities of rational mankind. Teilhard de Chardin gave the concept a more 'idealist' flavour, as if there were a sphere of mentality.

10 See Vernadsky (1929, 1986) for French and English translations respectively.

11 For a Russian analysis of Vernadsky's ideas, see Lapo (1982), and for a brief but informative biographical article Vinogradov (1963). It may be noted that James Lovelock, whose work is discussed later in this chapter, has stated that he erred by failing to give due recognition to Vernadsky in the first edition of *Gaia* (Lovelock, 1982: viii). He does not say whether this was due to oversight or unfamiliarity with Vernadsky's ideas.

12 See particularly Hutchinson (1965, 1970).

13 Victor, Wictor or Wiktor.

14 They drew on Ronov (1968).

15 E.g. predominance of ferric rather than ferrous compounds.

16 See, for example, Oparin (1953), Cairns-Smith (1982), Dyson (1985) and Eigen and Winkler-Oswatitch (1992). A journal – *Origins of Life: an International Journal Devoted to the Scientific Study of the Origin of Life* – was founded in 1974.

17 According to recent work, there are hints of photosynthetic activity on the earth as far back as 3.8 thousand million years ago. This idea is suggested by an increased C^{12}/C^{13} ratio in sedimentary organic matter from that time, which is taken to be an indication of photosynthetic activity. See Schidlowski (1988).

18 Garrels and Lerman (1984) calculated the amount of sulphate stored in the sulphate reservoir over time by two independent methods: (1) from sulphur isotope data; and (2) from carbon isotope data. The two curves matched each other closely. The overall coupling equation for the carbon and sulphur deposits was taken to be:

$$4FeS_2 + CaCO_3 + 7CaMg(CO_3) + 7SiO_2 + 15H_2O = 15CH_2O \text{ (organic carbon compounds)} + 8CaSO_4 + 2Fe_2O_3 + 7MgSiO_3$$

19 I call Lovelock an independent scientist – an amateur if you like – as he earns his living chiefly from his writing and inventing, rather than from some university, business corporation or whatever (although he does hold a visiting chair in cybernetics at Reading University). He is, however, highly regarded, and does scientific work on his computer

at his home in Cornwall. Previously, he worked for the National Aeronautics and Space Administration (NASA), and he is a Fellow of the Royal Society. Such persons are rare in the late twentieth century.

20 Incidentally, the name was suggested by Lovelock's neighbour, the well-known novelist, William Golding.

21 In contrast with the earth, Venus, with its dense, hot atmosphere, appears to have no plate-tectonic activity. Anderson (Director of the Seismological Laboratory at the California Institute of Technology [Caltech]) merely put forward the idea as an aside. His principal concern in his paper of 1984 was a critique of the assumptions of the orthodox plate-tectonics paradigm.

22 As we know, the clock model has provided a favourite way of thinking about the earth for centuries (see p. 44). It is generally associated with a mechanical view of the earth. Garrels' thinking seems to fall squarely within this tradition.

23 See note 17.

24 Kant's cosmology envisaged endless cycles. Planets and comets might eventually spiral into the sun, the heat of which would then be increased until everything resolved into a chaos of fragments, at which point everything would start all over again.

25 E.g. the 'tidal' theory, the 'binary-star' theory, the 'fission' theory, the 'Cepheid' theory, the 'electromagnetic' theory and the 'nova' theory.

26 Chamberlin, be it noted, did not have the idea that the existence of free oxygen might be due to the activities of living organisms. His suppositions about the formation of the atmosphere simply had its present composition in mind.

27 Chamberlin was, of course, writing in the middle of the First World War, and this may have influenced his choice of metaphors.

28 In these figures, the circles merely define tracts of ocean.

29 This theory, which was also propounded by the astronomer Sir James Jeans, suffered from the difficulty that a new-collision process might be expected to give highly elliptical orbits to the planets, rather than ones that were almost circular. The collision event also seemed to be improbable or peculiarly rare. The nebular 'condensation' model, deriving from the ideas of Kant and Laplace, had the problem of angular momentum, referred to above. Astronomers have tried to develop theories according to which momentum is transferred from the sun to the planets by the assistance of magnetic and electric fields set up by fields of radioactively generated charged particles. Such ideas are uncomfortably *ad hoc*. Consideration of them is not required here.

30 He was a man of considerable eminence in the USSR in his day, being a polar explorer in his younger days, a distinguished mathematician and, later, Director of the State Publishing House, Editor of the *Great Soviet Encyclopedia* and member of the Central Executive Committee of the USSR. Graham (1987: 386) describes him as a popular colourful character who sired two children by two different women on the same day. But he was bedridden with tuberculosis in his later years.

31 I am grateful to Daniel Alexandrov for this reference.

32 An empirical 'law' suggested by J.D. Titius but named after J.E. Bode, who published it in 1772 (Shapley and Howarth, 1929: 180–2). If the earth – sun distance is taken as 1, and the distance between Mercury and the sun is 0.4, then the distances from the sun of the other planets is given by the expression $0.4 + (0.3 \times 2^n)$, where n is 0 for Venus, 1 for the earth, etc. Opinion is divided as to whether this 'law' has any theoretical significance. The majority view is that it does not.

33 Shmidt (1958: 22) suspected that Laplace himself had been aware of the problem of angular momentum, which was why he had only sketched his theory in a qualitative fashion, tucking it away in one of the appendices to the *Exposition du systême du monde* (Laplace, 1796/Year 4). But Laplace's nineteenth-century followers were so awed by his abilities that they chose to overlook the momentum problem for many years.

34 In its English-language version, it made obeisance to the thoughts of Engels on scientific hypotheses, which may have led some readers to suppose that the work was just an exercise in dialectical materialism, But it certainly was not.

35 It may be argued, however, that dinosaurs did not really become extinct, but evolved into birds.

36 For an account of the ongoing controversy, see Glen (1994).

Chapter 13

1 Perhaps they still are. A newspaper report (Hills, 1995) of the recent tragedy of the Kobe earthquake (January 1995) began: 'Tokyo, Tuesday: When the Earth was young and the great gods Izanami and Izanagi ruled the universe, Awaji island was the first place they created, so Japanese legend has it.' And it concluded: 'It reminds Japan and the world how defenceless man is in the face of the terrible wrath of Izanami and Izanagi.' (I am indebted to Gillian Branagan for this reference.)

2 He died in 1994.

Bibliography

Abbreviations

American Journal of Science	*AJS*
Annales de Chimie et de Physique	*ACP*
Annales des Mines	*AM*
Annals of Science	*AS*
Astrophysical Journal: an International Review of Spectroscopy and Astronomical Physics	*AJ*
British Journal for the History of Science	*BJHS*
Bulletin de la Commission Géologique de Finland	*BCGF*
Bulletin de la Société Géologique de France	*BSGF*
Bulletin of the American Association of Petroleum Geologists	*BAAPG*
Bulletin of the Geological Society of America	*BGSA*
Bulletin of the Seismological Society of America	*BSSA*
Canadian Naturalist	*CN*
Earth Sciences History	*ESH*
Edinburgh New Philosophical Journal	*ENPJ*
Geofisica Pura e Applicata	*GPA*
Geological Journal	*GJ*
Geological Magazine	*GM*
Geologische Rundschau: Zeitschrift für allgemeine Geologie	*GRZAG*
Geophysical Supplements to the Monthly Notes of the Royal Astronomical Society	*GSMNRAS*
Histoire de l'Académie Royale des Sciences . . . Avec les Mémoires de Mathématique et de Physique pour la Même Année[1]	*HARS*
Historical Studies in the Physical Sciences	*HSPS*
History of Science	*HS*
Journal des Mines	*JM*
Journal of Geology	*JG*
Journal of the Society for the Bibliography of Natural History	*JSBNH*

[1]For references to this journal, the year of presentation of the memoir is given first, followed by the actual year of publication.

Mémoires de l'Institut des Sciences, Lettres et Arts. Sciences Mathématiques et Physiques[2]	*MIS*
Neues Jahrbuch für Mineralogie, Geognosie, Geologie und Petrefaktenkunde	*NJMGGP*
Philosophical Magazine	*PM*
Philosophical Transactions of the Royal Society, London	*PTRS*
Proceedings of the Geologists' Association	*PGA*
Proceedings of the Royal Society, London	*PRS*
Proceedings of the Yorkshire Geological Society	*PYGS*
Quarterly Journal of the Geological Society of London	*QJGS*
Report of the British Association for the Advancement of Science[3]	*RBAAS*
Scientific American	*SA*
Studies in History and Philosophy of Science	*SHPS*
Transactions of the Cambridge Philosophical Society	*TCPS*
Transactions of the Geological Society of Glasgow	*TGSG*
Transactions of the Geological Society of London	*TGSL*
Transactions of the Royal Society of Edinburgh	*TRSE*

Bibliographical entries

Adams, F.D. (1938/1954) *The Birth and Development of the Geological Sciences*, reprint, New York: Dover Publications.

Adhémar, J.-A. (1842/1860) *Révolutions de la mer: déluges périodiques*, Paris: Carilian-Goeury & V. Dalmont.

Agassiz, L.J.R. (1837) 'Des glaciers, des moraines, et les blocs erratiques', *Bibliothèque universelle des sciences, belles-lettres, et arts, faisant suite à la bibliothèque britannique rédigée à Genève. Parties des sciences*, 12: 369–94.

Agassiz, L.J.R. (1838) 'Upon glaciers, moraines, and erratic blocks', *ENPJ*, 24: 364–83.

Agassiz, L.J.R. (1840) *Etudes sur les glaciers*, Neuchâtel: Jent & Gassman.

Ager, D.V. (1973/1981) *The Nature of the Stratigraphical Record*, London: Macmillan.

Ager, D.V. (1993) *The New Catastrophism: the Importance of the Rare Event in Geological History*, Cambridge and New York: Cambridge University Press.

Agnew, D,C. (1989) 'Seismology: history', in *The Encyclopedia of Solid Earth Geophysics*, edited by D.E. James, 1198–202, New York: Van Nostrand & Reinhold.

Agricola, G. (1546) *De ortu et causis subterraneorum lib. V. De natura eorum quae effluent ex terra lib. IIII. De natura fossilium lib. X. De ueteribus & nouis metallis lib. II. Bermannus, siue de re metallica dialogus . . .* , Basle, Frobenius.

Agricola, G. (1556) *De re metallica . . .* , Basle, Frobenius.

Agricola, G. (1912/1950) *De re metallica . . .* , translated by H.C. Hoover and L.H. Hoover, reprint, New York: Dover Publications.

Airy, G.B. (1855) 'On the computation of the effect of mountain-masses as disturbing the apparent astronomical latitude of stations in geodetic surveys', *PTRS*, 145: 101–4.

[2] For references to this journal, the year of presentation of the memoir is given first, followed by the actual year of publication.

[3] For references to this journal, the date given is the year of publication. The paper was read at the British Association the year before.

Aiton, E.J. (1972) *The Vortex Theory of Planetary Motions*, London and New York: Macdonald/American Elsevier.
Albertus Magnus (1967) *Book of Minerals*, translated by D. Wyckoff, Oxford: Clarendon Press.
Albritton, C.C. (ed.) (1963) *The Fabric of Geology*, Stanford: Freeman Cooper.
Albritton, C.C. (1975) *Philosophy of Geohistory 1785–1970*, Stroudsburg: Hutchinson, Dowden & Ross.
Albritton, C.C. (1980) *The Abyss of Time: Changing Conceptions of the Earth's Antiquity after the Sixteenth Century*, San Francisco: Freeman, Cooper.
Albritton, C.C. (1989) *Catastrophic Episodes in Earth History*, London and New York: Chapman & Hall.
Albury, W.R. and Oldroyd, D.R. (1977) 'From renaissance mineral studies to historical geology in the light of Michel Foucault's *The Order of Things*', *BJHS*, 10: 187–215.
Aldrovandi, V. (or U.) (1648) *Mvsaevm metallicvm* . . . , Bononiae: Io. Babtistae Ferroni.
Allchin, D. (1994) 'James Hutton and phlogiston', *AS*, 51: 615–35.
Allègre, C. (1988) *The Behavior of the Earth: Continental Drift and Seafloor Mobility*, translated by D. Kurmes van Dam, Cambridge (Mass.) and London: Harvard University Press.
Allègre, C. (1992) *From Stone to Star: a View of Modern Geology*, translated by D. Kermes van Dam, Cambridge (Mass.) and London: Harvard University Press.
Allen, D.C. (1963) *The Legend of Noah: Renaissance Rationalism in Art, Science, and Letters*, Urbana: University of Illinois Press.
Ampferer, O. (1906) 'Uber das Bewegungsbild, von Faltengebirgen', *Jahrbuch der kaiserlich-königlichen geologischen Reichsanstalt*, 56: 539–622.
Anderson, D. (1984) 'The earth as a planet: paradigms and paradoxes', *Science*, 223: 347–55.
Anon. (1829) 'Geology', *The London Encyclopaedia*, 10: 77–112, London: Thomas Tegg.
Anon. (1901) *Aetna* . . . , translated by R. Ellis, Oxford: Clarendon Press.
Anon. (1955) *Minor Latin Poets*, translated by J.W. Duff and A.M. Duff, London: Heinemann; Cambridge (Mass.): Harvard University Press.
Anon. (1965) *Incerti auctoris Aetna*, edited by F.R.D. Goodyear, Cambridge: Cambridge University Press.
Arduino, G. (1760) 'Due lettere del Sig. Giovanni Arduino sopra varie sue osservazioni naturali. Al chiaris. Sig. Antonio Vallisnieri . . . Vicenza', *Nuova raccolta d'opusculi scientfici e filologici*, 6: 99–180.
Argand, E. (1916) 'Sur l'arc des alpes occidentales', *Eclogae geologicae helveticae*, 14: 145–91.
Argand, E. (1924) 'La tectonique de l'Asie', *Comptes rendus de la XIIIe Congrès International de Géologie*, 1 (Part 5): 171–372.
Argand, E. (1977) *Tectonics of Asia*, translated and edited by A.V. Carozzi, New York: Hafner.
Ariew, R. (1991) 'A new science of geology in the 17th century?', in *Revolution and Continuity*, edited by P. Barker and R. Ariew, 81–92, Washington, DC: Catholic University.
Aristotle (1939) *On the Heavens*, translated by W.K. Guthrie, London: Heinemann.
Aristotle (1952) *Meteorologica*, translated by H.P.D. Lee, London: Heinemann; Cambridge (Mass.): Harvard University Press.
Arkell, W.J. and Tomkeieff, S.I. (1953) *English Rock Terms: Chiefly as Used by Miners and Quarrymen*, London: Oxford University Press.
Arldt, T. (1917) *Handbuch der Paläogeographie*, 2 vols, Leipzig: Gebrüder Borntraeger.
Arrhenius, S. (1909) *The Life of the Universe as Conceived by Man from the Earliest Ages to the Present Time*, translated by H. Borns, 2 vols, London and New York: Harper.

Avicenna (1927) *De congelatione et conglutione lapidum, being sections of the Kitab-al-Shifa* . . . , translated by E.J. Holmyard and D.C. Mandeville, Paris: Geuthner.

Babbage, C. (1838) *Ninth Bridgewater Treatise: a Fragment*, 2nd edn, London: John Murray.
Ball, R. (1891) *The Cause of an Ice Age*, London: Kegan Paul, Trench, Trübner.
Barfield, A.O. (1957) *Saving the Appearances: a Study in Idolatory*, London: Faber & Faber.
Barr, J. (1985) 'Why the world was created in 4004 B.C.: Archbishop Ussher and biblical chronology', *Bulletin of the John Rylands University Library*, 67: 575–608.
Barrell, J. (1914) 'The strength of the earth's crust', *JG*, 22: 655–83.
Barrell, J. (1917) 'Rhythms and the measurements of geologic time', *BGSA*, 28: 745–904 and plates.
Barth, T.F.W. (1962) *Theoretical Petrology*, 2nd edn, New York: John Wiley.
Bartholomew, M. (1976) 'The non-progress of non-progression: two responses to Lyell's doctrine', *BJHS*, 9: 166–74.
Bascom, W. (1961) *A Hole in the Bottom of the Sea: the Story of the Mohole Project*. London: Weidenfeld & Nicolson.
Becher, J.J. (1664) *Oedipus chymicus* . . . , Amsterdam: Elizeum Weyerstraten.
Becher, J.J. (1669/1703/1738) *Physica subterranea profundam subterraneorum genesin, è principiis hucusque ignotis, ostendens. . . . Operam navavit & specimen Becherianum, fundamentorum, documentorum, experimentorum, subjunxit Georg. Ernestus Stahl*, Leipzig: Joh. Ludov. Gleditschium.
Belousov, V.V. (1962) *Basic Problems in Geotectonics*, edited by J.C. Maxwell and translated by P.T. Broneer, New York: McGraw-Hill (1st Russian edn, 1956).
Belousov, V.V. (1968) 'Modern conceptions about the structure and development of the earth's crust and the upper mantle of continents', *QJGS*, 122: 293–314.
Benioff, H. (1954) 'Orogenesis and deep crustal structure – additional evidence from seismology', *BGSA*, 65: 385–400.
Berger, A.L. (1977) 'Long-term variation of the earth's orbital elements', *Celestial Mechanics*, 15: 53–74.
Bergman, T.O. (1766) *Physik beskrifning öfver jord-klotet, pa cosmographiska sällskapets vägner författed*, Uppsala: no publisher stated (2nd edn, 2 vols, 1773–4; German translations 1769, 1780, 1791).
Bergman, T.O. (1777) 'Disquisitio chemica de terra gemmarum', *Nova Acta Regiae Societatis Scientiarum Upsaliensis*, 3: 137–70.
Bergson, H. (1911) *Creative Evolution*, translated by A. Mitchell, London and New York: Macmillan (1st French edn 1907).
Bernhard, C.J. (1985) *Through France with Berzelius: Live Scholars and Dead Volcanoes*, Oxford and New York: Pergamon Press.
Berry, W.B.N. (1968) *Growth of a Prehistoric Time Scale Based on Organic Evolution*, San Francisco and London: W.H. Freeman.
Berthon, S. and Robinson, A. (1991) *The Shape of the World: the Mapping and Discovery of the Earth*, London: George Philip.
Bertrand, M. (1886–7) 'La chaîne des Alpes, et la formation du continent européen', *BSGF*, 15 (3rd series): 423–47.
Bertrand, M. (1892) 'Physique du globe – sur la déformation de l'écorce terrestre', *Comptes Rendus Hebdomadaires de l'Académie des Sciences, Paris*, 114: 402–6.
Berzelius, J.J. (1811) 'Essai sur la nomenclature chimique', *Journal de Physique*, 73: 253–86.

Berzelius, J.J. (1814) *An Attempt to Establish a Pure Scientific System of Mineralogy, by the Application of the Electro-chemical Theory and the Chemical Proportions*, translated by J. Black, London: Robert Baldwin; Edinburgh: William Blackwood.

Bina, P.A. (1751) *Ragionamento sopra la cagione de' terremuoti, ed in particolare di quello della terra di gualdo di nocere nell'Umbria seguito l'anno 1751*. Perugia: Constantini and Maurizj.

Birch, C. (1990) *On Purpose*, Sydney: New South Wales University Press.

Birch, T. (1756–7) *A History of the Royal Society of London* . . . , 4 vols, London: A. Millar.

Biringuccio, V. (1540) *De la pirotechnia* . . . , Venice: V. Rofinello.

Birkett, K.R. and Oldroyd, D.R. (1991) 'Robert Hooke, physico-mythology, knowledge of the world of the ancients, and knowledge of the ancient world', in *The Uses of Antiquity in the Scientific Revolution*, edited by S. Gaukroger, 145–70, Dordrecht: Kluwer Academic Publishers.

Bischof, G. (1854–5) *Elements of Chemical and Physical Geology*, 2 vols, London: Cavendish Society.

Blake, J.F. (1893) 'On the felsites and conglomerates between Bethesda and Llanllyfni', *QJGS*, 49: 441–66.

Bolt, B.A. (1976) *Nuclear Explosions and Earthquakes: the Parted Veil*, San Francisco: W.H. Freeman.

Bolt, B.A. (1982) *Inside the Earth: Evidence from Earthquakes*, San Francisco: W.H. Freeman.

Boltwood, B.B. (1907) 'On the ultimate disintegration products of the radioactive elements. Part II. The disintegration products of uranium', *AJS*, 23 (4th series): 77–88.

Bonney, T.G. and Raisin, C.A. (1894) 'On the relations of some of the older fragmental rocks in North-western Caernarvonshire', *QJGS*, 50: 578–602.

Botting, D. (1973) *Humboldt and the Cosmos*, New York: Harper & Row; London: Sphere Books.

Bourguet, L. (1729) *Lettres philosophiques sur la formation des sels et des crystaux . . . avec un mémoire sur la théorie de la terre*, Amsterdam: François Honoré.

Bowen, N.L. (1922) 'The reaction principle in petrogenesis', *JG*, 30: 177–98.

Bowen, N.L. (1948) 'The granite problem and the method of multiple prejudices', in *Origin of Granite* . . . , edited by J. Gilluly, 79–90, New York: The Geological Society of America (Memoir No. 28).

Bowen, N.L. ([1928] 1956) *The Evolution of Igneous Rocks*, with a new introduction by J.F. Shairer, reprint, New York: Dover Publications.

Bowie, W. (1927) *Isostasy*, New York: E.P. Dutton.

Bowler, P.J. (1976) *Fossils and Progress*, New York: Science History Publications.

Bowler, P.J. (1983) *The Eclipse of Darwinism: Anti-Darwinian Evolution Theories in the Decades around 1900*, Baltimore and London: Johns Hopkins University Press.

Breislak, S. (1802) *Physische und lithologische Reisen durch Campanien nebst mineralogischen Beobachtungen über die Gegend von Rom*, translated by A. Reuss, 2 vols, Leipzig: Wilhelm Rein (1st Italian edn 1801).

Breithaupt, J.F.A. (1836–41) *Vollständiges Handbuch der Mineralogie*, 3 vols, Dresden and Leipzig: Arnold.

Briggs, J.C. (1987) *Biogeography and Plate Tectonics*, Amsterdam, Oxford, New York and Tokyo: Elsevier Science Publishing.

Brock, W.H. (1979) 'Chemical geology or geological chemistry?', in *Images of the Earth*, edited by L.L. Jordanova and R.S. Porter, 147–70, Chalfont St Giles: British Society for the History of Science.

Brøgger, W.C. (1890) 'Die Mineralien der Syenit-pegmatit-gänge der südnorwegischen Augit- und Nephelinsyenite . . .', *Zeitschrift für Krystallographie und Mineralogie*, 16: 1–663.

Brøgger, W.C. (1894) 'The basic eruptive rocks of Gran', *QJGS*, 50: 15–37.

Brongniart, A. (1813) 'D'une classification minéralogique des roches mélangées', *JM*, 34: 5–48.

Brongniart, A. (1827) *Classification et caractères minéralogiques des roches homogènes et hétérogènes.* Paris: F.-G. Levrault.

Brongniart, A. ([1829]) *Theoretical Table of the Most General European Succession and Disposition of the Strata and Rocks which Compose the Crust of the Earth . . .*, London: J.J. Griffin; Glasgow: R. Griffen.

Browne, J. (1983) *The Secular Ark: Studies in the History of Biogeography*, New Haven and London: Yale University Press.

Brush, S.G. (1979) 'Nineteenth-century debates about the inside of the earth', *AS*, 36: 224–54.

Brush, S.G. (1980) 'Discovery of the earth's core', *American Journal of Physics*, 48: 705–24.

Buch, L. von (1797) *Versuch einer mineralogischen Beschreibung von Landeck*, Breslau: Hirschberg & Lissa.

Buch, L. von (1798) 'Von der Uebergangsformation mit einer Anwendung auf Schlesien', *Jarbücher der Berg- und Hüttenkunde*, 2: 249–73.

Buch, L. von (1802–3) 'Observation sur les volcans d'Auvergne', *JM*, 13: 249–56.

Buch, L. von (1810a) *A Mineralogical Description of the Environs of Landeck*, translated by Charles Anderson, Edinburgh and London: Constable & Constable, Hunter, Parker & Hunter.

Buch, L. von (1810b) 'Etwas über locale und allgemeine Gebirgsformationen', *Magazin der Gesellschaft Naturforschender Freunde zu Berlin . . .*, 4: 69–74

Buch, L. von (1824) 'Ueber den geognostischen Systeme von Deutschland . . ., *Mineralogische Taschenbuch für das Jahr 1824, von K.C. Ritter von Leonhard*, 501–6 and map, Frankfurt-on-Main.

Buch, L. von (1825) *Physicalische Beschreibung der canarischen Inseln* (with atlas), Berlin: Königlichen Akademie der Wisssenschaften.

Buch, L. von (1836a) 'Ueber Erhebungscratere und Vulcane', *Annalen der Physik und Chimie*, 37: 169–90.

Buch, L. von (1836b) 'On volcanoes and craters of elevation', *ENPJ*, 21: 189–209.

Buchanan, R.C. and Maples, C.G. (1992) 'R.C. Moore and concepts of sea-level change in the midcontinent', in *Eustacy: the historical Ups and Downs of a Major Geological Concept*, edited by R.H. Dott, 73–81. Boulder: Geological Society of America (Memoir 180).

Buckland, W. (1820) *Vindiciae geologicae: or the Connexion of Geology with Religion Explained* . . . , Oxford: University Press.

Buckland, W. (1823) *Reliquiae diluvianae; or, Observations on the Organic Remains Contained in Caves, Fissures, and Diluvial Gravel, and on the Geological Phenomena, Attesting the Universal Deluge*, London: John Murray.

Buckland, W. and Conybeare, W.D. (1824) 'Observations on the south-western coast district of England', *TGSL*, 1 (new series): 210–316.

Buffetaut, E. (1987) *A Short History of Vertebrate Palaeontology*, London, Sydney and Wolfeboro: Croom Helm.

Buffon, G.-L. Leclerc, Comte de (1778) *Les époques de la nature*, 2 vols, Paris: Imprimerie Royale.

Buffon, G.-L. Leclerc, Comte de (1783–8) *Histoire naturelle des minéraux*, 5 vols, Paris: Imprimerie Royale.

Buffon, G.-L. Leclerc, Comte de (1797–1807) *Buffon's Natural History . . .*, London: H.D. Symonds.

Buffon, G.-L. Leclerc, Comte de (1962) *Les époques de la nature*, edited by Jacques Roger, Paris: Editions du Muséum.

Buffon, G.-L. Leclerc, Comte de *et al.* (1749–1804) *Histoire Naturelle . . .*, 44 vols, Paris: Imprimerie Royale.

Bullard, E.C. (1975) 'The emergence of plate tectonics: a personal view', *Annual Review of Earth and Planetary Science*, 3: 1–30.

Bullard, E.C., Maxwell, A.E. and Revelle, R. (1956) 'Heat flow through the deep sea floor', *Advances in Geophysics*, 3: 153–81.

Bullard, E.C., Everett, J.E. and Smith, A.G. (1965) 'The fit of the continents around the Atlantic', *PTRS*, Series A, 258: 41–51.

Bunsen, R.W. (1851) 'Ueber die Processe der vulkanischen Gesteins-bildungen Islands', *Annalen der Physik und Chimie*, 83: 197–212.

Bunsen, R.W. (1853) 'Recherches sur la formation des roches volcaniques en Islande', *ACP*, 38 (3rd series): 215–89.

Burchfield, J.D. (1975) *Lord Kelvin and the Age of the Earth*, New York: Science History Publications.

Burke, J.G. (1966) *Origins of the Science of Crystals*, Berkeley and Los Angeles: University of California Press.

Cairns-Smith, A.G. (1982) *Genetic Takeover and the Mineral Origins of Life*, Cambridge: Cambridge University Press.

Cannon, S.F. (1978) *Science in Culture: the Early Victorian Period*, Folkestone: William Dawson; New York: Science History Publications.

Carey, S.W. (1955) 'The orocline concept in geotectonics', *Papers and Proceedings of the Royal Society of Tasmania*, 89: 255–88.

Carey, S.W. ([1958] 1959) 'A tectonic approach to continental drift', in *Continental Drift . . . being a Symposium on the Present Status of the Continental Drift Hypothesis, Held . . . in 1956*, edited by S.W. Carey, 177–355, reprint, Hobart: Geology Department, University of Tasmania.

Carey, S.W. (1988) *Theories of the Earth and Universe: a History of Dogma in the Earth Sciences*, Stanford: Stanford University Press.

Carozzi, A.V. (1968) *Telliamed or Conversations Between an Indian Philosopher and a French Missionary on the Diminution of the Sea by Benoît de Maillet*, translated and edited by A.V. Carozzi, Urbana, Chicago and London: University of Illinois Press.

Carozzi, A.V. (1970) 'New historical data on the origin of the theory of continental drift', *BGSA*, 81: 283–6.

Carozzi, A.V. (1977) 'Preface and editor's introduction', in *Tectonics of Asia*, edited by E. Argand, translated by A.V. Carozzi, xi–xxvi, New York: Hafner.

Carozzi, A.V. (1988) 'Review of Laudan (1987)', *Eos*, 69: 637–45.

Carozzi, A.V. (1989) 'Forty years of thinking in front of the Alps: Saussure's (1796) unpublished theory of the earth', *ESH*, 8: 123–40.

Carozzi, A.V. (1992) 'De Maillet's Telliamed (1748): the diminution of the sea or the fall portion of a complete cosmic eustatic cycle', in *Eustasy: the Historical Ups and Downs of a Major Geological Concept*, edited by R.H. Dott, 17–24, Boulder: Geological Society of America (Memoir No. 180)

Carozzi, A.V. and Carozzi, M. (1991) 'Pallas' theory of the earth in German (1778). Translation and reevaluation. Reaction by a contemporary: H.-B. de Saussure', *Archives des Sciences (Geneva)*, 44: 1–105.

Cesalpinus, A. (1596) *De metallicis libri tres. . . .*, Rome: A. Zanneth.

Chamberlin, T.C. (1897) 'On the method of multiple working hypotheses', *JG*, 5: 837–48.

Chamberlin, T.C. (1898a) 'The ulterior basis of time divisions and the classification of geologic history', *JG*, 6: 445–62.

Chamberlin, T.C. (1898b) 'The influence of great epochs of limestone formation upon the constitution of the atmosphere', *JG*, 6: 609–21.

Chamberlin, T.C. (1899) 'An attempt to frame a working hypothesis of the cause of glacial periods on an atmospheric basis', *JG*, 7: 545–84, 667–85, 751–87.

Chamberlin, T.C. (1900) 'An attempt to test the nebular hypothesis by the relations of masses and momenta', *JG*, 8: 58–73.

Chamberlin, T.C. (1909) 'Diastrophism as the ultimate basis of correlation', *JG*, 17: 685–90.

Chamberlin, T.C. (1916) *The Origin of the Earth*, Chicago: University of Chicago Press.

Chamberlin, T.C. and Salisbury, R.D. (1906) *Geology: Earth History . . . Vol. II. Genesis – Paleozoic*, London: John Murray.

Charpentier, J. de (1834) 'Annonce d'un des principaux résultats des recherches de M. Venetz, sur l'état actuel et passé des glaciers du Vallais', *Verhandlungen der Schweizerischen Gesellschaft für die gesammten Naturwissenschaften, Luzern*, 23–4.

Charpentier, J. de (1836) 'Account of one of the most important results of the investigations of M. Venetz, regarding the present and earlier conditions of the glaciers of the Canton Vallais', *ENPJ*, 21: 210–20.

Charpentier, J. de (1841) *Essai sur les glaciers et sur le terrain erratique du bassin du Rhône*, Lausanne: Ducloux.

Chorley, R.J., Dunn, A.J. and Beckinsale, R.P. (1964) *The History of the Study of Landforms; or the Development of Geomorphology*, London: Methuen.

Cicero, M.T. (1979) *De natura deorum . . .*, translated by H. Rackham, London: Heinemann; Cambridge (Mass.): Harvard University Press.

Cloud, P. (ed.) (1970) *Adventures in Earth History: Being a Volume of Significant Writings from Original Sources . . .*, San Francisco: W. H. Freeman and Company.

Coquand, H. (1843) 'Observations concernant un changement relatif de niveau dans la mer crétacée', *Comptes rendus hebdomadaires des séances de l'Académie des sciences*, 17: 183–6.

Collins, H.M. (1981) *Changing Order: Replication and Induction in Scientific Practice*, London, Beverly Hills and New Delhi: Sage Publications.

Conybeare, W.D. and Phillips, W. (1822) *Outlines of the Geology of England and Wales . . .*, Part I, London: William Phillips.

Cope, J.C.W., Ingham, J.K. and Rawson, P.F. (1992) *Atlas of Palaeogeography and Lithofacies*, London: Geological Society.

Coppens, Y. (1994) 'East side story: the origin of humankind', *SA*, 270: 62–9.

Cordier, P.L.A. (1827) 'Essai sur la température de l'intérieur de la terre, *AM*, 2: 53–138.

Cotta, B. von (1866) *Rocks Classified and Described: a Treatise on Lithology*, translated by P.H. Lawrence, London: Longmans, Green.

Cox, A., Doell, R.R. and Dalrymple, G.B. (1963) 'Geomagnetic polarity epochs and Pleistocene geochonometry', *Nature*, 198: 1049–51.

Cox, L.R. (1942) 'New light on William Smith and his work', *PYGS*, 25: 1–99.

Cox, R.C. (ed.) (1982) *Robert Mallet 1810–1881*, Dublin: Institution of Engineers of Ireland.

Craig, G.Y., McIntyre, D.B. and Waterston, C.D. (eds) (1978) *James Hutton's Theory of the Earth the Lost Drawings*. Edinburgh: Scottish Academic Press.

Craw, R. (1992) 'Margins of cladistics: identity, difference and place in the emergence of phylogenetic systems 1864–1975' in *Trees of Life* . . . , edited by P. Griffith, 65–107, Dordrecht, Boston and London: Kluwer Academic Publications.

Croizat, L. (1958) *Panbiogeography* . . . , Caracas: L. Croizat.

Croizat, L. (1962[4]) *Space, Time, Form: the Biological Synthesis*, Caracas: L. Croizat.

Croizat, L., Nelson, G. and Rosen, D.E. (1974) 'Centers of origin and related concepts', *Systematic Zoology*, 23: 265–87.

Croll, J. (1864) 'On the physical cause of the change of climate during geological epochs', *PM*, 27: 121–37.

Croll, J. (1867a) 'On the eccentricity of the earth's orbit, and its physical relations to the glacial epoch', *PM*, 33: 119–31.

Croll, J. (1867b) 'On the change in the obliquity of the ecliptic, its influence on the climate of the polar regions and on the level of the sea', *PM*, 33: 426–45.

Croll, J. (1875) *Climate and Time in their Geological Relations: a Theory of Secular Changes of the Earth's Climate*, London: Daldy, Isbister.

Crook, T. (1933) *History of the Theory of Ore Deposits* . . . , London: Thomas Murby; New York: D. van Nostrand.

Cumming, D.A. (1981) 'Geological maps in preparation: John Macculloch on [the] western islands [of Scotland]', *Archives of Natural History*, 10: 255–71.

Cuvier, G.L.C.F.D. de (1813) *Essay on the Theory of the Earth . . . with . . . an Account of Cuvier's Geological Discoveries*, translated by R. Kerr, Edinburgh: William Blackwood; London: Robert Kerr.

Cuvier, G.L.C.F.D. de and Brongniart, A. (1808) 'Essai sur la géographie minéralogique des environs de Paris', *Annales du Muséum d'histoire naturelle*, 11: 293–326.

Cuvier, G.L.C.F.D. de and Brongniart, A. (1811) *Essai sur la géographie minéralogique des environs de Paris* . . . , Paris: Baudouin.

Cuvier, G.L.C.F.D. de and Brongniart, A. (1822) *Description géologique des environs de Paris . . . nouvelle édition* . . . , Paris and Amsterdam: G. Dufour & E. d'Ocagne.

D'Alembert, J. le R. (1749) *Recherches sur la précession des équinoxes*, Paris: David l'Ainé.

D'Alembert, J. le R. (1754/1759) 'Recherches sur la précession des équinoxes . . .', *HARS*, 116–19, 413–28.

Dalrymple, G.B. (1991) *The Age of the Earth*, Stanford: Stanford University Press.

Daly, R.A. (1903) 'The mechanics of igneous intrusion', *AJS*, 16 (4th series): 107–26.

Daly, R.A. (1914) *Igneous Rocks and their Origin*, New York and London: McGraw-Hill.

Daly, R.A. (1926) *Our Mobile Earth*, New York and London: Charles Scribner's Sons.

Daly, R.A. (1940) *Strength and Structure of the Earth*, New York: Prentice-Hall.

Dana, J.D. (1837) *A System of Mineralogy* . . . , New Haven: Durrie & Peck & Herrick & Noyes.

[4]This is the date given on the title page. The book did not, in fact, appear until 1964.

Dana, J.D. (1843) 'On the analogies between modern igneous rocks and the so-called primary formations, and the metamorphic changes produced by heat in the associated sediments', *AJS*, 45: 104–29.

Dana, J.D. (1846) 'On the volcanoes of the moon', *AJS*, 2 (2nd series): 335–55.

Dana, J.D. (1847a) 'On the origin of continents', *AJS*, 3 (2nd series): 94–100.

Dana, J.D. (1847b) 'Geological results of the earth's contraction in consequence of cooling', *AJS*, 3 (2nd series): 176–88.

Dana, J.D. (1847c) 'A general review of the geological effects of the earth's cooling from a state of igneous fusion', *AJS*, 4 (2nd series): 88–92.

Dana, J.D. (1847d) 'Origin of the grand outline features of the earth's crust', *AJS*, 4 (2nd series): 381–98.

Dana, J.D. (1849) *Narrative of the United States Exploring Expedition during the years 1838, 1839, 1840, 1841, 1842 . . . Vol. X, Geology*, Philadelphia: C. Sherman.

Dana, J.D. (1866) 'Observations on the origin of some of the earth's features', *AJS*, 42 (2nd series): 205–11, 252–3.

Dana, J.D. (1873a) 'On some results of the earth's contraction from cooling, including a discussion of the origin of mountains, and the nature of the earth's interior', *AJS*, 5 (3rd series): 423–43.

Dana, J.D. (1873b) 'On some results of the earth's contraction from cooling . . . Part V. Formation of the continental and oceanic depressions', *AJS*, 6 (3rd series): 161–72.

Dana, J.D. (1883) *New Text-book of Geology*, 4th edn, New York and Chicago: Ivison, Blakeman, Taylor.

Dana, J.D. (1890) 'Areas of continental progress in North America, and the influence of the conditions of these areas on the work carried forward within them', *BGSA*, 1: 36–48.

Dante Alighieri (1950) *The Divine Comedy of Dante Alighieri: the Carlyle–Okey–Wicksteed Translation*, New York: Random House.

Darwin, C.R. (1842) *The Structure and Distribution of Coral Reefs* . . . , London: Smith Elder.

Darwin, C.R. (1844) *Geological Observations on the Volcanic Islands* . . . , London: Smith Elder.

Darwin, C.R. (1846) *Geological Observations on South America* . . . , London: Smith Elder.

Darwin, C.R. (1859) *On the Origin of Species by Means of Natural Selection* . . . , London: John Murray.

Darwin, G.H. (1903) 'Radio-activity and the age of the sun', *Nature*, 68: 496.

Daubrée, G.-A. (1857) 'Observations sur le métamorphisme et recherches expérimentales sur quelques-uns des agents qui ont pu le produire', *AM*, 12 (5th series): 289–326.

Daubrée, G.-A. (1859) 'Etudes et expériences synthétiques sur le métamorphisme et sur la formation des roches cristallines', *AM*, 16 (5th series): 155–218, 393–476.

Daubrée, G.-A. (1860) *Etudes et expériences synthétiques sur le métamorphisme et sur la formation des roches cristallines*, Paris: Imprimerie Royale.

D'Aubuisson de Voisins, J.F. (1803/Year 11) 'Extrait d'une lettre de J.F. Daubuisson, à A.J. Brochant', *JM*, 13: 113–22.

D'Aubuisson de Voisins, J.F. (1807) 'Observations sur la chaleur souterraine, faites aux mines de Poullaouen et du Huelgout en Bretagne', *JM*, 21: 119–30.

Davis, W.M. (1889a) 'A pirate river', *Science*, 13: 108–9.

Davis, W.M. (1889b) 'The rivers and valleys of Pennsylvania', *National Geographic Magazine*, 1: 183–253.

Davis, W.M. (1889c) 'Topographical development of the Triassic formation of the Connecticut Valley', *AJS*, 37 (3rd series): 423–34.

Davy, H. (1980) *Humphry Davy on Geology . . .* , edited by R. Siegfried and R.H. Dott, Madison: University of Wisconsin Press.
Dean, D.R. (1979) 'The word geology', *AS*, 36: 35–43.
Dean, D.R. (1981) 'The age of the earth controversy: beginnings to Hutton', *AS*, 38: 435–46.
Dean, D.R. (1991) 'Robert Mallett and the founding of seismology', *AS*, 48: 39–67.
Dean, D.R. (1992) *James Hutton and the History of Geology*. Ithaca: Cornell University Press.
Debus, A.G. (1965) *The English Paracelsians*, London: Oldbourne Book Company.
Debus, A.G. (1991) *The French Paracelsians . . .* , Cambridge: Cambridge University Press.
De la Beche, H.T. (1834) *Researches in Theoretical Geology*, London: Charles Knight.
De la Beche, H.T. (1846) 'On the formation of the rocks of South Wales and south-western England', *Memoirs of the Geological Survey of Great Britain*, 1: 1–296.
Deluc, J.-A. (1790–1) 'Letters to Dr. James Hutton', *Monthly Review*, 2: 206–27, 582–601; 3: 573–86; 5: 564–86.
Descartes, R. (1644/1656) *Principia philosophiae*, Amsterdam: Apud Ludovicum Elzeverium.
Desmarest, N. (1771/1774) 'Mémoire sur l'origine & la nature du basalte à grandes colonnes polygones, déterminées par l'histoire naturelle de cette pierre, observée en Auvergne', *HARS*, 705–75.
Desmarest, N. (1773/1777) 'Mémoire sur le basalte où l'on traite du basalte des anciens; & où l'on expose l'histoire naturelle des différentes espèces de pierres auxquelles on a donnée, en différens temps, le nom de basalte', *HARS*, 599–670.
Desmarest, N. (1804/1806) 'Mémoire sur la détermination de trois époques de la nature par les produits des volcans, et sur l'usage qu'on peut faire de ces époques dans l'étude des volcans', *MIS*, 6:219–89.
Desmond, A.J. (1975) *Hot-blooded Dinosaurs*, London: Blond & Briggs.
Dewey, J. and Byerly, P. (1969) 'The early history of seismometry (to 1900)', *BSSA*, 59: 183–227.
Dietz, R.S. (1961) 'Continental ocean basin evolution by spreading of the sea floor', *Nature*, 190: 854–7.
Donovan, A.L. and Prentiss, J. (1980) 'James Hutton's Dissertatio physico-medica inauguralis de sanguine et circulatione microcosmi . . . Leyden Academy, 1749, with English translation . . . , *Transactions of the American Philosophical Society* (Philadelphia), 70 (Part 6).
D'Orbigny, A.D. (1842) 'Carte générale de la République de Bolivia', in *Voyage dans l'Amérique Méridionale*, by A.D. d'Orbigny (vol. 8 [plates], 1847), Paris: P. Bertrand, and Strasbourg: V. Levrault (map dated 1842 being a geologically coloured version of topographical map engraved by L. Bouffard, 1839).
Dott, R.H. (1974) 'The geosynclinal concept', in *Modern and Ancient Geosynclinal Sedimentation*, edited by R.H. Dott and R.H. Shaver, 1–13, Tulsa: Society of Economic Palaeontologists and Mineralogists (Special Publication No. 19).
Dott, R.H. (1978) 'Tectonics and sedimentation a century later', *Earth-science Reviews*, 14: 1–34.
Dott, R.H. (1979) 'The geosyncline – first major geological concept "made in America" ', in *Two Hundred Years of Geology in America*, edited by R.H. Dott, 239–64, Hanover [New Hampshire]: University Press of New England.
Dott, R.H. (ed.) (1992) *Eustacy: the Historical Ups and Downs of a Major Geological Concept*, Boulder: Geological Society of America (Memoir No. 180).
Douglas, N. and Douglas, N. (1990) *Vanuatu – a Guide*, 2nd edn, Alstonville (New South Wales): Pacific Profiles.
Dufrénoy, P.A. and Elie de Beaumont, L (1841) *Carte Géologique de la France*, Paris: l'Imprimérie Royale.

Duhem, P. (1958) *Le Système du monde: histoire des doctrines cosmologiques de Platon à Copernic. Tome IX. Cinquième partie: la physique Parisienne au XIVe siècle*, Paris: Hermann.

Du Rietz, G.E. (1940) 'Problems of bipolar plant distribution', *Acta phytogeographica suecica*, 13: 215–82.

Durocher, J.M.E. (1857a) 'Essai de pétrologie comparée, ou recherches sur la composition chimique et minéralogique des roches ignées . . .', *AM*, 11 (5th series): 217–59, 678–81.

Durocher, J.M.E. (1857b) 'Recherches sur les roches ignées, sur les phénomènes de leur émission et sur leur classification', *Comptes rendus hebdomadaires des séances de l'Académie des sciences*, 44: 325–30.

Dutka, J. (1993) 'Eratosthenes' measurement of the earth reconsidered', *Archive for the History of Exact Sciences*, 46: 55–66.

Du Toit, A. (1927) *A Geological Comparison of South America with South Africa*, Washington: Carnegie Institution.

Du Toit, A. (1937) *Our Wandering Continents: an Hypothesis of Continental Drifting*, Edinburgh: Oliver & Boyd.

Dutton, C.E. (1871) 'The causes of regional elevations and subsidences', *Proceedings of the American Philosophical Society*, 12: 70–2.

Dutton, C.E. (1880) *Report on the Geology of the High Plateaus of Utah, with Atlas*, Washington, DC: Government Printing Office.

Dutton, C.E. (1880–1) 'The physical geology of the Grand Cañon district', *United States Geological Survey. Second Annual Report*, 47–166, Washington, DC: Government Printing Office.

Dutton, C.E. (1882a) *Tertiary History of the Grand Cañon District*, Washington, DC: Government Printing Office.

Dutton, C.E. (1882b) *Atlas to Accompany the Monograph on the Tertiary History of the Grand Cañon District*, Washington, DC: Government Printing Office.

Dutton, C.E. (1892) 'On some of the greater problems of physical geology', *Bulletin of the Philosophical Society of Washington*, 11: 51–64 (read 27 April 1889).

Dyson, F. (1985) *Origins of Life*, Cambridge: Cambridge University Press.

Edmonds, J.M. (1975) 'The geological lecture-courses given in Yorkshire by William Smith and John Phillips, 1824–1825', *PYGS*, 40: 373–412.

Egyed, L. (1956) 'The change of the earth's dimensions determined from palaeogeographical data', *GPA*, 33: 42–8.

Egyed, L. (1957) 'A new dynamic conception of the internal constitution of the earth', *Geologische Rundschau: internationale Zeitschrift für Geologie*, 46: 101–21.

Egyed, L. (1960) 'Some remarks on continental drift', *GPA*, 45: 115–16.

Eigen, M. and Winkler-Oswatitch, R. (1992) *Steps Towards Life: a Perspective on Evolution*, translated by P. Woolley, Oxford, New York and Tokyo: Oxford University Press.

Elie de Beaumont, L. (1829–30) 'Recherches sur quelques-unes des révolutions de la surface du globe . . . , *Annales des Sciences Naturelles*, 18: 284–416; 19: 5–99, 177–240.

Elie de Beaumont, L. (1831) 'Researches on some of the revolutions on the surface of the globe; presenting various examples of the coincidence between the elevation of beds in certain systems of mountains, and the sudden changes which have produced the lines of demarcation observable in certain stages of the sedimentary deposits', *PM*, 10 (new series): 241–64.

Elie de Beaumont, L. (1852) *Notice sur les systèmes des montagnes*, 3 vols, Paris: P. Bertrand.

Ellenberger, F. (1988) *Histoire de la géologie. Tome 1. Des anciens à la première moitié du XVIIe siècle*, Paris: Technique et Documentation (Lavoisier).

Ellenberger, F. (1994) *Histoire de la géologie. Tome 2. La grande éclosion et ses prémices 1660–1810*, Paris, London and New York: Technique et Documentation (Lavoisier).

Emerton, N. 1984. *The Scientific Reinterpretation of Form*, Ithaca: Cornell University Press.

Ercker, L. (1951) *Lazarus Ercker's Treatise on Ores and Assaying*, translated by A.G. Sisco and C.S. Smith, Chicago and London: University of Chicago Press (1st German edn, Prague, 1574, 2nd, Frankfurt-on-Main, 1580).

Escher von der Linth, A. (1839) 'Erläuterung der Ansichten einiger Contact-Verhältnisse zwischen krystallinischen Feldspathgestein und Kalk im Berner Oberlande', *Neue Denkschrifte der Allgemeine Schweizerischen Gesellschaft für gesammten Naturwissenschaft*, 3: 1–13 and plates.

Escher von der Linth, A. (1841) 'Geologische und mineralogische Section', *Verhandlungen der Schweizerischen Naturforschenden Gesellschaft bei ihrer Versammlung zu Zürich, den 2., 3. und 4 August 1841*, 53–73 (at 54–63). Zurich: von Zürcher and Furrer.

Escher von der Linth, A. (1846) 'Gebirgskunde [des Kantons Glarus]', in *Der Kanton Glarus . . .*, edited by O. Heer and J.J. Blumer-Heer, 51–90. St Gallen and Bern: Huber.

Evans, J.W., Turner, H.H. and Wright, W.B. (1923) 'Discussion on Wegener's hypothesis of continental drift', *RBAAS*, 364–5.

Eve, A.S. (1939) *Rutherford: Being the Life and Letters of the Rt Hon. Lord Rutherford, O.M.*, Cambridge: Cambridge University Press.

Ewing, J.A. (1881) 'On a new seismograph', *PRS*, 31: 440–6.

Eyles, V.A. (1972) 'Mineralogical maps as forerunners of modern geological maps', *Cartographic Journal*, 9: 133–5.

Faill, R.T. (1985) 'Evolving tectonic concepts of the central and southern Appalachians', in *Geologists and Ideas: a History of North American Geology*, edited by E.T. Drake and W.M Jordan, 19–46, Boulder: Geological Society of America.

Faul, H. and Faul, C. (1983) *It Began with a Stone: a History of Geology from the Stone Age to the Age of Plate Tectonics*, New York: J. Wiley.

Ferrari, G. (ed.) (1990) *Gli strumenti sismici storici: Italia e contesto Europea. Historical Seismic Instruments: Italy and the European Framework*, Bologna: Storia–Geofisici–Ambiente.

Ferrari, G. (ed.) (1992) *Two Hundred Years of Seismic Instruments in Italy 1731–1940*, Bologna: Storia–Geofisici–Ambiente.

Feyerabend, P.K. (1970) 'Classical empiricism', in *The Methodological Heritage of Newton*, edited by R.E. Butts and J.W. Davis, 150–70, Oxford: Basil Blackwell.

Fischer, A.G. (1981) 'Climatic oscillations in the biosphere', in *Biotic Crises in Ecological and Evolutionary Time*, 103–31, New York: Academic Press.

Fisher, O. (1881/1891) *Physics of the Earth's Crust*, London: Macmillan.

Flügel, H.W. (1984) 'A. Wegener – O. Ampferer – R. Schwinner: the first chapter of the "new global tectonics" ', *ESH*, 3: 178–86.

Forbes, J.D. (1844) 'On the theory and construction of a seismometer, or instrument for measuring earthquake shocks, and other concussions', *TRSE*, 15: 219–28.

Foucault, M. (1966) *Les mots et les choses: une archéologie des sciences humaines*, Paris: Gallimard.

Foucault, M. (1970) *The Order of Things: an Archaeology of the Human Sciences*, London: Tavistock Publications.

Fouqué, F. and Michel-Lévy, A. (1882) *Synthèse des minéraux et des roches*, Paris: G. Masson, Libraire de l'Académie de Médecine.

Fourier, J. (1820) 'Extrait d'un mémoire sur le refroidissement séculaire du globe terrestre', *ACP*, 13: 418–38.

Fourier, J. (1824) 'Remarques générales sur les températures du globe terrestre et des espaces planétaires, *ACP*, 27: 136–67.

Frankel, H. (1979) 'The career of continental drift theory: an application of Imre Lakatos's analysis of scientific growth to the rise of drift theory', *SHPS*, 10: 21–66.

Frankel, H. (1980) 'Hess's development of his seafloor spreading hypothesis', in *Scientific Discovery: Case Studies*, edited by T. Nickles, 345–66, Dordrecht: D. Reidel.

Frankel, H. (1982) 'The development, reception, and acceptance of the Vine–Matthews–Morley hypothesis', *HSPS*, 13: 1–39.

Frazer, J. (1919) *Folk-lore in the Old Testament: Studies in Comparative Religion and Law*, 3 vols, London: Methuen.

Fritscher, B. (1991) *Vulkanismus und Geochemie . . .*, Stuttgart: Steiner.

Fuchs, J.N. von (1839) 'Chemical views regarding the formation of rocks, which seem to afford new arguments in favour of neptunism', *ENPJ*, 26: 182–94.

Füchsel, G.C. (1761a) 'Historia terrae et maris ex historia Thuringiae, per montium descriptionem eruta a Georgio Christiano Füchsel', *Actorum Academiae Electoralis Moguntinae scientiarum utilium quae Erfordiae est*, 2: 44–208.

Füchsel, G.C. (1761b) 'Eiusdem usus historiae suae terrae et maris', *Actorum Academiae Electoralis Moguntinae scientiarum utilium quae Erfordiae est*, 2: 209–54.

[Füchsel, G.C.] (1773) *Entwurf zu der ältesten Erd- und Menschengeschichte, nebst einem Versuch, den Ursprung der Sprache zu finden*, Frankfurt and Leipzig.

Galitzin [or Golitsyn], B.B. (1906) *Uber eine Abänderung des Zöllner'schen Horizontalpendels*, St Petersburg: Königliche Akademie der Wissenschaften.

Garrels, R.M. and Lerman, A. (1981) 'Phanerozoic cycles of sedimentary carbon and sulfur', *Proceedings of the National Academy of Science . . . Physical Series*, 78: 4652–6.

Garrels, R.M. and Lerman, A. (1984) 'Coupling of sedimentary sulfur and carbon cycles – an improved model', *AJS*, 284: 989–1007.

Garrels, R.M. and MacKenzie, F.T. (1971) *Evolution of Sedimentary Rocks*, New York: W.W. Norton.

Garrels, R.M. and Perry, E.A. (1974) 'Cycling of carbon, sulfur, and oxygen through geologic time', in *The Sea . . . Volume 5. Marine Chemistry*, edited by E.D. Goldberg, 303–36, New York: John Wiley.

Gassendi, P. (1678) *Abrégé de la philosophie de Gassendi . . . Par F. Bernier*, Lyon: Anisson.

Gayrard-Valy, Y. (1994) *The Story of Fossils: in Search of Vanished Worlds*, translated by I.M. Paris, London: Thames & Hudson; New York: Harry N. Abrams (1st French edn, 1987).

Geikie, A. ([1897, 1905] 1962) *The Founders of Geology*, reprint, New York: Dover Publications (1st edn, 1897).

Geikie, J. (1874) *The Great Ice Age and its Relation to the Antiquity of Man*, London: Edward Stanford.

Geikie, J. (1914) *The Antiquity of Man in Europe*, Edinburgh: Oliver & Boyd; London: Gurney & Jackson.

Geoffroy, E.F. (1716) 'Sur l'origine des pierres', *HARS*, 8–16.

Gesner, C. (1565) *De rerum fossilium, lapidum et gemmarum . . .*, Zurich: no publisher stated.

Gilbert, G.K. (1875) 'Report upon the geology of portions of Nevada, Utah, California, and Arizona, examined in the years 1871 and 1872 . . .' in *Report upon Geographical and Geological Explorations and Surveys West of the One Hundredth Meridian* . . . , edited by G.M. Wheeler, 17–187, Washington, DC: Government Printing Office.

Gilbert, G.K. (1877) *Report on the Geology of the Henry Mountains*, Washington, DC: Government Printing Office.

Gillispie, C.C. ([1951] 1959) *Genesis and Geology: a Study in the Relations of Scientific Thought, Natural Theology, and Social Opinion in Great Britain, 1790–1850*, reprint, New York: Harper & Row.

Gilluly, J. (ed.) (1948) *Origin of Granite* . . . , New York: Geological Society of America (Memoir No. 28).

Gläser, F.G. (1775) *Versuch einer mineralogische Beschreibung der gefürsteten Grafschaft Henneberg chursächsischen Antheils* . . . , Leipzig: Siegfried Crusius.

Glen, W. (1982) *The Road to Jaramillo* . . . , Stanford: Stanford University Press.

Glen, W. (ed.) (1994) *The Mass-extinction Debates* . . . , Stanford: Stanford University Press.

Gohau, G. (1990) *A History of Geology*, revised and translated by A.V. and M. Carozzi, New Brunswick and London: Rutgers University Press.

Goodchild, J.G. (1892) 'Note on a granite junction in the Ross of Mull', *GM*, 9 (decade 3): 447–51.

Gorshkov, G.S. (1958) 'On some theoretical problems in vulcanology', *Bulletin Vulcanologique*, 19 (2nd series): 103–13.

Gosselet, J. (1880) 'Sur la structure générale du bassin houiller franco-belge', *BSGF*, 8: 505–11.

Gould, S.J. (1965) 'Is uniformitarianism necessary?', *AJS*, 263: 223–8.

Gould, S.J. (1987) 'Charles Lyell, historian of time's cycle', in *Time's Arrow Time's Cycle: Myth and Metaphor in the Discovery of Geological Time*, by S.J. Gould 98–179, Cambridge (Mass.) and London: Harvard University Press.

Gould, S.J. (1989) *Wonderful Life: the Burgess Shale and the Nature of History*, New York and London: W.W. Norton.

Gould, S.J. (1993) 'Fall in the house of Ussher', in *Eight Little Piggies: Reflections in Natural History*, 181–93, by S.J. Gould, London: Penguin Books.

Grabau, A.W. (1906) 'Types of sedimentary overlap', *BGSA*, 17: 567–636.

Grabau, A.W. (1936) 'Oscillation or pulsation', *International Geological Congress. Report of the XVI Session. Washington 1933*, 1: 539–53.

Grabau, A.W. ([1940] 1978) *The Rhythm of the Ages: Earth History in the Light of the Pulsation and Polar Control Theories*, introduction by A.V. Carozzi, reprint, Huntington (NY): Robert E. Krieger Publishing.

Graham, L.R. (1987) *Science, Philosophy, and Human Behavior in the Soviet Union*, New York: Columbia University Press.

Gray, T. (1881) 'On instruments for measuring and recording earthquake motion', *PM*, 12 (5th series): 199–212.

Greene, M.T. (1982) *Geology in the Nineteenth Century: Changing Views of a Changing World*, Ithaca and London: Cornell University Press.

Greene, M.T. (1992) *Natural Knowledge in Preclassical Antiquity*, Baltimore: Johns Hopkins University Press.

Greenly, E. and Williams, H. (1930) *Methods in Geological Surveying*, London: Thomas Murby; New York: D. van Nostrand.

Greenough, G.B. (1819) *A Geological Map of England and Wales by G.B.G. Accompanied by an Explanatory Memoir*, London: Longman, Hurst, Rees, Orme & Brown for the Geological Society.

Greenough, G.B. (1820) *Memoir of a Geological Map of England . . .*, London: Longman, Hurst, Rees, Orme & Brown.

Gregor, B. (1992) 'Some ideas on the rock cycle: 1788–1988', *Geochimica et Cosmichimica Acta*, 56: 2993–3000.

Griffith, R. (1838) *Geological Map of Ireland to Accompany the report of the Railway Commissioners 1837*, Dublin.

Griggs, D. (1939) 'A theory of mountain building', *AJS*, 237: 611–50.

Grinewald, J. (1988) 'Sketch for a history of the idea of the biosphere', in *Gaia . . .*, edited by P. Bunyard and E. Goldsmith, 1–32, Camelford: Wadebridge Ecological Centre.

Grout, F.F. (1945) 'Scale models of structures related to batholiths', *AJS*, 243A: 260–84.

Guettard, J.-E. (1746/1751) 'Mémoire et carte minéralogique sur la nature & la situation des terrains qui traversent la France & l'Angleterre', *HARS*, 363–92 and plate.

Guettard, J.-E. (1752/1756) 'Mémoire sur quelques montagnes de la France qui ont été des volcans', *HARS*, 27–59.

Guettard, J.-E. (1779) *Mémoires sur la minéralogie du Dauphiné*, 2 vols, Paris: l'Imprimerie de Clousier.

Guettard, J.-E. (1784) *Carte minéralogique de France . . .*, Paris: l'Imprimerie Royale.

Guettard, J.-E. and Monnet, A.-G. (1780) *Atlas et description minéralogique de la France . . .*, Paris: Didot l'Ainé.

Gümbel, C.W. von (1866a) 'On the occurrence of Eozoön in the Primary rocks of eastern Bavaria', *QJGS*, 22 (Part 2): 23–24.

Gümbel, C.W. von (1866b) 'Ueber des Vorkommen von *Eozoön* in dem ostbayerischen Urgebirge', *Sitzungsberichte der königliche bayer Akademie der Wissenschaften zu Munich*, 1: 25–70 and plates.

Gümbel, C.W. von (1868) 'On the Laurentian rocks of Bavaria', *CN*, 3 (2nd series): 81–101 and plate.

Guntau, M. (1967) *Der Aktualismus in den geologischen Wissenschaften . . .*, Leipzig: Deutscher Verlag für Grundstoffindustrie.

Guntau, M. (1984) *Abraham Gottlob Werner*, Leipzig: Teubner.

Gutenberg, B (1914) 'Uber Erdbeben VIIA . . . Folgerungen über die Konstitution des Erdkörpers', *Nachrichten aus der Gesellschaft der Wissenschaften zu Göttingen. Mathematisch-physikalische Klasse*, 166–218.

Gutenberg, B. (1925) *Der Aufbau der Erde mit 23 Abbildungen*, Berlin: Gebrüder Borntraeger.

Gutenberg, B. (1926) 'Untersuchungen zur Frage, bis zu welcher Tiefe die Erde kristallin ist', *Zeitschrift für Geophysik*, 2: 24–9.

Gutenberg, B. (1928) 'Der Aggregatzustand im Erdinnern', *Müller-Pouillets Lehrbuch der Geophysik*, 5: 669–71.

Gutenberg, B. (1948) 'The layer of relatively low wave velocity at a depth of 80 kilometers', *BSSA*, 38: 121–48.

Gutenberg, B. (1954) 'Low velocity layers in the earth's mantle', *BGSA*, 65: 337–48.

Gutenberg, B. (1955) 'Channel waves in the earth's crust', *Geophysics*, 20: 283–94.

Gutenberg, B. (1957) 'The "boundary" of the earth's inner core', *Transactions of the American Geophysical Union*, 38: 750–3.

Gutenberg, B. (1931) 'On supposed discontinuities in the mantle', *BSSA*, 21: 216–22.

Gutenberg, B. and Richter, C.F. (1939) 'On supposed discontinuities in the mantle of the earth: new evidence for a change in physical conditions at depths near 100 km', *BSSA*, 29: 531–7.

Guthrie, W.K.C. (1962) *A History of Greek Philosophy. Volume 1. The Earliest Presocratics and the Pythagoreans*, Cambridge and New York: Cambridge University Press.

Haarmann, E. (1930) *Die Oszillations-Theorie: eine Erklärung der Krustenbewegungen von Erde und Mond*, Stuttgart: Ferdinand Enke.

Haber, F.C. (1959) *The Age of the World: Moses to Darwin*, Baltimore: Johns Hopkins University Press.

Hacking, I. (1983) *Representing and Intervening . . .*, Cambridge, London, New York, New Rochelle, Melbourne and Sydney: Cambridge University Press.

Hall, J. (Sir) (1805) 'Experiments on whinstone and lava', *TRSE*, 5: 43–76 (read 1798).

Hall, J. (Sir) (1812) 'Account of a series of experiments showing the effects of compression in modifying the action of heat', *TRSE*, 6: 71–186.

Hall, J. (Sir) (1826) 'On the consolidation of strata of the earth', *TRSE*, 10: 314–29.

Hall, J. (1859) 'Introduction', *Palaeontology, Volume III* [of *Natural History of New York*, Part 6] . . . *Part I: Text*, Albany: van Benthuysen.

Hall, J. ([1857] 1883) 'Contributions to the geological history of the North American continent', *Proceedings of the American Association for the Advancement of Science*, 31: 29–71 (Presidential Address for 1857).

Hallam, A. (1973) *A Revolution in the Earth Sciences: From Continental Drift to Plate Tectonics*, Oxford: Clarendon Press.

Hallam, A. (1983) *Great Geological Controversies*, Oxford: Oxford University Press.

Hallam, A. (1992) 'Eduard Suess and European thought on Phanerozoic eustacy', in *Eustacy: the Historical Ups and Downs of a Major Geological Concept*, edited by R.H. Dott, 25–29, Boulder: Geological Society of America (Memoir No. 180).

Halstead, L.B. (1982) *Hunting the Past: Fossils, Rocks, Tracks and Trails: the Search for the Origin of Life*, London: Hamish Hamilton.

Hamilton, W. and Myers, W.B. (1967) *The Nature of Batholiths*, Washington, DC: Government Printing Office. (Geological Survey Professional Paper 554–C).

Haq, B.U., Hardenbol, J. and Vail, P.R. (1987) 'Chronology of fluctuating sea levels since the Triassic', *Science*, 235: 1156–67.

Harker, A. (1909) *The Natural History of the Igneous Rocks*, London: Methuen.

Harker, A. (1932) *Metamorphism: a Study of the Transformations of Rock Masses*, London: Methuen.

Haug, E. (1900) 'Les géosynclinaux et les aires continentales: contributions à l'étude des transgressions et régressions marines', *BSGF*, 28 (3rd series): 617–710.

Haug, E. (1907–11) *Traité de géologie*, 2 vols, Paris: Librairie Armand Colin.

Haüy, R.-J. (1784) *Essai d'une théorie sur la structure des crystaux, appliquée à plusieurs genres de substances crystallisées*, Paris: Chez Gogué et Née de la Rochelle.

Haüy, R.-J. (1801) *Traité de minéralogie*, 5 vols. Paris: Conseil des Mines.

Haüy, R.-J. (1822) *Traité de crystallographie . . .*, 3 vols, Paris: Bachelier et Hazard.

Hays, J.D., Imbrie, J. and Shackleton, N.J. (1976) 'Variations in the earth's orbit: pacemakers of the ice ages', *Science*, 194: 1121–32.

Heckel, P.H. (1984) 'Changing concepts of midcontinent Pennsylvanian cyclothems, North America', *Neuvième congrès international de stratigraphie et de géologie du Carbonifère. Washington and Champaign–Urbana May 17–26, 1979. Compte rendu*, 3: 535–53.

Hedberg, H.D. (1969a) 'The influence of Torbern Bergman (1735–1784) on stratigraphy: a resumé', in *Toward a History of Geology*, edited by C.J. Schneer, 186–91, Cambridge (Mass.): MIT Press.

Hedberg, H.D. (1969b) 'The influence of Torbern Bergman (1735–1784) on stratigraphy', *Acta Universitatis Stockholmiensis: Stockholm Contributions in Geology*, 20: 19–47.

Heezen, B.C. (1959) 'Géologie sous-marine et déplacements des continents', in *La topographie et la géologie des profondeurs océaniques (Colloques internationaux du Centre National de la Recherche Scientifique . . .)*, 83: 295–302, Paris: Centre National de la Recherche Scientifique.

Heezen, B.C. (1960) 'The rift in the ocean floor', *SA*, 203: 99–110.

Heim, A. (1878) *Untersuchungen über den Mechanismus der Gebirgsbildung: im Anschluss an die geologische Monographie der Tödi-Windgällen-Gruppe*, 2 vols and atlas, Basle: Benno Schwab.

Helmont, J.B. van (1662) *Oriatrike, or, Physick Refined* . . . , translated by J. Chandler, London: Lodowick Loyd.

Hennig, W. (1966) *Phylogenetic Systematics*, translated by D.D. Davis and R. Zanger, Urbana: University of Illinois Press (1st German edn, 1950).

Henslow, J.S. (1822) 'Geographical description of Anglesea', *TCPS*, 1: 359–452 and plate.

Herodotus (1954) *The Histories*, translated by A. de Selincourt, Harmondsworth: Penguin Books.

Herries Davies, G.L. ([1969]) *The Earth in Decay: a History of British Geomorphology 1578–1878*, London: Macdonald Technical and Scientific.

Herries Davies, G.L. (1983) *Sheets of Many Colours: the Mapping of Irish Rocks 1750–1890*, Dublin: Royal Dublin Society.

Herries Davies, G.L. (1989) 'A science receives its character', in *Two Centuries of Earth Science 1650–1850*, edited by G.L. Herries Davies and A.R. Orme, 1–28, Los Angeles: University of California.

Herries Davies, G. L. (1995) *North from the Hook: 150 years of the Geological Survey of Ireland*, Dublin: Geological Survey of Ireland.

Herries Davies, G.L. and Mollan, R.C. (eds) (1980) *Richard Griffith 1784–1878* . . . , Dublin: Royal Dublin Society.

Herschel, J. ([1830] 1835) 'On the astronomical causes which may influence geological phænomena', *TGSL*, 3, part 2 (2nd series): 301–420 and plates.

Hesiod (1914) *The Homeric Hymns and Homerica*, translated by H.G. Evelyn-White, London: Heinemann.

Hess, H.H. (1946) 'Drowned ancient islands of the Pacific basin', *AJS*, 244: 772–91.

Hess, H.H. (1962) 'History of ocean basins', in *Petrologic Studies*, edited by A.E.J. Engel, H.L. James and B.F. Leonard, 599–620, Washington, DC: Geological Society of America.

Heyne, B. (1814) *Tracts, Historical and Statistical, on India; with Journals of Several Tours through Various Parts of the Peninsula* . . . , London: Baldwin & Black, Parry.

Hills, B. (1995) 'The Japan quake. World's turn to fear this ancient rage', *Sydney Morning Herald*, 18 January: 9.

Hochstetter, F.W. von (1864) *Geologie von Neu-Seeland* . . . , Vienna: Kaiserlich-königlichen Hof und Staatsdruckerei.

Hoff, K.E.A. von (1826–35) 'Verzeichniss von Erdbeben, vulcanischen Ausbrüchen und merdwürdigen meteorologischen Erscheinungen seit dem Jahre 1821', *Annalen der Physik und Chemie*, 7: 159–70, 289–304; 9: 589–600; 12: 555–84; 15: 363–83; 18: 38–56; 21: 202–18; 25: 59–90; 29: 415–46, 447–51; 34: 85–108.

Hoffmann, R. (1869) 'Chemische Untersuchung des Eozoongesteins von Raspenau in Böhmen', *Journal für praktische Chemie*, 101: 129–37.

Holmes, A. (1911) 'The association of lead with uranium in rock-minerals, and its application to the measurement of geological time', *PRS*, Series A, 85: 248–56.

Holmes, A. (1928) 'Radioactivity and continental drift', *GM*, 65: 236–8.

Holmes, A. (1928–31) 'Radioactivity and earth movements', *TGSG*, 18: 559–606.

Holmes, A. (1929) 'A review of the continental drift hypothesis', *Mining Magazine*, 40: 205–9.

Holmes, A. (1944/1965) *Principles of Physical Geology*, London: Thomas Nelson and Sons (revised edn 1965).

Holmquist, P.J. (1916) 'Swedish archaean structures and their meaning', *Bulletin of the Geological Institution of the University of Upsalla*, 15: 125–48.

Hooke, R. (1705) *Posthumous Works* . . . , edited by R. Waller, London: S. Smith & B. Walford.

Hooker, J.D. (1853) *The Botany of the Antarctic Voyage of H.M. Ships Erebus and Terror in the Years 1839–1843. Part II. Flora Novae-Zelandiae*, London: Lovell Reeve.

Hooykaas, R. (1963) *Natural Law and Divine Miracle: the Principle of Uniformity in Geology, Biology and Theology*, Leiden: E.J. Brill.

Hooykaas, R. (1970) *Catastrophism in Geology, its Scientific Character in Relation to Actualism and Uniformitarianism*, Amsterdam: North Holland.

Hopkins, W. (1838) 'Researches in physical geology', *TCPS*, 6: 1–84.

Hopkins, W. (1839) 'Researches in physical geology', *PTRS*, 129: 381–85.

Hopkins, W. (1840) 'Researches in physical geology – second series. On precession and nutation, assuming the interior of the earth to be fluid and homogeneous', *PTRS*, 130: 193–208.

Hopkins, W. (1842) 'Researches in physical geology – third series. On the thickness and constitution of the earth's crust', *PTRS*, 132: 43–56.

Howell, B.F. (1991) 'How misconceptions on heat flow may have delayed discovery of plate tectonics', *ESH*, 10: 44–55.

Hoyle, F. (1950) *The Nature of the Universe: a Series of Broadcast Lectures*, Oxford: Basil Blackwell.

Hsü, K.J. (1973) 'The odyssey of geosyncline', in *Evolving Concepts in Sedimentology*, edited by R.N. Ginsburg, 66–92, Baltimore: Johns Hopkins University Press.

Hsü, K.J. (1992) *Challenger at Sea: a Ship that Revolutionized Earth Science*, Princeton: Princeton University Press.

Huggett, R. (1989) *Cataclysms and Earth History: the Development of Diluvialism*, Oxford: Clarendon Press.

Hull, D.L. (1988) *Science as a Process* . . . , Chicago and London: University of Chicago Press.

Humboldt, F.H.A. von (Year 9/1801) 'Esquisse (1) d'un tableau géologique de l'Amérique Méridionale', *Journal de Physique*, 53: 30–59.

Humboldt, F.H.A. (1823a) *Essai géognostique sur le gisement des roches* . . . , Paris: F.G. Levrault.

Humboldt, F.H.A. (1823b) *A Geognostic Essay on the Superposition of Rocks* . . . , London: Longman, Hurst, Rees, Orme, Brown & Green.

Humboldt, F.H.A. (1845–62) *Kosmos* . . . , 5 vols, Stuttgart and Tübingen: J.G. Cotta.

Humboldt, F.H.A. (1880–3) *Cosmos* . . . , translated by E.C. Otté, B.H. Paul and W.S. Dallas, 5 vols, London: George Bell & Sons.

Humboldt, F.H.A. von and Bonpland, A. (1814–29) *Personal Narrative of Travels to the Equinoctial Regions of the New Continent, During the Years 1799–1804* . . . , translated by

H.M. Williams, 7 vols, London: Longman, Hurst, Rees, Orme & Brown (1st French edn, 3 vols, 1814–25).

Hunt, T.S. (1858) 'On the chemistry of the primeval earth . . .', *AJS*, 25 (2nd series): 102–3.

Hunt, T.S. (1859) 'On some points in chemical geology', *QJGS*, 15: 488–96.

Hunt, T.S. (1867) 'On the chemistry of the primeval earth', *CN*, 3 (new series): 225–34.

Hunt, T.S. (1871) 'The geognosy of the Appalachian mountains and the origin of crystalline rocks', *American Naturalist*, 5: 451–509.

Hunt, T.S. (1875) *Chemical and Geological Essays*, Boston: James Osgood.

Hunt, T.S. (1885) 'The origin of crystalline rocks', *Transactions of the Royal Society of Canada . . . Section III. Mathematical, Physical and Chemical Sciences*, 2: 1–67.

Huntington, E. (1925) *The Character of Races as Influenced by Physical Environment, Natural Selection and Historical Development*, New York and London: Charles Scribner's Sons.

Hutchinson, G.E. (1965) *The Ecological Theater and the Evolutionary Play*, New Haven and London: Yale University Press.

Hutchinson, G.E. (1970) 'The biosphere', *SA*, 223 (September): 44–53.

Hutton, D.H.W. (1977) 'A structural cross-section from the aureole of the main Donegal granite', *GJ*, 12: 99–111.

Hutton, J. (1749) *Dissertatio physico-medica inauguralis de sanguine et circulatione microcosmi . . .*, Leyden: Wilhelmum Boot.

Hutton, J. (1788) 'Theory of the earth; or an investigation of the laws observable in the composition, dissolution, and restoration of land upon the globe', *TRSE*, 1: 209–304.

Hutton, J. (1795–1899) *Theory of the Earth, with Proofs and Illustrations*, 2 vols, London: Cadell, Junior and Davies; Edinburgh: William Creech. Vol. 3, London: Geological Society.

Hutton, J. ([1785] 1987) *Abstract of a Dissertation read in the Royal Society of Edinburgh, upon the seventh of March, and fourth of April, M,DCC,LXXXV, Concerning the System of the Earth, its Duration, and Stability*, reprinted as *The 1785 Abstract of James Hutton's Theory of the Earth*, introduced by G.Y. Craig, Edinburgh: Scottish Academic Press.

Ihde, D. (1979) *Technics and Praxis*, Dordrecht: D. Reidel.

Ihde, D. (1990) *Technology and the Life World: From Garden to Earth*, Bloomington and Indianapolis: Indiana University Press.

Iliffe, R. (1993) ' "Applatisseur du monde et de Cassini": Maupertuis, precision measurement, and the shape of the earth', *HSPS*, 31: 335–75.

Imbrie, J. and Imbrie, K.P. (1979) *Ice Ages: Solving the Mystery*, London and Basingstoke: Macmillan.

Irons, J.C. (1896) *Autobiographical Sketch of James Croll . . . with Memoir of his Life and Work*, London: Edward Stanford.

Irvine, T.N. (1989) 'A global convection framework: concepts of symmetry, stratification, and system in the earth's dynamic structure', *Economic Geology*, 84: 2059–114.

Irving, E. (1956) 'Palaeomagnetic and palaeoclimatological aspects of polar wandering', *GPA*, 33: 23–41.

Jacobs, J.A., Russell, R.D. and Wilson, J.T. (1959) *Physics and Geology*, New York: McGraw-Hill.

Jaki, S. (1974) *Science and Creation*, Edinburgh and London: Scottish Academic Press.

Jameson, R. (1805) *Treatise on the External Characters of Minerals*, Edinburgh: Bell & Bradfute; Guthrie & Tate, & W. Blackwood.

Jameson, R. (1808) *System of Mineralogy . . . [Vol. III]. Elements of Geognosy*, Edinburgh: Archibald Constable.
Jameson, R. (1809–10) 'On colouring geognostical maps', *Memoirs of the Wernerian Natural History Society*, 1: 149–61 (read 1808).
Jamieson, T.F. (1865) 'On the history of the last geological changes in Scotland', *QJGS*, 21: 161–203.
Jamieson, T.F. (1882) 'On the cause of the depression and re-elevation of land during the glacial period, *Geological Magazine*, 9: 400–7.
Jeffreys, H. (1924/1929/1952/1959/1970/1976) *The Earth . . .*, Cambridge: Cambridge University Press.
Jeffreys, H. (1927) 'On the earth's thermal history and some related geological phenomena', *Gerlands Beiträge zur Geophysik*, 18: 1–18.
Jeffreys, H. (1928a) 'The rigidity of the earth's central core', *GSMNRAS*, 1: 371–83.
Jeffreys, H. (1937–40 [1939]) 'The times of the core waves', *GSMNRAS*, 4: 537–61, 594–615.
Jevons, F.R. (1964) 'Paracelsus' two-way astrology. I. What Paracelsus meant by "stars." II. Man's relation to the stars', *BJHS*, 2: 139–47, 148–55.
Johns, R.K. (1976) *History and Role of Government Geological Surveys in Australia*, Adelaide: South Australia Government Printer.
Johnson, M.E. (1992) 'A.W. Grabau's embryonic sequence stratigraphy and eustatic curve', in *Eustacy: the Ups and Downs of a Major Historical Concept*, edited by R.H. Dott, 43–45, Boulder: Geological Society of America (Memoir No. 180).
Johnson, M.E. and McKerrow, W.S. (1991) 'Sea level and faunal changes during the latest Llandovery and earliest Ludlow (Silurian)', *Historical Biology*, 5: 153–69.
Joly, J. (1909) *Radioactivity and Geology: an Account of the Influence of Radioactive Energy on Terrestrial History*, London: Archibald Constable.
Joly, J. (1925) *The Surface-history of the Earth*, Oxford: Clarendon Press.
Jones, J. (1985) 'James Hutton's agricultural researches and his life as a farmer', *AS*, 42: 573–601.
Jordan, P. (1971) *The Expanding Earth: Some Consequences of Dirac's Gravitation Hypothesis*, translated and edited by A. Beer, Oxford and New York: Pergamon Press (1st German edn, 1966).
Judd, J.W. (1897) 'William Smith's manuscript maps', *GM*, 4 (decade 4): 439–47.

Kant, I. (1755) *Algemeine Naturgeschichte und Theorie des Himmels . . .*, Königsberg: J.F. Peterman.
Kant, I. (1981) *Universal Natural History and Theory of the Heavens*, translated by S.L. Jaki, Edinburgh: Scottish Academic Press.
Kay, M. ([1951] 1963) *North American Geosynclines*, reprint, New York: Geological Society of America (Memoir No. 48).
Keferstein, C. (1831) 'Notice sur Füchsel et ses ouvrages', *Journal de Géologie*, 2: 191–7.
Kelvin, W. Thomson, Lord (1862) 'On the age of the sun's heat', *Macmillan's Magazine*, 5: 388–93.
Kelvin, W. Thomson, Lord (1863) 'On the secular cooling of the earth', *PM*, 25 (4th series): 1–14.
Kelvin, W. Thomson, Lord (1864) 'On the secular cooling of the earth', *TRSE*, 23: 157–70.
Kelvin, W. Thomson, Lord (1871) 'On geological time', *TGSG*, 3: 1–28.
Kendrick, T.D. (1956) *The Lisbon Earthquake*, London: Methuen.

King, W. and Rowney, T.H. (1881) *An Old Chapter of the Geological Record. . . . With . . . an Annotated History of the Controversy on . . . "Eozoön canadense" . . .*, London: John van Voorst.

Kircher, A. (1678) *Mundus subterraneus, in XII libros digestus . . .*, 3rd edn, 2 vols, Amsterdam: Joannem Janssonius & Elizeum Weyerstraten (1st edn, 1664–5).

Kirk, G.S. (1970) *Myth: its Meaning and Functions in Ancient and Other Cultures*, Cambridge: Cambridge University Press.

Kirk, G.S. and Raven, J.E. (1957) *The Presocratic Philosophers: a Critical History with a Selection of Texts*, Cambridge: Cambridge University Press.

Kirwan, R. (1799) *Geological Essays*, London: T. Bensley for D. Bremner.

Koehne, W. (1915) 'Die Entwickelungsgeschichte der geologischen Landesaufnahmen in Deutschland', *GRZAG*,6: 178–92.

Kuhn, T.S. (1962) *The Structure of Scientific Revolutions*, Chicago and London: University of Chicago Press.

Lakatos, I. (1970) 'Falsification and the methodology of scientific research programmes', in *Criticism and the Growth of Knowledge*, edited by I. Lakatos and A. Musgrave, 91–195, London and New York: Cambridge University Press.

Lamarck, J.-B. (1797/Year 5) *Mémoirs de physique et d'histoire naturelle . . .*, Paris: J.-B. Lamarck, Agasse & Maradin.

Lamarck, J.-B. (1802/Year 10) *Hydrogéologie . . .*, Paris: J.-B. Lamarck, Agasse & Maillard.

Landes, K.K. (1952) 'Our shrinking globe', *BGSA*, 63: 225–39.

Langenheim, R.J. and Nelson, W.J. (1992) 'The cyclothem concept in the Illinois basin: a review', in *Eustacy: the Historical Ups and Downs of a Major Geological Concept*, edited by R.H. Dott, 55–71, Boulder: Geological Society of America (Memoir No. 180).

Laplace, P.S. de (1796/Year 4) *Exposition du système du monde*, 2 vols, Paris: Circle-Social.

Lapo, A.V. (1982/1987) *Traces of Bygone Biospheres*, translated by V.A. Purto, Moscow: Mir Publishers.

Lapworth, C. (1879) 'On the tripartite classification of the lower Palaeozoic rocks', *GM*, 6 (decade 2): 1–15.

Lapworth, C. (1883–4) 'On the structure and metamorphism of the rocks of the Durness–Eriboll district', *PGA*, 8: 438–42.

Latour, B. (1987) *Science in Action . . .*, Milton Keynes: Open University Press.

Laudan, L. (1977) *Progress and its Problems*, Berkeley: University of California Press.

Laudan, R. (1976) 'William Smith: stratigraphy without palaeontology', *Centaurus*, 20: 210–26.

Laudan, R. (1980) 'The recent revolution in geology and Kuhn's theory of scientific change', in *Paradigms and Revolutions . . .*, edited by G. Gutting, 284–96, Notre Dame and London: University of Notre Dame Press.

Laudan, R. (1987) *From Mineralogy to Geology . . .*, Chicago and London: University of Chicago Press.

Lavoisier, A.L. (1789a) *Traité élémentaire de chimie . . .*, 2 vols, Paris: Cuchet.

Lavoisier, A.L. (1789b [1792]) 'Observations générales, sur les couches modernes horizontales, qui ont été déposées par la mer . . .', *Histoires de l'Académie des Sciences. Année M. DCC. LXXXIX . . .*, 351–71 and plates.

Lawrence, P. (1977) 'Heaven and earth – the relation of the nebular hypothesis to geology', in *Cosmology, History, and Theology*, edited by W. Yourgrau and A.D. Breck, 253–81, New York and London: Plenum Press.

Lees, G.M. (1953) 'The evolution of a shrinking earth', *QJGS*, 109: 217–57.

Le Fèvre, N. (1664) *A Compendious Body of Chymistry . . .*, 2 vols, London: T. Davies & T. Sadler.

Le Grand, H.E. (1988) *Drifting Continents and Shifting Theories . . .*, Cambridge, New York, New Rochelle, Melbourne and Sydney: Cambridge University Press.

Lehmann, I. (1936) '*P'*', *Publications du Bureau Central Séismologique International. Série A. Travaux scientifiques*, 14: 87–115.

Lehmann, I. (1987) 'Seismology in the days of old', *Eos*, 68: 33–5.

Lehmann, J.G. (1753) *Abhandlungen von den Metall-Müttern und der Erzeugung der Metalle aus der Naturlehre und Bergwerkwissenschaft hergeleitet und mit chymischen Versuchen*, Berlin: Cristoph Gottlieb Nicolai.

Lehmann, J.G. (1756) *Versuch einer Geschichte von Flötz-Gebürgen . . .*, Berlin: Klüterschen Buchhandlung.

Lehmann, J.G. (1759) *Essai d'une histoire naturelle des couches de la terre . . .* [Volume 3 of *Traités de physique d'histoire naturelle, de minéralogie et de métallurgie*], Paris: J.-T. Hérissant.

Lehmann, J.G. (1884) *Untersuchungen über die Entstehung der Altkristallinischen Schiefergestein mit besondere Bezugnahme auf das Sächsische Granulitgebirge, Erzgebirge, Fichtelgebirge und Bairisch-Böhmisch Grenzgebirge*, Bonn: M. Hochgürtel.

Leibniz, G.W. von (1693) 'Protogaea autore GGL', *Acta Eruditorum Anno MDCXCIII*, 40–2. Leipzig.

Leibniz, G.W. (1749) *Protogaea, sive de prima facie telluris et antiquissimae historiae vestigiis in ipsis naturae monumentis . . .*, Göttingen: Sumptibus Ioh. Guil. Schmidii.

Leonardo da Vinci (1906) *Leonardo da Vinci's Note-books*, translated by E. McCurdy, London: Duckworth.

Le Roy, E. (1927) *L'Exigence idéaliste et le fait de l'évolution*, Paris: Boivin.

Lhwyd, E. (1699) 'Letter to John Ray (De fossilium marinorum & foliorum mineralium origine)', in *Lithophylacii britannici ichnographia . . .*, edited by E. Luidii, 128–39, London: Ex Officina M.C.

Lhwyd, E. (1945) 'Letter to John Ray . . . ', in *Early Science in Oxford: Vol. XIV. Life and Letters of Edward Lhwyd . . .*, edited by R.T. Gunther, 381–96, Oxford: Printed for the subscribers.

Linnaeus, C. (1770) *Systema Naturae . . . Tomus III*, Vienna: Ionnis Thomae Nob. de Trattnern.

Lister, M. (1671) 'A letter of Mr. Martin Lister, written at York August 25 1671 . . . [A]dding some notes upon D. Swammerdam's book of insects, and on that of M. Steno concerning petrify'd shells', *PTRS*, 6: 2281–4.

Loewinson-Lessing, F.Y. (1954) *A Historical Survey of Petrology*, translated from the Russian edition of 1936 by S.I. Tomkeieff, Edinburgh and London: Oliver & Boyd.

Logan, R.K. (1986) *The Alphabet Effect: the Impact of the Phonetic Alphabet on the Development of Western Civilization*, New York: St Martin's Press.

Lossen, K.A. (1869) 'Metamorphische Schichten aus der paläozoischen Schichtenfolge des Ostharzes', *Zeitschrift der Deutschen Geologischen Gesellschaft*, 21: 281–340.

Love, A.E.H. (1911) *Some Problems of Geodynamics*, Cambridge: Cambridge University Press.

Lovelock, J.E. (1979/1982/1987) *Gaia: a New Look at Life on Earth*, Oxford and New York: Oxford University Press.

Lovelock, J.E. (1988/1989/1991) *The Ages of Gaia: a Biography of Our Living Earth*, Oxford: Oxford University Press.

Lucretius (1951) *On the Nature of the Universe*, translated by R.E. Latham, Harmondsworth: Penguin.

Ludwig, C. (1856) 'Diffusion zwischen ungleich erwärmten Orten gleich zusammengesetzter Lösungen', *Sitzungsberichte der mathematisch-naturwissenschaftlichen Classe der Kaiserlichen Akademie der Wissenschaften* (Vienna) 20: 539.

Lugeon, M. (1940) 'Emile Argand', *Bulletin de la Société Neuchâteloise des Sciences Naturelles*, 65: 25–53.

Lugg, A. (1978) 'Overdetermined problems in science', *SHPS*, 9: 1–18.

Lyell, C. (1830–3) *Principles of Geology, being an Attempt to Explain the Former Changes of the Earth's Surface, by Reference to Causes Now in Operation*, 3 vols, London: John Murray.

Lyell, C. (1845) *Travels in North America, in the Years 1841–2* . . . , 2 vols, New York: Wiley & Putnam.

Lyell, C. (1863) *Geological Evidences of the Antiquity of Man* . . . , London: John Murray.

Lyell, C. (1970) *Sir Charles Lyell's Scientific Journals on the Species Question*, edited by L.G. Wilson, New Haven and London: Yale University Press.

Lyell, K.M. (ed.) (1881) *Life, Letters and Journals of Sir Charles Lyell* . . . , 2 vols, London: John Murray.

McCartney, P.J. (1977) *Henry De la Beche: Observations on an Observer*, Cardiff: National Museum of Wales.

McConnell, A. (1986) *Geophysics and Geomagnetism: Catalogue of the Science Museum Collection*, London: Her Majesty's Stationery Office.

Macculloch, J. (1836) *A Geological Map of Scotland by Dr. John Macculloch* . . . , London.

Macelwane, J.B. and Sohon, F.W. (1936) *Introduction to Theoretical Seismology. Part I. Geodynamics*, New York: John Wiley & Sons; London: Chapman & Hall.

MacKenzie, F.T. (1989) 'Memorial of Robert Minard Garrels August 24, 1916–March 8, 1988', *American Mineralogist*, 74: 497–9.

MacLaren, C. (1842) 'The glacial theory of Prof. Agassiz', *AJS*, 42: 346–65.

Maclure, W. (1809) 'Observations on the geology of the United States, explanatory of a geological map', *Transactions of the American Philosophical Society*, 6: 411–27.

Maclure, W. (1817) *Observations on the Geology of the United States of America* . . . , Philadelphia: Abraham Small.

McMahon, D.M. (1992) 'The Turin papyrus map: the oldest known map of geological significance', *ESH*, 11: 9–12.

Maillet, B. de (1748) *Telliamed, ou entretiens d'un philosophe indien avec un missionnaire françois sur la diminution de la mer, la formation de la terre, l'origine de l'homme*, edited by J. A. G[uer], 2 vols, Amsterdam: chez l'Honoré & Fils.

Mallet, R. (1848) 'On the dynamics of earthquakes', *Transactions of the Royal Irish Academy*, 21: 50–106.

Mallet, R. (1859) 'Fourth report on the facts and theory of earthquake phenomen', *RBAAS*, 1–136.

Mallet, R. (1862) *Great Neapolitan Earthquake of 1857: the First Principles of Observational Seismology as Developed in the Report to the Royal Society* . . . , 2 vols and maps, London: Chapman & Hall.

Marcou, J. (1857–60) *Lettres sur les roches de Jura et leur distribution géographique dans les deux hémisphères*, Paris: Friedrich Klincksieck.

Marggraf, A.S. (1762) *Opuscules chymiques . . .*, 2 vols, Paris: Vincent.

Mark, K. (1992) 'From geosynclinal to geosyncline', *ESH*, 11: 68–9.

Marsh, B.D. (1982) 'On the mechanics of igneous diapirism, stoping, and zone melting', *AJS*, 282: 808–55.

Marvin, U. (1973) *Continental Drift . . .*, Washington, DC: Smithsonian Institution Press.

Mason, R.G. and Raff, A.D. (1961) 'Magnetic survey off the west coast of North America, 32°N. latitude to 52°N. latitude', *BGSA*, 72: 1267–70 and plate.

Mather, K.F. and Mason, S.L. (1939) *A Source Book in Geology*, New York and London: McGraw-Hill Book Company.

Matthew, W.D. (1915) *Climate and Evolution*, New York: New York Academy of Sciences.

Melville, C. and Wood, R.M. (1987) 'Robert Mallet, first modern seismologist', in *Mallet's Macroseismic Survey on the Neopolitan Earthquake of 16th December, 1857*, edited by E. Guidoboni and G. Ferrari, 17–48, Bologna: Instituto Nazionale di Geofisica.

Menard, H.W. (1986) *The Ocean of Truth . . .*, Princeton: Princeton University Press.

Merriam, D.F. (1992) 'Linnaeus' 1747 geological profile (cross section) of Kinne-Kulle, central Sweden', *The Compass*, 69: 320–5.

Merrill, G.P. (1920) *Contributions to a History of American State Geological and Natural History Surveys*, Washington, DC: Government Printing Office (Smithsonian Institution Bulletin 109).

Merrill, G.P. ([1924] 1969) *The First One Hundred Years of American Geology*, reprint, New York and London: Hafner.

Meyerhoff, A.A. and Meyerhoff, H.A. (1972a) 'The new global tectonics: major inconsistencies', *BAAPG*, 56: 269–336.

Meyerhoff, A.A. and Meyerhoff, H.A. (1972b) 'The new global tectonics: age of linear magnetic anomalies of ocean basins', *BAAPG*, 56: 337–59.

Michell, J. (1759–60) 'Conjectures concerning the cause, and observations upon the phaenomena of earthquakes; particularly that of the great earthquake of the first of November, 1755 . . .', *PTRS*, 51: 566–634.

Michel-Lévy, A. (1893–4) 'Contributions à l'étude du granite de Flamanville et des granites français en général, *Bulletin des Services de la Carte Géologique de la France*, 5 (no. 36): 317–57.

Milankovich, M. (1920) *Théorie mathématique des phénomènes thermiques produits par la radiation solaire*, Paris: Gauthier-Villars.

Milankovich, M. (1938) 'Astronomische Mittel zur Erforschung der erdgeschichtlichen Klimate', in *Handbuch der Geophysik*, edited by B. Gutenberg, 9: 593–698, Berlin: Gebrüder Borntraeger.

Milne, J. (1883) 'Earth pulsations', *Nature*, 28: 367–70.

Milne, J. (1886) *Earthquakes and Other Earth Movements*, London: Kegan Paul, Trench, Trübner.

Milton, R. (1992) *The Facts of Life: Shattering the Myths of Darwinism*, London: Fourth Estate.

Mohorovicic, A. (1910) 'Das Beben vom 8. X. 1909', *Jahrbuch des meteorologischen Observatoriums in Zagreb (Agram) für das Jahr 1909*, 9 (Part 4, Section 1): 1–63 and charts (in Serbo-Croat).

Mohs, F. (1820) *The Characters of the Classes, Orders, Genera, and Species; or, the Characteristics of the Natural History System of Mineralogy . . .*, Edinburgh: W. & C. Tait.

Mohs, F. (1825) *Treatise on Mineralogy* . . . , translated by W. Haidinger, 3 vols, Edinburgh: Constable; London: Hurst, Robinson (1st German edn, 1822–4).

Montefiore, H. (1985) *The Probability of God*, London: SCM Press.

Moore, R.C. (1931) 'Pennsylvanian cycles in the northern mid-continent region', *Bulletin of the Illinois Geological Survey*, 60: 247–57.

Moore, R.C. (1935) 'Late Paleozoic crustal movements of Europe and North America', *BAAPG*, 19: 1253–307.

Moore, R.C. (1936) 'Stratigraphic evidence bearing on problems of continental tectonics', *BGSA*, 47: 1785–1808.

Morello, N. (1979) *La macchina della terra* . . . , Turin: Loescher Editore.

Morley, L.W. and Larochelle, A. (1964) 'Paleomagnetism as a means of dating geological events', in *Geochronology in Canada*, edited by F.F. Osborne, 39–51, Toronto: University of Toronto Press.

Moro, A.-L. (1740) *De' crostacei e degli altri marini corpi che si truovano su' monti libri due*, Venice: Stefano Monti.

Moulton, F.R. (1900) 'An attempt to test the nebular hypothesis by an appeal to the laws of dynamics', *AJ*, 17: 103–30.

Moulton, F.R. (1905) 'On the evolution of the solar system', *AJ*, 22: 165–81.

Mrazec, L. (1907) 'Despre cute cu sâmbure de străpungere [On folds with piercing cores]', *Buletinul Societăţii de Ştiinţe din Bucureşti-România*, 16: 6–7 (in Romanian).

Mrazec, L. (1914–15) 'Les Plis diapirs', *Comptes Rendus des Séances de l'Institut Géologique de Roumanie*, 6: 1–45.

Müller, D.W., McKenzie, J.A. and Weissert, H. (eds) (1991) *Controversies in Modern Geology*, London: Academic Press.

Murchison, R.I. (1839) *The Silurian System, Founded on Geological Researches in the Counties of Salop, Hereford, Radnor, Montgomery, Caermarthen, Brecon, Pembroke, Monmouth, Gloucester, Worcester, and Stafford* . . . , 3 vols, London: John Murray.

Murchison, R.I. (1854) *Siluria: the History of the Oldest Known Rocks Containing Organic Remains* . . . , London: John Murray.

Murchison, R.I. and Geikie, A. (1861) *First Sketch of a New Geological Map of Scotland* . . . , Edinburgh: W. & A.K. Johnston & W. Blackwood & Sons; London: E. Stanford.

Murphy, J.B. and Nance, R.D. (1991) 'Supercontinental model for the contrasting character of late proterozoic orogenic belts', *Geology*, 19: 469–72.

Murphy, J.B. and Nance, R.D. (1992) 'Mountain belts and the supercontinental cycle', *SA*, 266 (April): 34–41.

Naumann, C.F. (1826) *Lehrbuch der Mineralogie*, Leipzig: Engelmann (cited in Tomkeieff, 1983: 358).

Necker, L.A. ([1808]) [*Geological Map of Scotland*] [coloured on to : T. Kitchen (1778) *North Britain or Scotland* . . . , London: Wm Fadden] republished by the Geological Society, n.d.

Needham, J. and Wang Ling (1959) *Science and Civilization in China. Volume 3. Mathematics and the Sciences of the Heavens and the Earth*, Cambridge: Cambridge University Press.

Nelson, G. and Platnick, N.I. (1981) *Sytematics and Biogeography: Cladistics and Vicariance*, New York: Columbia University Press.

Nelson, G. and Rosen, D.E. (eds) (1981) *Vicariance Biogeography: a Critique* . . . , New York: Columbia University Press.

Neumayr, M. (1886–7) *Erdgeschichte*, 2 vols, Leipzig: Verlag des Bibliographischen Institut.

Newton, I. (1687) *Philosophiae Naturalis Principia Mathematica*, London: The Royal Society.

Newton, I. (1934) *[M]athematical Principles of Natural Philosophy and his System of the World*, translated by A. Motte and edited by F. Cajori, Cambridge: Cambridge University Press.

Nicol, W. (1829) 'On a method of so far increasing the divergency of the two rays in calcareous spar that only one image may be seen at a time', *ENPJ*, 6: 83–4.

Nicolson, M.H. (1959) *Mountain Gloom and Mountain Glory: the Development of the Aesthetics of the Infinite*, Ithaca: Cornell University Press.

O'Brien, C.F. (1970) '*Eozoön canadense*: the dawn animal of Canada', *Isis*, 61: 206–23.

Oldham, R.D. (1899) 'Report on the great earthquake of 12 June 1897', *Memoirs of the Geological Survey of India*, 29: 1–379.

Oldham, R.D. (1900) 'On the propagation of earthquake motion to great distances', *PTRS*, Series A, 194: 135–74.

Oldham, R.D. (1906) 'The constitution of the earth as revealed by earthquakes', *QJGS*, 62: 456–75.

Oldham, R.D. (1913) 'Radium and the evolution of the earth's crust', *Nature*, 91: 635.

Oldham, R.D. (1914) 'The constitution of the interior of the earth as revealed by earthquakes', *Nature*, 92: 684–5.

Oldroyd, D.R. (1972a) 'Robert Hooke's methodology of science as exemplified in his Discourse of Earthquakes', *BJHS*, 6: 109–30.

Oldroyd, D.R. (1972b) 'Nineteenth-century controversies concerning the Mesozoic/Tertiary boundary in New Zealand', *AS*, 29: 39–57.

Oldroyd, D.R. (1974a) 'A note on the status of A.F. Cronstedt's simple earths and his analytical methods', *Isis*, 65: 506–12.

Oldroyd, D.R. (1974b) 'Some phlogistic mineralogical schemes, illustrative of the evolution of the concept of "earth" in the 17th and 18th centuries', *AS*, 31: 269–305.

Oldroyd, D.R. (1974c) 'Some Neo-Platonic and Stoic influences on mineralogy in the sixteenth and seventeenth centuries', *Ambix*, 21: 128–56.

Oldroyd, D.R. (1974d) 'Mechanical mineralogy', *Ambix*, 21: 157–78.

Oldroyd, D.R. (1976–7) 'The doctrine of peoperty-conferring principles in chemistry: origins and antecedents', *Organon*, 12–13: 139–55.

Oldroyd, D.R. (1979) 'Historicism and the rise of historical geology', *HS*, 17: 191–213, 227–57.

Oldroyd, D.R. (1989) 'Hooke vs Wallis: geological controversy in the seventeenth century', in *Robert Hooke: New Studies*, edited by M. Hunter and S. Schaffer, 207–33, Woodbridge: Boydell & Brewer.

Oldroyd, D.R. (1990a) *The Highlands Controversy: Constructing Geological Knowledge through Fieldwork in Nineteenth-century Britain*, Chicago and London: University of Chicago Press.

Oldroyd, D.R. (1990b) 'Newton, the inverse square law, and Kepler's laws', *School Science Review*, 71: 107–13.

Oldroyd, D.R. (1993) 'The Archaean Controversy in Britain part III – the rocks of Anglesey and Caernarvonshire', *AS*, 50: 523–84.

Oldroyd, D.R. (1994) 'James Hutton: the founder of modern geology?', *BJHS*, 27: 213–19.

Oldroyd, D.R. and Howes, J.B. (1978) 'The first published version of Leibniz's *Protogæa*', *JSBNH*, 9: 56–60.

Omalius d'Halloy, J.J. d' (1808) 'Essai sur la géologie du nord de la France', *JM*, 24: 123–58, 271–318, 345–92.

Oparin, A.I. ([1938] 1953) *The Origin of Life*, translated by S. Morgulis, reprint, New York: Dover Publications.

Opdyke, N.D. (1962) 'Palaeoclimatology and continental drift', in *Continental Drift*, edited by S.K. Runcorn, 41–65, New York and London: Academic Press.

Oreskes, N. (1988) 'The rejection of continental drift', *HSPS*, 18: 311–48.

Oreskes, N. (1990) 'American geological practice', PhD dissertation, Stanford University.

Ospovat, A.M. (1980) 'The importance of regional geology in the geological theories of Abraham Gottlob Werner – a contrary opinion', *AS*, 37: 433–40.

Ospovat, D. (1981) *The Development of Darwin's Theory* . . . , Cambridge, London, New York, New Rochelle, Melbourne and Sydney: Cambridge University Press.

Ovid (1955) *Metamorphoses*, translated by M.A. Innes, Harmondsworth: Penguin Books.

Owen, H.G. (1976) 'Continental displacement and expansion of the earth during the Mesozoic and Cenozoic', *PTRS*, Series A, 281: 223–91.

Owen, H.G. (1981) 'Constant dimensions or an expanding earth?', in *The Evolving Earth*, edited by L.R.M. Cocks, 179–92, London: British Museum (Natural History); London, Cambridge, New York and Melbourne: Cambridge University Press.

Page, L. (1969) 'Diluvialism and its critics in Great Britain in the early nineteenth century', in *Toward a History of Geology*, edited by C.J. Schneer, 257–71, Cambridge (Mass.) and London: MIT Press.

Paisley, P.B. and Oldroyd, D.R. (1979) 'Science in the Silver Age: *Ætna*, a classical theory of volcanic activity', *Centaurus*, 23: 1–20.

Palassou, P.B. (1781) *Essai sur la minéralogie des Monts-Pyrénées* . . . , Paris: Didot le Jeune.

Pallas, P.S. (1777/1778) 'Observations sur la formation des montagnes; & les changemens arrivés au globe . . . , *Acta Academia Scientiarum Imperialis Petropolitanae*, 1 (Part 1): 21–64, St Petersburg: Academy of Sciences.

Pallas, P.S. (1778) *Betrachtungen über die Beschaffenheit der Gebürge und Veränderungen der Erdkugel, besonders in Beziehung auf das Russische Reich* . . . , Frankfurt and Leipzig.

Palmieri, L. (1871) 'The electro-magnetic seismograph [translated by B.O. Duncan]', *Annual Report of . . . the Smithsonian Institution . . . for the Year 1870*, 469–71, Washington, DC: Smithsonian Institution.

Palmieri, L. (1872) 'Description of instruments used for observing the earthquake shocks which precede and accompany the eruptions of Vesuvius', *The Engineer*, 33 (7 June): 407.

Paracelsus (1894) *The Hermetic and Alchemical Writings* . . . , translated by A.E. Waite, 2 vols, London: John Elliott.

Penck, A. and Brückner, E. (1901–9) *Die Alpen im Eiszeitalter*, 3 vols, Tauchnitz: Leipzig.

Phillips, J. (1844) *Memoirs of William Smith* . . . , London: John Murray.

Pitcher, W.S. and Read, H.H. (1960) 'The aureole of the main Donegal granite', *QJGS*, 116: 1–36 and plates.

Pitman, W.C. and Heirtzler, J.R. (1966) 'Magnetic anomalies over the Pacific–Antarctic ridge', *Science*, 154: 1164–71.

Plato (1975) *Timaeus Critias Cleitophon Menexenus Epistles*, translated by R.G. Bury, London: Heinemann; Cambridge (Mass): Harvard University Press.

Plato (1977) *Euthyphro Apology Crito Phaedo Phaedrus*, translated by H.N. Fowler, London: Heinemann; Cambridge (Mass.): Harvard University Press.

Playfair, J. ([1802] 1964) *Illustrations of the Huttonian Theory of the Earth*, introduction by G.H. White, reprint, New York: Dover Publications.

Playfair, J. (1805) 'Biographical account of James Hutton, M.D. F.R.S.Ed.', *TRSE*, 5: 39–99.

Pliny (1962) *Natural History*, vol. 10, books 36–37, translated by D.E. Eicholz, London: Heinemann; Cambridge (Mass.): Harvard University Press.

Pliny the Younger (1978) *Pliny: a Self Portrait in Letters*, translated by B. Radice, London: The Folio Society.

P[lot], R. (1677) *The Natural History of Oxfordshire . . .*, Oxford: The Theatre.

Polanyi, M. (1958) *Personal Knowledge: Towards a Post-critical Philosophy*, London: Routledge.

Polanyi, M. (1966) *The Tacit Dimension*, New York: Doubleday.

Popper, K.R. (1958–9) 'Back to the Pre-Socratics', *Proceedings of the Aristotelian Society*, 59 (new series): 1–24.

Porter, R.S. (1976) 'Charles Lyell and the principles of the history of geology', *BJHS*, 9: 91–103.

Porter, R.S. (1977) *The Making of Geology: Earth Science in Britain 1660–1815*, Cambridge, London, New York and Melbourne: Cambridge University Press.

Pott, J.H. (1753) *Lithgéognosie . . .*, 2 vols, Paris: J.T. Hérissant (1st German edn, 1746).

Powell, J.W. (1875) *Exploration of the Colorado River of the West and its Tributaries . . . in 1869, 1870, 1871 and 1872 . . .*, Washington, DC: Government Printing Office.

Pratt, J.H. (1855) 'On the attraction of the Himalayan mountains and of the elevated regions beyond them, upon the plumb-line in India', *PTRS*, Series B, 145: 53–100.

Prévost, L.C. (1839–40) 'Opinion sur la théorie des soulèvements', *BSGF*, 10: 430; 11: 183–203.

Pyne, S.J. (1982) *Dutton's Point: an Intellectual History of the Grand Canyon*, Grand Canyon Natural History Association (Monograph 5).

Raleigh, W. (1614) *The History of the World*, London: W. Burre.

Ramsay, A.C. (1862) 'On the glacial origin of certain lakes . . .', *QJGS*, 18: 185–204.

Rappaport, R. (1968) 'Lavoisier's geological activities', *Isis*, 58: 375–84.

Rappaport, R. (1969) 'The geological atlas of Guettard, Lavoisier, and Monnet: conflicting views of the nature of geology', in *Toward a History of Geology*, edited by C.J. Schneer, 272–87, Cambridge (Mass.) and London: MIT Press.

Rappaport, R. (1973) 'Lavoisier's theory of the earth', *BJHS*, 6: 247–60.

Ray, J. (1692a) *Miscellaneous Discourses Concerning the Dissolution and Changes of the World . . .*, London: Samuel Smith.

Ray, J. (1692b) *The Wisdom of God Manifested in the Works of Creation . . .*, 2nd edn, London: Samuel Smith.

Ray, J. (1693) *Three Physico-Theological Discourses . . .*, London: Samuel Smith.

Ray, J. (1713) *Three Physico-theological Discourses, Concerning I. The Primitive Chaos and Creation of the World. II. The General Deluge, its Causes and Effects. III. The Dissolution of the World, and Future Conflagration . . .*, London: William Innys.

Rayleigh, Lord (J.W. Strutt) (1887) 'On waves propagated along the plane surface of an elastic solid', *Proceedings of the London Mathematical Society*, 17: 4–11.

Read, H.H. (1957) *The Granite Controversy* . . . , London: Thomas Murby.

Réaumur, R.A.F. de (1721/1723) 'Sur la nature et la formation des cailloux', *HARS*, 255–76 and plates.

Réaumur, R.A.F. de (1723/1725) 'Sur la rondeur que semblent affecter certaines espèces de pierres, & entr'autres sur celle qu'affectent les cailloux', *HARS*, 273–84.

Réaumur, R.A.F. de (1730/1732) 'De la nature de la terre en général, & du caractère des différentes espèces de terres', *HARS*, 243–83.

Rebeur-Paschwitz, E.L.A. von (1889) 'The earthquake of Tokio, April 18, 1889', *Nature*, 40: 294–5.

Reinhardt, O. and Oldroyd, D.R. (1983) 'Kant's theory of earthquakes and volcanic action', *AS*, 40: 247–72.

Revelle, R. and Maxwell, A.E. (1952) 'Heat flow through the floor of the north Pacific Ocean', *Nature*, 170: 199–200.

Richthofen, F.P.W. von (1868) *Principles of the Natural System of Volcanic Rocks*, San Francisco: Towne & Bacon.

Richthofen, F.P.W. von (1877–1912) *China. Ergebnisse eigener Reisen und darauf gegründeter Studien*, 5 vols, Berlin: D. Reimer.

Ristoro d'Arezzo (1862) *Della composizione del mondo, testo Italiano del 1282* . . . (*Biblioteca rara*, edited by G. Daelli, Part 54), Milan.

Roger, J. (1962) 'Buffon: les époques de la nature', *Mémoires du Museum National d'Histoire Naturelle. Séries C, Sciences de la Terre*, 10.

Roger, J. (1989) *Buffon: un philosophe au Jardin du Roi*, Paris: Fayard.

Rogers, G.S. (1918) 'Intrusive origin of the Gulf Coast salt domes', *Economic Geology*, 13: 447–85.

Rogers, H.D. (1858) 'On the laws of of structure of the more disturbed zones of the earth's crust', in *The Geology of Pennsylvania*, 2: 885–916, London and Edinburgh: W. Blackwood & Sons; Philadelphia: J.P. Lippincott.

Rogers, H.D. and Rogers, W.B. (1843) 'On the physical structure of the Appalachian chain, as exemplifying the laws which have regulated the elevation of great mountain chains generally', *Reports of the First, Second, and Third Meetings of the Association of American Geologists and Naturalists* . . . , 1: 474–531, Boston: Gould, Kendall & Lincoln.

Rohault, J. (1723) *Rohault's System of Natural Philosophy* . . . , translated by J. Clarke, 2 vols, London: Knapton.

Romé de l'Isle, J.-B.L. (1772) *Essai de cristallographie, ou description des figures géometriques, propres à différens corps du règne minéral* . . . , Paris: Didot Jeune.

Romé de l'Isle, J.-B.L. (1784) *Des Caractères Extérierurs des Minéraux* . . . , Paris: The Author.

Ronov, A.B. (1968) 'Probable changes in the composition of sea water during the course of geological time', *Sedimentology*, 10: 25–43.

Rosenbusch, K.H.F. (1882) 'Ueber das Wesen der körnigen und porphyrischen Struktur bei Massengesteinen', *NJMGGP*, 2: 1–17.

Rosenbusch, K.H.F. (1890) 'Ueber die chemischen Beziehungen der Eruptiv-gestein', *Mineralogische und petrographische Mittheilungen*, 11 (new series): 144–78, 438.

Rosenbusch, K.H.F. (1896) *Mikroskopische Physiographie der Mineralien und Gesteine*, Stuttgart: E. Schweizerbart.

Rossi, P. (1984) *The Dark Abyss of Time: the History of the Earth and the History of Nations from Hooke to Vico*, translated by L.G. Cochrane, Chicago and London: University of Chicago Press (1st Italian edn, 1979).

Rothé, E. (1924) 'Sur la propagation des ondes séismiques au voisinage de l'épicentre. . . . Ondes P et P′, exposé d'après les travaux de A. Mohorovicic', *Publications du Bureau Central Séismologique International. Séries A. Travaux scientifiques*, 1: 17–59.

Rudwick, M.J.S. (1962) 'Hutton and Werner compared: George Greenough's geological tour of Scotland in 1805', *BJHS*, 1: 117–35.

Rudwick, M.J.S. (1969) 'Lyell on Etna, and the antiquity of the earth', in *Toward a History of Geology*, edited by C.J. Schneer, 288–304, Cambridge (Mass) and London: MIT Press.

Rudwick, M.J.S. (1970) 'The strategy of Lyell's *Principles of Geology*', *Isis*, 61: 4–33.

Rudwick, M.J.S. (1972) *The Meaning of Fossils* . . . , London and New York: Macdonald & American Elsevier.

Rudwick, M.J.S. (1974) 'Poulett Scrope on the volcanoes of Auvergne: Lyellian time and political economy', *BJHS*, 7: 205–42.

Rudwick, M.J.S. (1976) 'The emergence of a visual language for geological science', *HSPS*, 14: 149–95.

Rudwick, M.J.S. (1985) *The Great Devonian Controversy* . . . , Chicago and London: University of Chicago Press.

Rudwick, M.J.S. (1992) *Scenes from Deep Time: Early Pictorial Representations of the Prehistoric World*, Chicago and London: University of Chicago Press.

Runcorn, S.K. (1955) 'Rock magnetism – geophysical aspects', *Advances in Physics*, 4: 244–91.

Runcorn, S.K. (1956) 'Paleomagnetic comparisons between Europe and North America', *Proceedings of the Geological Society of Canada*, 8: 77–85.

Runcorn, S.K. (1962) 'Palaeomagnetic evidence for continental drift and its geophysical cause', in *Continental Drift*, edited by S.K. Runcorn, 1–40, New York and London: Academic Press.

Rupke, N.A. (1983) *The Great Chain of History: William Buckland and the English School of Geology*, Oxford and New York: Clarendon Press.

Rutherford, E. and Barnes, H.T. (1903) 'Heating effect of the radium emanation', *Nature*, 68: 622; 69: 126.

Santillana, G. de and Dechend, H. von (1977) *Hamlet's Mill: an Essay on Myth and the Frame of Time*, Boston: David R. Godine (1st edn, 1969).

Sarjeant, W.A.S. (1980) *Geologists and the History of Geology: an International Bibliography from the Origins to 1978*, New York: Arno.

Sarjeant, W.A.S. (1987) *Geologists and the History of Geology . . . Supplement, 1979–1984*, Malabar (Florida): R.E. Krieger.

Sarton, G. (1919) 'La synthèse géologique de 1775 à 1918', *Isis*, 2: 357–94.

Sartorius von Waltershausen, W. (1853) *Ueber die vulkanischen Gesteine in Sicilien und Island und ihre submarine Umbildung*, Göttingen: Dieterich.

Scheele, C.W. (1901) *Chemical Essays*, translated by J. Murray, London: Scott Greenwood.

Schidlowski, M. (1988) 'A 3,800-million-year isotopic record of life from carbon in sedimentary rocks', *Nature*, 333: 313–18.

Schneer, C.J. (ed.) (1969) *Toward a History of Geology* . . . , Cambridge (Mass.) and London: MIT Press.

[Schröter, J.S.] (1775) 'Georg Christian Füchsel [review of Füchsel's *Historia terrae et maris*]', *Journal für Liebhaber des Steinreichs*, 2: 54–63.

Schuchert, C. (1916) 'The problem of continental fracturing and diastrophism[5] in Oceania', *AJS*, 42 (4th series): 92–105.

Schuchert, C. (1923) 'Sites and nature of the North American geosynclines', *BGSA*, 20: 92–105.

Schuchert, C. (1932) 'Gondwana land bridges', *BGSA*, 43: 875–916.

Schwarzbach, M. (1986) *Alfred Werner: the Father of Continental Drift*, translated by Carla Love, Madison: Science Tech Publishers; Berlin: Springer-Verlag.

Schwinner, R. (1915) 'Analogien in Bau der Ostalpen', *Centralblatt für Mineralogie, Geologie und Paläontologie . . .*, 16: 52–62.

Schwinner, R. (1920) 'Vulkanismus und Gebirsbildung: ein Versuch', *Zeitschrift für Vulkanologie*, 5: 175–230.

Scott, H.I. (1977) *The Development of Landform Studies in Australia*, Artarmon (Sydney): Bellbird Publishing House; Farnborough: Teakfield.

Scrope, G.P. (1825) *Considerations on Volcanos . . . Leading to the Establishment of a New Theory of the Earth*, London: William Phillips.

Scrope, G.P. (1827) *Memoir on the Geology of Central France . . .*, 2 vols, London: Longman.

Scrope, G.P. (1859) 'On lamination and cleavage occasioned by the mutual friction of the particles of rocks while in irregular motion', *QJGS*, 15: 84–6.

Scrope, G.P. (1862) *Volcanos. The Character of their Phenomena, their Share in the Structure and Composition of the Surface of the Globe, and their Relation to the Internal Forces . . .*, 2nd edn, London: Longman, Green, Longmans & Roberts.

Secord, J.E. (1986) *Controversy in Victorian Geology: the Cambrian–Silurian Dispute*, Princeton: Princeton University Press.

Sederholm, J.J. (1907) 'Om granit och gneiss deras uppkomst, uppträdande och utbreding inom urberget i Fennoskandia', *BCGF* (Helsinki), no. 23 (with English summary).

Sederholm, J.J. (1923) 'On migmatites and associated Pre-cambrian rocks of southwestern Finland. Part I. The Pellige region', *BCGF* (Helsinki), no. 58.

Sederholm, J.J. (1926) 'On migmatites and associated Pre-cambrian rocks of southwestern Finland. Part II. The region around the Barösundsfjärd w[est] of Helsingfors and neighbouring areas', *BCGF* (Helsinki), no. 77.

Sederholm, J.J. (1934) 'On migmatites and associated Pre-cambrian rocks of southwestern Finland. Part III. The Åland islands', *BCGF* (Helsinki), no. 107.

Sedgwick, A. (1846) 'The geology of the Lake District in four letters addressed to W. Wordsworth, Esq.', in W. Wordsworth, *A Complete Guide to the Lakes . . .*, 3rd edn, Kendal: J. Hudson; London: Longman & Whittaker.

Seneca (1910) *Quaestiones naturales*, translated by John Clarke with notes by A. Geikie, London: Macmillan.

Shapley, H. and Howarth, H.H. (eds) (1929) *A Source Book in Astronomy*, New York and London: McGraw-Hill Book Company.

Shepard, C.U. (1835) *Treatise on Mineralogy . . .*, 2 vols, New Haven: H. Howe, Herrick & Noyes.

Shepherd, E.S., Rankin, G.A. and Wright, F.E. (1909) 'The binary systems of alumina with silica, lime and magnesia', *AJS*, 28 (4th series): 293–333.

Shmidt, O.Y. (1944) 'Meteoric theory of the origin of the planet earth', *Doklady Akademii*

[5]Crustal movements, faulting, etc.

Nauk SSSR [*Reports of the Academy of Science of the USSR*], 45: 245–49 (in Russian. See Shmidt [1958], p. 131).

Shmidt, O.Y. (1958) *A Theory of the Earth's Origin: Four Lectures*, translated by G.H. Hanna, Moscow: Foreign Languages Publishing House.

Sisco, A.G. and Smith, C.S. (1949) *Bergwerk- und Probierbüchlein. A Translation . . . of the Bergbüchlein, a Sixteenth-century Book on Mining Geology, . . . and of the Probierbüchlein, a Sixteenth-century Work on Assaying . . .*, New York: American Institute of Mining and Metallurgical Engineers.

Sloss, L.L. (1991) 'The tectonic factor in sea level change: a countervailing view', *Journal of Geophysical Research*, 96: 6609–17.

Smith, W. (1799) 'A table of the order of strata, and their embedded organic remains, in the vicinity of Bath; examined and proved prior to 1799', MS held in the archives of the Geological Society, London.

Smith, W. (1815a) *A Delineation of the Strata of England and Wales, with Part of Scotland . . .*, London: J. Cary.

Smith, W. (1815b) *A Memoir to the Map and Delineation of the Strata of England and Wales, with Parts of Scotland*, London: J. Carey.

Smith, W. (1819) *A Geological Section from London to Snowdon. Showing the Varieties of the Strata, and the Correct Attitude of the Hills . . . Coloured to Correspond with . . . Geological Maps of England and Wales*, London: No publisher stated.

Smith, W. (1824) *Geological Map of Cumberland* [entered on J. Carey, *A New Map of Cumberland Divided into Hundreds*].

Snider-Pellegrini, A. (1858) *La Création et ses mystères dévoilés*, Paris: A. Franck.

Socolow, A.A. (ed.) (1988) *The State Geological Surveys [of the United States]: a History*, Grand Forks: Association of American State Geologists.

Sorby, H.C. (1851) 'On the microscopical structure of the calcareous grit of the Yorkshire coast', *Proceedings of the Geological and Polytechnic Society of the West-riding of Yorkshire*, 3: 197–205.

Sorby, H.C. ([1856] 1857) 'On the microscopical structure of mica-schist', *RBAAS*, 78.

Soret, C. (1879–1880) 'Sur l'état d'équilibre que prend au point de vue de sa concentration une dissolution saline primitivement homogène dont deux parties sont portées à des températures différentes. *Bibliothèque universelle. Archives des Sciences Physiques et Naturelles* (Geneva) 3 (3rd series): 48–61; 4: 209–13.

Spengler, O. (1926) *The Decline of the West*, translated by C.F. Atkinson, London: Allen & Unwin.

Sproul, B. 1979. *Primal Myths: Creating the World*, San Francisco: Harper & Row.

Stafford, R.A. (1989) *Scientist and Empire: Sir Roderick Murchison, Scientific Exploration and Western Imperialism*, Cambridge, New York, Port Chester, Melbourne and Sydney: Cambridge University Press.

Stegagno, G. (1929) *Il Veronese Giovanni Arduino e il suo contributo al progresso della scienza geologica*, Verona: Tip. Operaia.

Steno, N. (1669) *De solido intra solidvm natvraliter contento dissertationis prodromvs . . .*, Florence: Ex Typographia Sub Signo Stellae.

Steno, N. (1968) *The Prodromus of Nicolaus Steno's Dissertation Concerning a Solid within a Solid*, New York and London: Hafner.

Stewart, J.A. (1990) *Drifting Continents and Colliding Paradigms . . .*, Bloomington and Indianapolis: Indiana University Press.

Stille, H. (1936a) 'Die Entwicklung des amerikanischen Kordillerensystems in Zeit und Raum', *Preussischer Akademie der Wissenschaft, physikalisch-mathematische Klasse, Sitzungsberichte*, 15: 134–55.

Stille, H. (1936b) 'The present tectonic state of the earth', *BAAPG*, 20: 849–80.

Stille, H. (1940) *Einführung in den Bau Amerikas*, Berlin: Gebrüder Borntraeger.

Strabo (1917) *The Geography*, translated by H.L. Jones, London: Heinemann; New York: G.P. Putnam's Sons.

Strachey, J. (1719) 'A curious description of the strata observ'd in the coal-mines of Mendip in Somersetshire,' *PTRS*, 30: 968–73.

Suess, E. (1875) *Die Entstehung der Alpen*, Vienna: Wilhelm Braumüller.

Suess, E. (1883–1909) *Das Antlitz der Erde*, 5 vols, Vienna and Prague: F. Tempsky; Leipzig: G. Freytag.

Suess, E. (1904–24) *The Face of the Earth (Das Antlitz der Erde)*, translated by H.B.C. Sollas, 5 vols, Oxford: Clarendon Press.

Sweet, J.M. and Waterston, C.D. (1967) 'Robert Jameson's approach to the Wernerian theory of the earth 1796', *AS*, 23: 81–95.

Targioni-Tozzetti, G. (1768/1777) *Relazioni d'alcuni viaggi fatti in diverse parti della Toscana* . . . , 2 vols, Florence: Gaetano Cambiagi.

Taylor, F.B. (1910) 'Bearing of the Tertiary mountain-belt on the origin of the earth's plan', *BGSA*, 21: 179–226.

Taylor, K.L. (1985) 'Early geoscience mapping, 1700–1830', *Proceedings of the Geoscience Information Society*, 15: 15–49.

Taylor, K.L. (1992) 'The historical rehabilitation of theories of the earth', *The Compass*, 69: 334–45.

Taylor, T.G. (1919) 'Climatic cycles and evolution', *Geographical Review*, 8: 289–328.

Teilhard de Chardin, P. (1969) *Christianity and Evolution*, translated by René Hague, New York: Harcourt Brace Jovanovich.

Termier, H. and Termier, G. (1952) *Histoire géologique de la biosphère: la vie et les sédiments dans les géographies successives*, Paris: Masson.

Termier, P. (1904) 'Les schistes cristallins des Alpes occidentales', *Congrès géologique international. Compte rendu de la IX session, Vienne, 1903*, 571–86, Vienna: Hollinck Frères.

Thackray, J.C. (1978) 'R.I. Murchison's *Silurian System*, *JSBNH*, 8: 421–33.

Tharp, M. and Frankel, H. (1986) 'Mappers of the deep', *Natural History*, 95: 49–62.

Thompson, S.J. (1988) *A Chronology of Geological Thinking from Antiquity to 1899*, Metuchen: Scarecrow Press.

Tinkler, K.J. (1985) *A Short History of Geomorphology*, London: Croom-Helm.

Tinkler, K.J. (ed.) (1989) *History of Geomorphology* . . . , Boston: Unwin/Hyman.

Tomkeieff, S.I. (1983) *Dictionary of Petrology*, edited by E.K. Walton, B.A.O. Randall, M.H. Battey and O. Tomkeieff, Chichester, New York, Brisbane, Toronto and Singapore: John Wiley & Sons.

Topley, W. (1885) 'Report upon national geological surveys: part I. Europe', *RBAAS*, 221–37.

Torrens, H.S. (1994) 'Patronage and problems: Banks and the earth sciences', in *Sir Joseph Banks: a Global Perspective*, edited by R.E.R. Banks, B. Elliott, J.G. Hawkes, D. King-Hele and G. Ll. Lucas, 49–75. Kew: Royal Botanical Gardens.

Toulmin, S. and Goodfield, J. ([1965] 1967) *The Discovery of Time*, reprint, Harmondsworth: Penguin Books.

Tournefort, J.-P. de (1702/1720) 'Description du labirinthe de Candie, avec quelques observations sur l'accroissement & sur la génération des pierres', *HARS*, 224–41 (2nd edn).

Toussaint Charpentier, J.F.W. von. (1778) *Mineralogische Geographie der chursächsischen Lande*, Leipzig: S.L. Crusius.

Tuge, H. (1968) *Historical Development of Science and Technology in Japan*, Tokyo: Kokusai Bunka Shinko.

Turner, A.J. (1974) 'Hooke's theory of the earth's axial displacement: some contrary opinion', *BJHS*, 7: 166–70.

Tuttle, O.F. and Bowen, N.L. (1958) *Origin of Granite in the Light of Experimental Studies in the System $NaAlSi_3O_8$ [albite] – $KAlSi_3O_8$ [orthoclase] – SiO_2 [quartz] – H_2O*, New York: Geological Society of America (Memoir No. 74).

Udden, J.A. (1912) *Geology and Mineral Resources of the Peoria Quadrangle*, US Washington, DC: Geological Survey (Bulletin 506).

Ulrich, E.O. (1911) 'Revision of the Paleozoic systems', *BGSA*, 22: 281–680.

Usher, J. (1658) *Annals of the World. Deduced from the Origin of Time, and Continued to the Beginning of the Emperour Vespasians Reign, and the Totall Destruction and Abolition of the Temple and Commonwealth of the Jews . . . Collected from All History . . .*, London: J. Crook (1st Latin edn, 1650).

Vaccari, E. (1993) *Giovanni Arduino (1714–1795): il contributo di uno scienziato Veneto al dibattito settentesco sulle scienze della terra*, Florence: Leo S. Olschki.

Vail, P.R. (1975) 'Eustatic cycles from seismic data for global stratigraphic analysis', *BAAPG*, 59: 2198–9.

Vail, P.R., Mitchum, R.M., Todd, R.G., Widmier, J.M., Thompson III, S., Sangree, J.B., Bubb, J.N. and Hatlelid, W.G. (1977) 'Seismic stratigraphy and global changes in sea level', in *Seismic Stratigraphy – Applications to Hydrocarbon Exploration*, edited by C.E. Payton, 47–212, Tulsa: American Association of Petroleum Geologists (Memoir No. 26).

Van der Gracht, W.A.J.M. van W. (ed.) (1928) *Theory of Continental Drift: a Symposium on the Origin and Movement of Land Masses both Inter-continental and Intra-continental, as Proposed by Alfred Wegener*, Delft: NV Technische Boekhandelen Drukkerej J. Waltman Jr.

Venetz, J. (1837) 'Ueber den gegenwärtigen und früheren Zustand der Walliser Gletscher und über die erratischen Blöcke oder die Bruchstücke alpinischer Felsarten . . .', *NJMGGP*, 8: 472–7.

Vening Meinesz, F.A. (1932) *Gravity Expeditions at Sea 1923–1930. Vol. I. The Expeditions, the Computations and the Results*, Delft: N.V. Technische Boekhandel en Drukkerij J. Waltman Jr.

Vening Meinesz, F.A. (1934) 'Gravity and the hypothesis of convection currents in the earth', *Proceedings of the Koninklijke Nederlandse Akademie van Wetenschappen*, 37: 37–45.

Vening Meinesz, F.A., Umbgrove, J.H.F. and Kuenen, Ph.H. (1934) *Gravity Expeditions at Sea. Vol. II. Report of the Gravity Expedition in the Atlantic of 1932 and Interpretation of the Results*, Delft: NV Technische Boekhandel en Drukkerij J. Waltman Jr.

Vernadsky, V.I. [W.] (1924) *La Géochimie*, Paris: Librairie Félix Alcan (first published in Russia in 1924).

Vernadsky, V.I. [W.] (1929) *La Biosphère*, Paris: Librairie Félix Alcan.

Vernadsky, V.I. [W.] (1944) 'Problems of biochemistry, II[6] The fundamental matter-energy difference between the living and the inert natural bodies of the biosphere', *Transactions of the Connecticut Academy of Arts and Sciences*, 35: 487–515 (translated by G. Vernadsky; first published in Russian in 1938).

Vernadsky, V.I. [W.] (1945) 'The biosphere and the noösphere', *American Scientist*, 33: 1–12.

Vernadsky, V.I. [W.] (1986) *The Biosphere*, translated anonymously, Oracle (Arizona) and London: Synergetic Press (first published in Russia in 1926).

Vine, F.J. and Matthews, D.H. (1963) 'Magnetic anomalies over oceanic ridges', *Nature*, 199: 947–9.

Vine, F.J. and Wilson, J.T. (1965) 'Magnetic anomalies over a young ocean ridge off Vancouver Island', *Science*, 150: 485–9.

Vinogradov, A.P. (1963) 'Development of V.I. Vernadskiy's ideas', *Soviet Soil Science*, 8: 727–32.

Virlet d'Aoust, T. (1846–7). 'Observations sur la métamorphisme normal et la probabilité de la non-existence de véritables roches primitives à la surface du globe, *BSGF*, 4 (2nd series): 498–505.

Vitaliano, D.B. (1976) *Legends of the Earth: their Geologic Origins*, Seceucas: Citadel Press.

Wadati, K. (1934–5) 'On the activity of deep-focus earthquakes in the Japan Islands and neighbourhoods', *Geophysical Magazine*, 8: 305–25.

Wallace, A.R. (1876) *The Geographical Distribution of Animals*, 2 vols, London: Macmillan.

Wallerius, J.G. (1747) *Mineralogia, eller mineral-riket, indelt och beskrifvit af Johan Gotschalk Wallerius*, Stockholm: L. Salvii.

Wallerius, J.G. (1753) *Minéralogie, ou description générale des substances du règne minéral, ouvrage traduit de l'allemand*, 2 vols, Paris: Durand, Pissot.

Walther, J. (1893–4) *Einleitung in die Geologie als historische Wissenschaft. Beobachtungen über die Bildung der Gesteine und ihrer organischen Einschlüsse*, 3 vols, Jena: Gustav Fischer.

Wanless, H.R. (1931) 'Pennsylvanian cycles in western Illinois', in *Papers Presented at the Quarter Centennial Celebration of the Illinois State Geological Survey* (Illinois Geological Survey, Bulletin 60), 179–93.

Wanless, H.R. and Shepard, F.P. (1936) 'Sea level and climatic changes related to late Paleozoic cycles', *BGSA*, 47: 1177–206.

Wanless, H.R. and Weller, J.M. (1932) 'Correlation and extent of Pennsylvanian cyclothems', *BGSA*, 43: 1003–16.

Webster, J. (1671) *Metallographia . . .*, London: Walter Kettilby.

Webster, T. (1814) 'On the freshwater formations of the Isle of Wight, with some observations on the strata over the chalk in the south-east part of England', *TGSL*, 2: 161–254 and plates.

Wegener, A.L. (1912a) 'Die Entstehung der Kontinente', *Dr. A. Petermanns geographische Mitteilungen aus Justus Perthes geographischer Anstalt*, 58: 185–95, 253–56, 305–09.

Wegener, A.L. (1912b) 'Die Enstehung der Kontinente', *GRZAG*, 3: 276–92.

Wegener, A.L. (1915) *Die Entstehung der Kontinente und Ozeane*, Braunschweig: Friedrich Vieweg & Sohn.

Wegener, A.L. (1924) *The Origin of Continents and Oceans*, translated from 3rd German edn by J.G.A. Skerl, London: Methuen.

[6]The first part of this paper was not translated into English.

Wegener, A.L. (1966) *The Origin of Continents and Oceans*, translated from 4th German edn (1929) by J Biram, New York: Dover Publications.

Wegmann, C.E. (1935) 'Zur deutungder Migmatite', *GRZAG*, 26: 305–50.

Wegmann, C.E. (1970) 'Argand, Émile', in *Dictionary of Scientific Biography*, edited by C.C. Gillispie, 1: 235–37, New York: Scribner.

Wegmann, E. (1969) 'Changing ideas about moving shorelines', in *Toward a History of Geology*, edited by C.J. Schneer, 386–414, Cambridge (Mass.) and London: MIT Press.

Weller, J.M. (1930) 'Cyclical sedimentation of the Pennsylvanian period and its significance', *JG*, 38: 97–135.

Weller, J.M. (1931) 'The conception of cyclical sedimentation during the Pennsylvanian period', *Bulletin of the Illinois Geological Survey*, 60: 163–77.

Wells, A.K. (1938) *Outline of Historical Geology*, London: George Allen & Unwin.

Werner, A.G. (1774) *Von den äusserlichen Kennzeichen der Fossilien*, Leipzig: Siegfried Crusius.

Werner, A.G. (1786) 'Kurze Klassifikation und Beschreibung der verschiedenen Gebirgsarten', *Abhandlungen der Böhemischen Gesellschaft der Wissenschaften, auf das Jahr 1786*, 272–97, Prague and Dresden: In der Walterischen Hofbuchhandlung. (Reproduced in facsimile in Werner [1971].)

Werner, A.G. (1788) 'Anzeigen, Auszüge und Recensionen bergmannischer und mineralogischer Schriften', *Bergmännisches Journal*, 2: 812–41.

Werner, A.G. (1962) *On the External Characters of Minerals*, translated by A.V. Carozzi, Urbana: University of Illinois Press.

Werner, A.G. (1971) *Short Classification and Description of the Various Rocks*, translated by A.M. Ospovat, New York: Hafner.

Wertenbaker, W. (1974) *The Floor of the Sea: Maurice Ewing and the Search to Understand the Earth*, Boston: Little, Brown.

Westbroek, P. (1988) 'The geological impact of life', in *Gaia* . . . , edited by P. Bunyard and E. Goldsmith, 99–107, Camelford: Wadebridge Ecological Centre.

Westbroek, P. (1991) *Life as a Geological Force: Dynamics of the Earth*, New York and London: W.W. Norton.

Wetherill, G. (1976) 'The role of large bodies in the formation of earth and moon', *Proceedings of the Seventh Lunar Scientific Conference . . . Volume 3. The Moon and Other Bodies*, 3245–57, New York: Pergamon Press.

[Whewell, W.] (1832) '[Review of] *Principles of geology* . . . by Charles Lyell . . . Vol. II', *The Quarterly Review*, 47: 103–32.

Whiston, W. (1696) *A New Theory of the Earth* . . . , London: R. Roberts for Benj. Tooke.

White, D.A., Roeder, D.H., Nelson, T.H. and Crowell, J.C. (1970) 'Subduction', *BGSA*, 81: 3431–32.

Wiechert, E. (1896) 'Ueber die Beschaffenheit des Erdinnern', *Schriften der Physikalisch-ökonomischen Gesellschaft zu Königsberg*, 37: 4–5.

Wiechert, E. (1897) 'Ueber die Massenverteing im Innern der Erde', *Nachtrichten von der Königlichen Gesellschaft der Wissenschaften zu Göttingen. Mathematische-physikalische Klasse aus dem Jahre 1897*, 221–43.

Wiechert, E. (1903) 'Ein astatisches Pendel hoher Empfindlichkeit zur mechanischen Registrierung von Erdbeben', *Physikalische Zeitschrift*, 4: 821–9.

Wiechert, E. and Zoeppritz, K. (1907) 'Über Erdbebenwellen . . . ', *Nachrichten von der Königlichen Gesellschaft der Wissenschaften zu Göttingen. Mathematisch-physikalische Klasse aus dem Jahre 1907*, 415–549, 529–49.

Willis, B. (1893) 'The mechanics of Appalachian structure', *Annual Report of the United States Geological Survey . . . 1891–1892*, Part 2, 211–82, Washington, DC: Government Printer.

Willis, B. (1910) 'Principles of paleogeography', *Science*, 31: 241–60.

Willis, B. (1932) 'Isthmian links', *BGSA*, 43: 917–52.

Wilson, J.T. (1951) 'On the growth of continents', *Papers and Proceedings of the Royal Society of Tasmania for the Year 1950*, 85: 85–111.

Wilson, J.T. (1954) 'The development and structure of the crust', in *The Earth as a Planet*, edited by G.P. Kuiper, 138–214. Chicago: The University of Chicago Press.

Wilson, J.T. (1965) 'A new class of faults and their bearing on continental drift', *Nature*, 207: 343–7.

Wilson, L.G. (1972) *Charles Lyell, the Years to 1841 . . .*, New Haven: Yale University Press.

Wood, R.M. (1985) *The Dark Side of the Earth*, London and Boston: Allen & Unwin.

Woodward, H.B. (1908) *The History of the Geological Society of London*. London: Longmans, Green.

Woodward, J. (1695) *An Attempt towards a Natural History of the Earth . . . With an Account of the Universal Deluge: and of the Effects that it had upon the Earth*, London: R. Wilkin.

Woodward, J. (1726) *The Natural History of the Earth, Illustrated, and Inlarged, and also Defended, and the Objections against it . . . Answered . . .*, translated by B. Holloway, London: Thomas Edlin.

Woodward, J. (1728) *Fossils of All Kinds, Digested into a Method Suitable to their Mutual Relation and Affinity . . .*, London: William Innys.

Wroe, M. (1994) 'Noah's "Ark found near Mt Ararat" ', *The Observer*, 16 January.

Yang, J.-Y. and Oldroyd, D.R. (1989) 'The introduction and development of continental drift theory and plate tectonics in China . . .', *AS*, 46: 21–43.

Yoder, H.S. (1993) 'Timetable of petrology', *Journal of Geological Education*, 41: 447–89.

Zittel, K.A. von (1901) *History of Geology and Palaeontology to the End of the Nineteenth Century*, translated by M. Ogilvie-Gordon, London: Walter Scott; Reprinted 1962, Weinheim: Cramer (1st German edn, 1899).

Zöllner, F. (1873) 'Ueber eine neue Methode zur Messung anziehender und abstossender Kräfte', *Annalen der Physik und Chemie*, 40: 131–4 (first published in Leipzig in 1869).

Index